# Antimicrobial Peptides

## Challenges and Future Perspectives

*Edited by*

## K. Ajesh and K. Sreejith

Department of Biotechnology and Microbiology, Kannur University, Kannur, Kerala, India

Academic Press is an imprint of Elsevier
125 London Wall, London EC2Y 5AS, United Kingdom
525 B Street, Suite 1650, San Diego, CA 92101, United States
50 Hampshire Street, 5th Floor, Cambridge, MA 02139, United States
The Boulevard, Langford Lane, Kidlington, Oxford OX5 1GB, United Kingdom

Copyright © 2023 Elsevier Inc. All rights reserved.

No part of this publication may be reproduced or transmitted in any form or by any means, electronic or mechanical, including photocopying, recording, or any information storage and retrieval system, without permission in writing from the publisher. Details on how to seek permission, further information about the Publisher's permissions policies and our arrangements with organizations such as the Copyright Clearance Center and the Copyright Licensing Agency, can be found at our website: www.elsevier.com/permissions.

This book and the individual contributions contained in it are protected under copyright by the Publisher (other than as may be noted herein).

**Notices**

Knowledge and best practice in this field are constantly changing. As new research and experience broaden our understanding, changes in research methods, professional practices, or medical treatment may become necessary.

Practitioners and researchers must always rely on their own experience and knowledge in evaluating and using any information, methods, compounds, or experiments described herein. In using such information or methods they should be mindful of their own safety and the safety of others, including parties for whom they have a professional responsibility.

To the fullest extent of the law, neither the Publisher nor the authors, contributors, or editors, assume any liability for any injury and/or damage to persons or property as a matter of products liability, negligence or otherwise, or from any use or operation of any methods, products, instructions, or ideas contained in the material herein.

ISBN: 978-0-323-85682-9

For Information on all Academic Press publications
visit our website at https://www.elsevier.com/books-and-journals

*Cover image legend*: Cyclic and linear forms of the antimicrobial lipopeptide 'Kannurin' derived using molecular dynamics simulation.
*Image courtesy*: Shabeer Ali Hassan.

*Publisher*: Stacy Masucci
*Acquisitions Editor*: Kattie Washington
*Editorial Project Manager*: Aera F. Gariguez
*Production Project Manager*: Punithavathy Govindaradjane
*Cover Designer*: Mark Rogers

Typeset by MPS Limited, Chennai, India

# Contents

| | | |
|---|---|---|
| List of contributors | | xiii |
| Preface | | xvii |

**1   Historical developments of antimicrobial peptide research** — 1
*Benu George, S. Pooja, T.V. Suchithra and Denoj Sebastian*
- 1.1   Introduction — 1
- 1.2   History and development of antimicrobial peptides — 2
- 1.3   Antimicrobial peptides as host innate defense barricade — 4
- 1.4   Peptide-based database: barn house for AMPs — 5
- 1.5   Current timeline of antimicrobial peptide approvals — 6
- 1.6   Chemical developments in AMPs — 7
- 1.7   Antimicrobial peptides modification for medical application — 9
- 1.8   Antimicrobial peptides modification for industrial applications — 10
- 1.9   An interdisciplinary upgrade to AMPs — 11
- 1.10  Conclusion — 12
- References — 13

**2   Biosynthesis of peptide antibiotics and innate immunity** — 17
*K. Ajesh and K. Sreejith*
- 2.1   Introduction — 17
- 2.2   Antimicrobial peptides in innate immunity — 18
- 2.3   Biosynthesis of nonribosomal and ribosomal peptides — 21
  - 2.3.1   Nonribosomal peptides — 22
  - 2.3.2   Ribosomally synthesized and posttranslationally modified peptides — 24
- 2.4   Summary and conclusion — 26
- References — 27

**3   Antimicrobial peptides: features and modes of action** — 33
*Feba Francis and Nitin Chaudhary*
- 3.1   Introduction — 33
- 3.2   Historical perspective — 34
- 3.3   Features of antimicrobial peptides — 35
  - 3.3.1   Diversity — 35
  - 3.3.2   Cationicity and amphipathicity — 35
  - 3.3.3   Structure — 36
- 3.4   Biosynthesis and regulation — 41
- 3.5   Some common families of antimicrobial peptides — 43

|  |  | 3.5.1 | Cathelicidins | 45 |
| --- | --- | --- | --- | --- |
|  |  | 3.5.2 | Defensins | 45 |
|  |  | 3.5.3 | Thionins | 46 |
|  |  | 3.5.4 | Antimicrobial peptides rich in specific amino acids | 46 |
|  | 3.6 | Relationship of structure with function | | 48 |
|  | 3.7 | Modes of action | | 48 |
|  |  | 3.7.1 | Membrane-mediated action | 49 |
|  |  | 3.7.2 | Membrane-independent/nonmembrane-disruptive mechanism | 51 |
|  | 3.8 | Multifaceted roles of antimicrobial peptides | | 51 |
|  |  | 3.8.1 | Anticancer antimicrobial peptides | 51 |
|  |  | 3.8.2 | Wound-healing antimicrobial peptides | 52 |
|  |  | 3.8.3 | Antidiabetogenic peptides | 52 |
|  |  | 3.8.4 | Antiinflammatory and immunomodulatory peptides | 52 |
|  |  | 3.8.5 | Spermicidal peptides | 53 |
|  | 3.9 | Limitations and challenges | | 53 |
|  |  | 3.9.1 | Stability | 53 |
|  |  | 3.9.2 | Toxicity | 54 |
|  |  | 3.9.3 | Salt sensitivity | 54 |
|  |  | 3.9.4 | Aggregation propensity | 54 |
|  | 3.10 | Conclusion | | 55 |
|  | References | | | 55 |
| **4** | **Purification and characterization of antimicrobial peptides** | | | **67** |
|  | *A.R. Sarika and Arunan Chandravarkar* | | | |
|  | 4.1 | Purification techniques | | 67 |
|  |  | 4.1.1 | Solid-phase extraction on C18 column | 68 |
|  |  | 4.1.2 | Ion-exchange chromatography | 68 |
|  |  | 4.1.3 | Gel permeation chromatography | 69 |
|  |  | 4.1.4 | Affinity chromatography | 70 |
|  |  | 4.1.5 | Membrane filtration | 70 |
|  |  | 4.1.6 | High-performance liquid chromatography | 71 |
|  | 4.2 | Characterization techniques | | 73 |
|  |  | 4.2.1 | Amino acid analysis | 73 |
|  |  | 4.2.2 | Sequencing—Edman procedure | 74 |
|  |  | 4.2.3 | Two dimensional—poly acrylamide gel electrophoresis | 74 |
|  |  | 4.2.4 | Mass spectrometry | 74 |
|  | References | | | 76 |
| **5** | **Antimicrobial lipopeptides of bacterial origin—the molecules of future antimicrobial chemotherapy** | | | **81** |
|  | *P. Prajosh, H. Shabeer Ali, Renu Tripathi and K. Sreejith* | | | |
|  | 5.1 | Introduction | | 81 |
|  | 5.2 | Lipopeptides | | 82 |
|  |  | 5.2.1 | Types of lipopeptides produced by different bacterial genera | 82 |

|  |  | 5.2.2 Structure—activity relationship of lipopeptides | 88 |
|---|---|---|---|
|  |  | 5.2.3 Mechanism of action of lipopeptides | 89 |
|  |  | 5.2.4 Antiadhesion and antibiofilm activities of lipopeptides | 90 |
|  |  | 5.2.5 Natural role of lipopeptides | 90 |
|  |  | 5.2.6 Lipopeptides in the treatment of multidrug-resistant infections | 90 |
|  | 5.3 | Conclusion | 91 |
|  | References |  | 91 |
| **6** | **Antimicrobial peptides of fungal origin** |  | **99** |
|  | *S. Shishupala* |  |  |
|  | 6.1 | Introduction | 99 |
|  | 6.2 | Fungi-producing antimicrobial peptides | 100 |
|  | 6.3 | Fungal peptides | 101 |
|  | 6.4 | Mode of action and biological activities | 102 |
|  | 6.5 | Mechanisms of synthesis | 103 |
|  | 6.6 | Detection methods of antimicrobial peptides | 103 |
|  | 6.7 | Peptide databases | 107 |
|  |  | 6.7.1 Peptaibol database | 108 |
|  | 6.8 | Biotechnological applications | 108 |
|  | 6.9 | Summary and conclusions | 110 |
|  | Acknowledgments |  | 110 |
|  | References |  | 111 |
| **7** | **Insect peptides with antimicrobial effects** |  | **117** |
|  | *Daljeet Singh Dhanjal, Chirag Chopra, Sonali Bhardwaj,* |  |  |
|  | *Parvarish Sharma, Eugenie Nepovimova, Reena Singh and Kamil Kuca* |  |  |
|  | 7.1 | Introduction | 117 |
|  | 7.2 | The need for antimicrobial peptides | 118 |
|  | 7.3 | Classification of insect peptides | 119 |
|  |  | 7.3.1 Attacins | 120 |
|  |  | 7.3.2 Cecropins | 120 |
|  |  | 7.3.3 Defensins | 126 |
|  |  | 7.3.4 Gloverins | 127 |
|  |  | 7.3.5 Lebocins | 127 |
|  |  | 7.3.6 Moricins | 128 |
|  | 7.4 | Mode of action | 128 |
|  | 7.5 | Concluding remarks | 129 |
|  | References |  | 130 |
| **8** | **Amphibian host defense peptides** |  | **139** |
|  | *A. Anju Krishnan, A.R. Sarika, K. Santhosh Kumar and* |  |  |
|  | *Arunan Chandravarkar* |  |  |
|  | 8.1 | Antimicrobial peptides: critical component of innate immune system | 139 |

|  |  |  |
|---|---|---|
| 8.2 | Antimicrobial peptide from amphibians | 140 |
| | 8.2.1 Antimicrobial peptides isolated from African frogs | 140 |
| | 8.2.2 Antimicrobial peptide isolated from amphibians in North America | 141 |
| | 8.2.3 Antimicrobial peptides isolated from amphibians in South America | 142 |
| | 8.2.4 Antimicrobial peptide isolated from amphibians in Australia | 144 |
| | 8.2.5 Antimicrobial peptide isolated from amphibians in Europe | 145 |
| | 8.2.6 Antimicrobial peptide isolated from amphibians in Asia | 145 |
| 8.3 | Conclusion | 149 |
| | References | 149 |

## 9 Plant-derived antimicrobial peptides — 157

*Jane Mary Lafayette Neves Gelinski, Bernadette Dora Gombossy de Melo Franco and Gustavo Graciano Fonseca*

|  |  |  |
|---|---|---|
| 9.1 | General characteristics of bioactive peptides derived from plants | 158 |
| 9.2 | Antimicrobial peptides derived from different plant families | 158 |
| | 9.2.1 Cyclotides | 158 |
| | 9.2.2 Thionins | 159 |
| | 9.2.3 Defensins | 159 |
| | 9.2.4 Snakins | 160 |
| | 9.2.5 Heveins and hevein-like peptides | 160 |
| 9.3 | Extraction and identification of plant antimicrobial peptides | 163 |
| 9.4 | Perspectives in technological and therapeutic applications | 164 |
| 9.5 | Concluding remarks | 165 |
| | References | 165 |

## 10 Mammalian antimicrobial peptides — 171

*M. Divya Lakshmanan, Swapna M. Nair and B.R. Swathi Prabhu*

|  |  |  |
|---|---|---|
| 10.1 | Introduction | 171 |
| 10.2 | History of antimicrobial peptides | 171 |
| 10.3 | Mammalian antimicrobial peptides as first-line defense against invading microbes | 172 |
| 10.4 | Classification of mammalian antimicrobial peptides | 173 |
| | 10.4.1 Classification of antimicrobial peptides based on amino acid sequence | 173 |
| | 10.4.2 Classification of antimicrobial peptides based on the structure | 175 |
| | 10.4.3 Classification of antimicrobial peptides based on the activity | 179 |
| 10.5 | Common mechanism of action of mammalian antimicrobial peptides | 182 |
| | 10.5.1 Membrane-targeting mechanism | 182 |

|  |  | 10.5.2 | Cell wall-targeting mechanism | 184 |
|  | 10.5.3 | Targeting intracellular processes | 184 |
|  | 10.5.4 | Immunomodulatory mechanism | 185 |

| | | |
|---|---|---|
| 10.6 | Clinical applications of antimicrobial peptides | 185 |
| 10.7 | Current and future prospects and challenges in developing antimicrobial peptides | 188 |
| References | | 189 |

## 11 Antimicrobial peptides from marine environment — 197

*M.S. Aishwarya, R.S. Rachanamol, A.R. Sarika, J. Selvin and A.P. Lipton*

| | | | |
|---|---|---|---|
| 11.1 | Introduction | | 197 |
| 11.2 | Antimicrobial peptides from marine invertebrates | | 198 |
| | 11.2.1 | Antimicrobial peptides from marine sponges | 198 |
| | 11.2.2 | Antimicrobial peptides from marine molluscs | 198 |
| | 11.2.3 | Antimicrobial peptides from ascidians | 201 |
| | 11.2.4 | Antimicrobial peptides from crustaceans | 201 |
| | 11.2.5 | Antimicrobial peptides from marine worms | 202 |
| | 11.2.6 | Antimicrobial peptides from Cnidaria | 203 |
| | 11.2.7 | Antimicrobial peptides from Echinodermata | 203 |
| 11.3 | Antimicrobial peptides from marine microorganisms | | 203 |
| | 11.3.1 | Antimicrobial peptides from marine bacteria | 203 |
| | 11.3.2 | Antimicrobial peptides from marine actinomycetes | 204 |
| | 11.3.3 | Antimicrobial peptides from marine fungi | 206 |
| 11.4 | Antimicrobial peptides from marine vertebrates | | 207 |
| | 11.4.1 | Antimicrobial peptides from marine fishes | 207 |
| 11.5 | Antimicrobial peptides from marine algae | | 207 |
| 11.6 | Conclusions | | 207 |
| References | | | 208 |

## 12 Peptides with antiviral activities — 219

*Anjali Jayasree Balakrishnan, Aswathi Kodenchery Somasundaran, Prajit Janardhanan and Rajendra Pilankatta*

| | | | |
|---|---|---|---|
| 12.1 | Introduction | | 219 |
| 12.2 | Viral life cycle | | 220 |
| 12.3 | Peptides as viral inhibitors | | 222 |
| 12.4 | Mechanism of inhibition | | 223 |
| | 12.4.1 | Viral attachment inhibitors | 223 |
| | 12.4.2 | Plasma membrane and viral fusion inhibitors | 224 |
| | 12.4.3 | Endosomal acidification inhibitors | 226 |
| | 12.4.4 | Replication and translation inhibitors | 227 |
| 12.5 | Peptides as therapeutics | | 228 |
| 12.6 | Challenges and future scope | | 228 |
| Acknowledgments | | | 229 |
| References | | | 230 |

## 13 Antimicrobial peptide antibiotics against multidrug-resistant ESKAPE pathogens — 237
*Guangshun Wang, Atul Verma and Scott Reiling*
- 13.1 Introduction — 237
- 13.2 Antibiotic resistance of ESKAPE pathogens — 238
  - 13.2.1 Direct drug interaction — 238
  - 13.2.2 Indirect drug resistance — 239
- 13.3 Strategies to combat the ESKAPE pathogens — 240
  - 13.3.1 Vaccines — 240
  - 13.3.2 Phage therapy — 240
  - 13.3.3 Antibiotic derivatives — 241
  - 13.3.4 Antimicrobial peptides — 241
- 13.4 Advantages and disadvantages of cationic antimicrobial peptides — 241
- 13.5 Antimicrobial peptides to stop ESKAPE pathogens — 243
  - 13.5.1 Structure-based design — 243
  - 13.5.2 Library-based search and peptide mimetics — 244
  - 13.5.3 Peptide conjugates — 244
  - 13.5.4 Combined treatment — 245
  - 13.5.5 Formulated antimicrobial peptides — 246
  - 13.5.6 Surface immobilized antimicrobial peptides — 246
- 13.6 Mechanisms of bacterial killing by antimicrobial peptides — 247
  - 13.6.1 Bacterial membranes — 247
  - 13.6.2 Cell wall — 248
  - 13.6.3 Bacterial ribosomes — 248
- 13.7 Efficacies in animal models and clinical use of antimicrobial peptides — 249
- 13.8 Concluding remarks — 249
- Acknowledgment — 250
- References — 250

## 14 Antimicrobial peptide resistance and scope of computational biology in antimicrobial peptide research — 261
*C.K.V. Ramesan, N.V. Vinod and Sinosh Skariyachan*
- 14.1 Introduction — 261
- 14.2 Antimicrobial peptide resistance in gram-positive bacteria — 263
  - 14.2.1 Bacterial cell surface—cell wall and cell membrane — 263
  - 14.2.2 Extracellular mechanism of antimicrobial peptide resistance — 264
  - 14.2.3 Inhibition of antimicrobial peptide activity by surface-associated polysaccharides — 265
- 14.3 Mechanisms of antimicrobial peptides resistance in gram-negative bacteria — 265
  - 14.3.1 Modifications in the bacterial outer membrane — 266
  - 14.3.2 Biofilm formation — 268
  - 14.3.3 Efflux pumps — 270

|  |  | 14.3.4 | Binding and sequestering cationic antimicrobial peptides | 271 |
|---|---|---|---|---|
|  |  | 14.3.5 | Proteolytic degradation of antimicrobial peptides | 273 |
|  |  | 14.3.6 | Modulation of cationic antimicrobial peptide expression | 273 |
|  | 14.4 | Scope of computational biology in antimicrobial peptide research | | 275 |
|  |  | 14.4.1 | Antimicrobial peptide databases | 278 |
|  |  | 14.4.2 | Detection of antimicrobial peptides and their resistance patterns by machine learning approach | 281 |
|  |  | 14.4.3 | Recent perspectives on the scope of computational biology in antimicrobial peptide research | 282 |
|  | 14.5 | Conclusion | | 283 |
|  | References | | | 284 |
| 15 | **Recent advances and challenges in peptide drug development** | | | **297** |
|  | *N.K. Hemanth Kumar, K. Poornachandra Rao, Rakesh Somashekaraiah, Shobha Jagannath and M.Y. Sreenivasa* | | | |
|  | 15.1 | Introduction | | 297 |
|  | 15.2 | Historical overview of peptide drug development | | 299 |
|  | 15.3 | Basic drawbacks of peptide drugs | | 301 |
|  | 15.4 | Present approaches toward the discovery of protein–protein modulators | | 301 |
|  |  | 15.4.1 | High-throughput screening | 301 |
|  |  | 15.4.2 | Fragment-based drug discovery | 302 |
|  |  | 15.4.3 | Structure-based design | 302 |
|  | 15.5 | Peptides and protein–protein interactions | | 303 |
|  |  | 15.5.1 | Potential developments for intrusive peptides | 303 |
|  |  | 15.5.2 | Computational and experimental methods for determining protein–protein interactions | 303 |
|  |  | 15.5.3 | Computer-assisted docking strategies | 304 |
|  |  | 15.5.4 | Structural-based predictions | 304 |
|  | 15.6 | Innovations and computational methods for peptide–protein interactions | | 304 |
|  |  | 15.6.1 | Selection of preliminary peptide scaffolds | 304 |
|  |  | 15.6.2 | Molecular docking for peptide–protein interactions | 305 |
|  |  | 15.6.3 | Docking methods at local and global levels | 305 |
|  |  | 15.6.4 | Template-based docking method | 306 |
|  | 15.7 | Conclusion | | 306 |
|  | References | | | 307 |
| 16 | **Future perspective of peptide antibiotic market** | | | **311** |
|  | *B. Arun, E.P. Rejeesh and N. Megha Rani* | | | |
|  | 16.1 | Introduction | | 311 |
|  | 16.2 | Global antimicrobial peptides market overview | | 311 |

| | | | |
|---|---|---|---|
| 16.3 | Applications of antimicrobial peptide | | 312 |
| | 16.3.1 Prospects in medicine | | 312 |
| | 16.3.2 Food industry | | 313 |
| | 16.3.3 Animal husbandry and aquaculture | | 313 |
| 16.4 | Important parameters of market analysis | | 314 |
| 16.5 | Drivers and restraints of the peptide antibiotics market | | 314 |
| 16.6 | Conclusion | | 314 |
| References | | | 319 |

**Index** **321**

# List of contributors

**M.S. Aishwarya** Centre for Marine Science & Technology (CMST), Manonmaniam Sundaranar University, Rajakkamangalam, Tamil Nadu, India

**K. Ajesh** Department of Biotechnology and Microbiology, Kannur University, Kannur, Kerala, India

**B. Arun** Department of Biotechnology and Microbiology, Kannur University, Kannur, Kerala, India

**Anjali Jayasree Balakrishnan** Department of Biochemistry and Molecular Biology, Central University of Kerala, Periye, Kasaragod, Kerala, India

**Sonali Bhardwaj** School of Bioengineering and Biosciences, Lovely Professional University, Phagwara, Punjab, India

**Arunan Chandravarkar** Kerala State Council for Science, Technology and Environment, Thiruvananthapuram, Kerala, India

**Nitin Chaudhary** Department of Biosciences and Bioengineering, Indian Institute of Technology Guwahati, Guwahati, Assam, India

**Chirag Chopra** School of Bioengineering and Biosciences, Lovely Professional University, Phagwara, Punjab, India

**Bernadette Dora Gombossy de Melo Franco** Food Research Center – FoRC, Sao Paulo University, USP, Sao Paulo, SP, Brazil

**Daljeet Singh Dhanjal** School of Bioengineering and Biosciences, Lovely Professional University, Phagwara, Punjab, India

**Gustavo Graciano Fonseca** Faculty of Natural Resource Sciences, School of Business and Science, University of Akureyri, Akureyri, Iceland

**Feba Francis** Department of Biosciences and Bioengineering, Indian Institute of Technology Guwahati, Guwahati, Assam, India

**Jane Mary Lafayette Neves Gelinski** Laboratory of Protein Chemistry and Biochemistry, Department of Cell Biology, University of Brasilia-Federal District, Brasilia, Federal District, Brazil

**Benu George** School of Biotechnology, National Institute of Technology Calicut, Kattangal, Kerala, India

**N.K. Hemanth Kumar** Department of Studies in Botany, University of Mysore, Mysuru, Karnataka, India; Department of Botany, Yuvarajas College, University of Mysore, Mysuru, Karnataka, India

**Shobha Jagannath** Department of Studies in Botany, University of Mysore, Mysuru, Karnataka, India

**Prajit Janardhanan** Department of Biochemistry and Molecular Biology, Central University of Kerala, Periye, Kasaragod, Kerala, India

**A. Anju Krishnan** Chemical Biology Lab, Rajiv Gandhi Centre for Biotechnology, Thiruvananthapuram, Kerala, India

**Kamil Kuca** Department of Chemistry, Faculty of Science, University of Hradec Kralove, Hradec Kralove, Czech Republic

**K. Santhosh Kumar** Chemical Biology Lab, Rajiv Gandhi Centre for Biotechnology, Thiruvananthapuram, Kerala, India

**M. Divya Lakshmanan** Yenepoya Research Centre, Yenepoya (Deemed to be University), Mangalore, Karnataka, India

**A.P. Lipton** Centre for Marine Science & Technology (CMST), Manonmaniam Sundaranar University, Rajakkamangalam, Tamil Nadu, India

**N. Megha Rani** Department of Pharmacology, Yenepoya Medical College, Mangalore, Karnataka, India

**Swapna M. Nair** Yenepoya Research Centre, Yenepoya (Deemed to be University), Mangalore, Karnataka, India

**Eugenie Nepovimova** Department of Chemistry, Faculty of Science, University of Hradec Kralove, Hradec Kralove, Czech Republic

**Rajendra Pilankatta** Department of Biochemistry and Molecular Biology, Central University of Kerala, Periye, Kasaragod, Kerala, India

**S. Pooja** School of Biotechnology, National Institute of Technology Calicut, Kattangal, Kerala, India

**K. Poornachandra Rao** Department of Studies in Microbiology, University of Mysore, Mysuru, Karnataka, India; Department of Analytical Food Microbiology, VIMTA Labs. Ltd., Hyderabad, Telangana, India

**P. Prajosh** Department of Biotechnology and Microbiology, Kannur University, Kannur, Kerala, India

**R.S. Rachanamol** Department of Microbiology, A.J. College of Science & Technology, Thonnakkal, Kerala, India

**C.K.V. Ramesan** Department of Microbiology, Sree Narayana College, Kannur, Kerala, India

**Scott Reiling** Department of Pathology and Microbiology, College of Medicine, University of Nebraska Medical Center, Omaha, NE, United States

**E.P. Rejeesh** Department of Pharmacology, Mount Zion Medical College, Adoor, Kerala, India

**A.R. Sarika** Kerala State Council for Science, Technology and Environment, Thiruvananthapuram, Kerala, India

**Denoj Sebastian** Department of Life Sciences, University of Calicut, Thenhipalam, Kerala, India

**J. Selvin** Department of Microbiology, Pondicherry University, Kalapet, Puducherry, India

**H. Shabeer Ali** Department of Biotechnology and Microbiology, Kannur University, Kannur, Kerala, India; Division of Molecular Parasitology and Immunology, CSIR—Central Drug Research Institute, Lucknow, Uttar Pradesh, India

**Parvarish Sharma** School of Pharmaceutical Sciences, Lovely Professional University, Phagwara, Punjab, India

**S. Shishupala** Department of Microbiology, Davangere University, Davangere, Karnataka, India

**Reena Singh** School of Bioengineering and Biosciences, Lovely Professional University, Phagwara, Punjab, India

**Sinosh Skariyachan** Department of Microbiology, St. Pius X College, Kasaragod, Kerala, India

**Rakesh Somashekaraiah** Department of Studies in Microbiology, University of Mysore, Mysuru, Karnataka, India

**Aswathi Kodenchery Somasundaran** Department of Biochemistry and Molecular Biology, Central University of Kerala, Periye, Kasaragod, Kerala, India

**K. Sreejith** Department of Biotechnology and Microbiology, Kannur University, Kannur, Kerala, India

**M.Y. Sreenivasa** Department of Studies in Microbiology, University of Mysore, Mysuru, Karnataka, India

**T.V. Suchithra** School of Biotechnology, National Institute of Technology Calicut, Kattangal, Kerala, India

**B.R. Swathi Prabhu** Yenepoya Research Centre, Yenepoya (Deemed to be University), Mangalore, Karnataka, India

**Renu Tripathi** Division of Molecular Parasitology and Immunology, CSIR—Central Drug Research Institute, Lucknow, Uttar Pradesh, India

**Atul Verma** Department of Pathology and Microbiology, College of Medicine, University of Nebraska Medical Center, Omaha, NE, United States

**N.V. Vinod** Department of Microbiology, St. Pius X College, Kasaragod, Kerala, India

**Guangshun Wang** Department of Pathology and Microbiology, College of Medicine, University of Nebraska Medical Center, Omaha, NE, United States

# Preface

Among the great discoveries in the field of medicine, none is as prized as penicillin, which is considered the world's first antibiotic. Mankind is indebted to the great Sir Alexander Fleming, who has saved millions of human beings from untimely deaths since then by this discovery. The discovery of penicillin was not merely accidental. It was the product of turning Sir Fleming's intelligence into the consequences of an incident of negligence made during his experiments.

Although many great scientists and antibiotics—natural as well as synthetic forms—have been born in the world since penicillin, it remains a fact that microorganisms can survive them through various strategies and the predator–prey game that has been going on from the beginning of life continues to this day between antibiotics and microbes. Hence, in this era of exponentially increasing incidences of multiple drug-resistant infections globally, there is an urgent need to explore new areas in antimicrobial therapy, and peptide antibiotics are unquestionably seen as a potential solution.

It is at this point that peptides with antimicrobial properties are emerging as alternatives to conventional antibiotics. Antimicrobial peptides, as they are having broad-spectrum antimicrobial activity and have been shown to cause only low levels of drug resistance, making them highly promising clinical antibiotics of the future.

In this milieu, *Antimicrobial Peptides: Challenges and Future Perspectives* provides information pertinent to the antimicrobial peptides discovered through global research. The book presents several chapters starting from the historical developments to the global peptide antibiotic market. It also adds an array of techniques for peptide isolation and peptides from a wide range of organisms, including bacteria, fungi, insects, amphibians, mammals, and plants, and their analysis. Additional sections give a thorough overview of marine antimicrobial peptides, antiviral peptides, and those targeting multidrug-resistant bacteria. With the much recent information in computational biology and data sciences, the authors treat the chapters on antimicrobial peptide resistance and future perspectives of peptide drug development in an effective manner. This book will be a valuable resource for researchers, teachers, and students in microbiology, molecular biology, biochemistry, and other allied disciplines.

In this venture, we are deeply indebted to all the authors for doing an outstanding job in structuring their chapters more informative and understandable to the readers. We also want to thank our friends and colleagues from Kannur

University, Kerala who have shared knowledge and concepts across the subject with us.

Finally, we would like to express appreciation to everyone at the Academic Press.

**K. Ajesh and K. Sreejith**

# Historical developments of antimicrobial peptide research

Benu George[1], S. Pooja[1], T.V. Suchithra[1] and Denoj Sebastian[2]
[1]School of Biotechnology, National Institute of Technology Calicut, Kattangal, Kerala, India, [2]Department of Life Sciences, University of Calicut, Thenhipalam, Kerala, India

## 1.1 Introduction

Several pathogenic strains have been well identified over the years. Humankind has also witnessed a friendly as well as a foe nature of pathogens as they are an integral part of the ecological niche. Various therapies based on synthetic sources have been proved beneficial to control these pathogenic invasions. But, would the story end there? As aware of the adapting capability of microbes over the years. The emergence of "superbugs" that are multidrug resistant has raised concerns about the use of synthetic antibiotics in the long run [1]. Thus improved variants of fluoroquinolones and imipenem have been serving as broad-spectrum drugs, but indeed, bacteria can become resistant to the variants by the alteration of known resistance mechanisms [2]. In the absence of an immune response, human beings and other species manage to stay uninfected because of the effectiveness of nonspecific defenses. The research identifies that organisms use cationic peptides as a nonspecific defense against infections [3]. Thus an urge for new antibiotics or antibacterial resources has directed scientists to explore peptide defense mechanism. Discovered in 1939, antimicrobial peptides (AMPs) or host defense peptides exhibit broad-spectrum and dynamic antimicrobial efficacy against bacteria, fungi, and viruses [4]. These peptides are crucial components of the innate immune system that endures against forging attacks and forms the first-line defense [5]. The antimicrobial mechanism that AMPs offer is unlike traditional antibiotics that make it possible for being functional against microbes that are even drug resistant [6].

The broad range activity for AMPs is attributed to their reduced toxicity and diminished target cell resistance development [7]. The AMPs occur in a varied range of $\alpha$-helices, $\beta$-strands, loops, and extended structures. Such diversified secondary structures have been highly beneficial for AMPs' broad-spectrum antimicrobial activity [8]. Researchers have observed that rapid diffusion and extracellular secretion of AMPs facilitate instant defense response against pathogenic microbes [9]. Alternatively, the differences in the lipid composition of cell membranes in prokaryotes and eukaryotes signify the targets for AMPs. Thus the specificity of AMPs incited against the target cell is extremely dependent on the interaction of peptides preferentially with the microbial cells, which enable AMPs to be target-specific without affecting the host cells [10]. The net charge and the hydrophobicity of

AMPs form a vital component in the cellular association of peptides for target-specific antimicrobial activity [11]. However, to focus our discussion, this chapter reviews the developmental path of AMPs, their ancient origins, and evolution along with their functional development and emergence of the various database that has supported the development of AMPs.

## 1.2 History and development of antimicrobial peptides

It is fascinating to witness the evolution where every life form is always prayed by microbial infection. The adaptive immune system manages well to protect against infection. Nevertheless, the question is, how do plants and insects endure infections? They lack an adaptive immune system, yet it is all accomplished. Historically, the early report on the production of eukaryotic AMPs was a discovery from plants [12]. In 1896 the eukaryotic AMPs were primarily the research focus and by the mid-20thcentury cecropins from moths and magainins from frogs were discovered. Research on AMPs has flourished over the year 2000, mostly associated with all eukaryotic organisms. By the 1920s, Alexander Fleming identified lysozyme, which was considered the first peptide with antimicrobial activity [13]. However, lysozyme enzymatically destructs the bacterial cell wall; thus the discovery of lysozyme classifies a diverse category of AMPs, due to the mechanism of action [14]. Later in 1928, Fleming, Howard Florey, and Ernst Chain discovered penicillin [15], with the advent of therapeutic use of penicillin and streptomycin in 1943, the "Golden Age of antibiotics," resulted in a relapse of curiosity in the therapeutic potential of natural host antibiotics [16].

Recognition of AMPs since 1939 emerged with gramicidins isolated from *Bacillus brevis* that were potent against a broad class of Gram-positive bacteria. As research flourished, gramicidins showed effectiveness to cure infected wounds on the skin of guinea pigs, demonstrating a clinical use [17], and thus first AMPs to be manufactured as antibiotics were established [18]. In the year 1942, the antimicrobial substance from the wheat endosperm (*Triticum aestivum*) was established as a peptide that inhibited the phytopathogens like *Pseudomonas solanacearum* and *Xanthomonas campestris* [19]. Far along, a peptide, purothionin (mid-1970s) [20], from the family of thionins was discovered [12]. However, with the dusk of age antibiotics due to the emergence of multidrug-resistant microbial pathogens, that is "superbugs" an awakening of host defense molecules was prompted [21]. At this point in history, the true origin of research into AMPs [22] emerged. And in the commencement of the 1950s and 1960s, it was learned that human neutrophils possess a cationic protein to kill bacteria via oxygen-independent mechanisms [23,24].

An increase in bacterial resistance to antibiotics led to the discovery of new therapeutic molecules that focus not only on drug prospection but also on its improved immune potential from the existing antibiotics. Chemical modifications of AMPs improve their stability, efficacy, as well as physical properties. Various analogous modification experiments resulted in designing peptides based on sequence and amphipathicity. In 1997 various natural peptides were investigated, and by observing the most predominant amino acids among 20 N-terminus residues of 80 different peptides, a model alpha-helical antibacterial peptide was synthesized [25].

Such chemically synthesized peptides are privileged and can undertake desired modification, specificity, stability, and toxicity. Thus the synthesized peptide in a hydrophobic environment adopts an alpha-helix, which many other alpha-helical peptides fail. Modification of AMPs offers a broad scope for developing cytotoxic peptide into a biocompatible peptide. Chemical modification of pardaxin achieved in 1999 enabled the peptide to retain its antimicrobial activity. Pardaxin displays lytic activity with both microbial and mammalian cells, but the addition of D-amino acid residues into the peptide alters the $\alpha$-helix structure of pardaxin into $\beta$-structure [26]. Thus the $\beta$-structured pardaxin fails to display hemolytic activity as well as retains the antimicrobial activity.

It has been recognized that the chemical synthesis of peptides is a preferably beneficial method for preparing AMPs. In 2001 an investigation observed that a medium-sized peptide synthesis can be achieved by a standard way of solid-phase peptide synthesis method. Large and complex AMPs synthesis requires a combination of solid-phase peptide synthesis and in-solution fragment condensation or ligation techniques [27]. A report on 2016 signifies the importance of cyclization-based modifications in linear peptides and their effectiveness in improving membrane permeating ability of AMPs. Several linear peptides were cyclized and their stability, hydrophobicity, amphipathicity, and charge index were significantly enhanced, which resulted in increased membrane permeation of bacterial cells [28]. The reduced conformational flexibility of cyclic peptides enhances membrane disruption, thereby making AMPs an effective alternative to antibiotics. A site-elective modification of dehydroalanine and dehydrobutyrine by employing photoredox catalysis to natural peptide was achieved in 2018. Such site-selective change is meant to be a promising strategy since potential targets are often products of sophisticated biological posttranslational machinery, making it difficult to modify by means of common bio-orthogonal chemistry bioengineering approaches [29]. A study in 2020 evaluated the need for chemical modifications for a peptide, KR-12 minimalized antimicrobial fragment of human host peptide LL-37. The original activity of LL-37 was related to wound healing and biofilm activity. Further investigations on specific active fragments of peptides revealed that KR-12 exhibit similar antimicrobial potential as that of LL-37. Key positions of the fragment were Gln5 and Asp9, when substituted with Lys or Ala increased the broad-spectrum activity up to eightfold. Reversing the sequence of KR-12 and introducing cyclic dimers were found to increase the antifungal activity 4- to 16-fold. The addition of N-terminal cysteine and N-benzymidazolinone groups facilitate peptide cyclization and increases the ligation nature, which in turn improves the hydrophobicity and strengthens the disulfide bonds, resulting in increased cytotoxic activity [30].

**Table 1.1** FDA-approved synthetic and modified peptides.

| Drug name | Year | Peptide type | Category |
|---|---|---|---|
| SCENESSE | 2019 | Synthetic | Receptor binding |
| GIAPREZA | 2017 | Synthetic | Receptor binding |
| TYMLOS | 2017 | Synthetic | Receptor binding |
| PARSABIV | 2017 | Synthetic | Receptor binding |
| OZEMPIC | 2017 | Chemically modified | Receptor binding |
| ADLYXIN | 2016 | Synthetic | Receptor binding |
| KYPROLIS | 2012 | Modified tetrapeptidyl | Inhibitor |
| INCIVEK | 2011 | Chemically modified | Inhibitor |
| FIRAZYR | 2011 | Synthetic | Receptor binding |
| EGRIFTA | 2010 | Synthetic | Receptor binding |
| VICTOZA | 2010 | Synthetic | Receptor binding |
| FIRMAGON | 2008 | Synthetic | Receptor binding |
| BYETTA | 2005 | Synthetic | Receptor binding |
| PRIALT | 2004 | Synthetic | Inhibitor |
| PLENAXIS | 2003 | Synthetic | Inhibitor |
| VELCADE | 2003 | Chemically modified | Inhibitor |
| ANGIOMAX | 2000 | Synthetic | Inhibitor |
| CETROTIDE | 2000 | Synthetic | Receptor binding |
| TRELSTAR | 2000 | Synthetic | Receptor binding |

Significantly chemical synthesized or modified AMPs are of great interest and a survey conducted in 2019 on small therapeutic peptides (less than 50 amino acids) approved by the Food and Drug Administration (FDA) in the previous two decades (Table 1.1) mostly consisted of receptor binding and inhibitor category [31]. Modifications can achieve precision in achieving the desired function for example, KYPROLIS approved in 2012 function as an inhibitor for multiple myeloma treatment. Development of such synthetic AMP or modified analogs helps in improving biological activities and metabolic stability of AMPs, which leads to arise "smart antibiotics" that aim to improve the therapies and industrial role of AMPs.

## 1.3 Antimicrobial peptides as host innate defense barricade

As we know the evolutionary adaptability to withstand a foreign invasion from external sources is established by all life forms [32]. In higher organisms, AMPs develop as an element of innate immunity to protect against infection occurring in the host, which in contrast, for bacteria AMPs are triggered to survive in the same ecological niche [33]. AMPs extraordinarily exhibit a broad range of antimicrobial activity against organisms such as Gram-positive, Gram-negative bacteria, fungi, viruses, and unicellular protozoa [34,35]. Recently, several AMPs research has promising display results of pathogen clearance by modulating host innate immune response [36].

AMPs are a natural product of ribosomal translation of mRNA or nonribosomal peptide synthesis. Bacteria mainly form the nonribosomal peptides but the ribosomal AMPs synthesis is encoded genetically, which is found in all species, including bacteria [37]. The nonribosomal AMPs are well known as antibiotics (e.g., polymyxins and gramicidin S) and the ribosomal AMPs have been commonly known for their part in innate immunity [32,37,38].

The mammalian AMPs are typically found within the skin and mucosal epithelial cell secretions and neutrophil granules. [38–40]. The AMPs are mostly encoded in clusters and co-expressed, which results in the accumulation of numerous AMPs at a single site [41]. Fascinatingly, most AMPs are inactive precursor products that require a proteolytic cleavage to become active [42], so their expression depends on regulation as well as the surplus presence of suitable proteases [41]. In multicellular organisms, AMPs are stockpiled in greater concentrations as granules, which are inactive and released at the site of inflammation. Surprisingly, the AMPs in higher mammals are also an expression that is associated with pathogen-associated molecular patterns or cytokines responses [34,41].

## 1.4 Peptide-based database: barn house for AMPs

Several databases make it possible to learn about naturally occurring AMPs; these databases cover more than 23,253 peptides [43]. The sequences for AMPs are easily retrieved from UniProt database (http://www.uniprot.org) that contains peptide sequences in a large number with broad origins and functions. However, due to the comprehensive classification of peptides available, the AMPs databases are designed and modified to gather, and filter the information available for example, BACTIBASE for bacteriocins peptide sources and PhytAMP for plant-based peptide sources [44]. The database mostly classifies AMPs into two main groups: general and specific databases. The general databases are a depository of all types of AMPs, irrespective of source and type. Specific databases are conserved to a specific number of peptides belonging to a particular source or family. The search option of the database utilizes sequence alignment tools to recognize likenesses between a given template and the deposited AMPs. However, the results are not promising due to low sequence similarity [45]. These databases also provide physicochemical profiles that help in evaluating the AMPs features.

Recently developed database, LAMP2 has an overall of unique 23,253 AMPs [43]. The unique designed AMPs database cross-link various databases and provide options to select individual databases as per the users' need on the subject of interest. As the database updates, it would include details such as structure, collection, composition, source, and function activities of AMPs [43]. Similarly, the THPdb (http://crdd.osdd.net/raghava/thpdb/) represents a comprehensive database which is based on the Food Drug Administration approved or investigational therapeutic

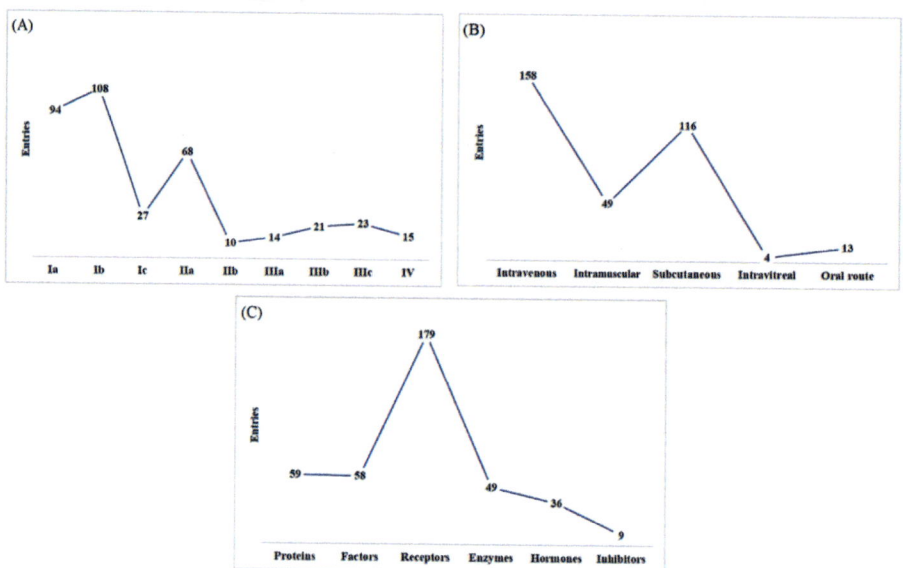

**Figure 1.1** Statistical distribution of AMPs in THPdb (A) based on function, (B) based on the route of administration, and (C) based on the target of AMPs. AMPs, Antimicrobial peptides.

peptides [46]. The THPdb entries of AMPs are categorized into various classes based on function (Table 1.1). The entries in THPdb information are compiled from research papers, patents, the pharmaceutical company, drug bank, USFDA, and others. The database statically (Fig. 1.1B) shows that most of the entries of AMPs are intravenous, subcutaneous, and intramuscular administration. Only 5% of developed AMPs classes are orally administered. It can be observed that the AMPs entry in THPdb is mostly associated with receptor targets (47.6%) followed by protein and factors as targets (Fig. 1.1C). The entries of AMPs base on function (Fig. 1.1A) is classified as, Group Ia—replacing a protein that is deficient or abnormal; Group Ib—augmenting an existing pathway; Group Ic—providing a novel function or activity; Group IIa—interfering with a molecule or organism; Group IIb—delivering other compounds or proteins; Group IIIa—protecting against a deleterious foreign agent; Group IIIb—treating an autoimmune disease; Group IIIc—treating cancer; and Group IV—protein diagnostics.

## 1.5 Current timeline of antimicrobial peptide approvals

The peptide discovery in recent years has been trending among researchers. The research evolves from academic groups to create new peptide-focused companies and businesses to solve much of pharma drug alternatives [47,48]. Intestinally, of 208 new drugs approved by the USFDA, 15 drugs were peptides, or peptide-

# Historical developments of antimicrobial peptide research

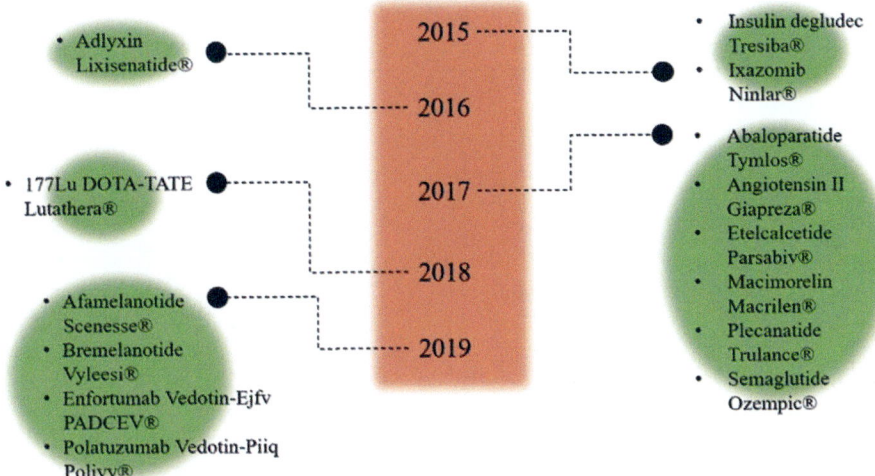

**Figure 1.2** Recent peptide-based drugs approved by the FDA (2015–19).

containing molecules (Fig. 1.2) and the rest belonged principle were 150 new chemical entities and 58 biologics in the last 5 years (2015–19) [49,50]. The unique place shared by peptide-approved drugs is with small molecules in completion. The peptide approvals by the FDA are an indication of pharmaceutical relevance as an active that display the diversity of the peptide realm.

## 1.6 Chemical developments in AMPs

The stability of AMPs can be improved with the help of certain chemical modifications. Posttranslational modifications of natural peptides have significantly improved the bioactivity of peptides to a greater extent. Some of the common chemical modifications include terminal capping, D-amino acid modifications, halogenation, hydroxylation, oxidation, phosphorylation, glycosylation, sulfation, reduction, disulfide bridges, thioether bridges, and cyclization. AMPs can be conjugated (Table 1.2) with polymers such as polyethylene glycol, chitosan, hyaluronic acid, and hyperbranched glycerol (HPG) to develop drug conjugates with improved efficacy. A study conducted in 2017 showed the immunomodulatory effect of HPG-amine conjugates on red blood cells, which showed a tremendous therapeutic advantage over synthetic ones, owing to its lesser toxicity and antibiofilm activity [60]. Since then, HPG-based AMPs are screened for their activities. Despite having several advantages of conjugated/modified AMPs, there are certain demerits to it. Right from reduction of antimicrobial activity to loss of specificity of targets several issues have to be rectified to make the new-gen AMPs into the market. Considering the bioavailability and toxicity issues, certain posttranslational modifications for linear peptides have to be done before conjugating them with polymers [61].

Table 1.2 Chemical modifications of antimicrobial peptides (AMPs) and their significance.

| Conjugating agent | AMP | Source of AMP | Advantage | Reference |
|---|---|---|---|---|
| Polyethylene glycol | Aurein | • *Litoria aurea*<br>• *Litoria raniformis* | • Decreases cytotoxicity | [51] |
| | KYE 28 | • Long peptide from helix D of Heparin cofactor II | • Decrease in hemolysis<br>• Improved selectivity in blood-bacteria mixture | [52] |
| Hyperbranched polyglycerol | Aurein | • *Litoria aurea*<br>• *Litoria raniformis* | • Improved immunomodulatory activities<br>• Less toxicity | [51] |
| Hyaluronic acid | Nisin | • *Lactococcus lactis* | • Improves the antimicrobial activity<br>• Useful for several medical applications such as wound dressing, contact lenses, and as a cosmetic preservative | [53] |
| Liposomes by combining cholesterol, soybean lecithin, and dioleoyl-sn-glycero-3-phosphoethanolamine | pH-responsive peptide, [D-]H$_6$L$_9$ | • Cell-penetrating peptide | • pH reduction of [D-Lip] from 7.4 to 6.3 increased the cellular uptake and tumor uptake efficiency | [54] |
| Poly lactic-co-glycolic acid electrospun fibers loaded with epithelial growth factors | Magainin 2 | • *Xenopus laevis* | • Lowers bacterial adhesion, reduces toxicity, and improves wound healing properties | [55] |
| Ag nanoparticles | Cathelicidin LL37 | • Human skin | • LL37-AgNp composite exhibited antibiofilm activity<br>• Improves cell proliferating activity<br>• Ability to prevent infection in burn care. | [56] |
| Au nanoparticles | Indolicidin | • Neutrophil blood cells of cow | • Reduced the iNOS mRNA expression levels of candidiasis patients<br>• Reduced cytotoxic activity | [57] |
| Carboxylated graphene quantum dots magnetic nanoparticles | Cecropin P1<br>Ceraginins | • Pig intestine<br>• Synthetic mimic of endogenous antimicrobial peptide | • Biosensing applications<br>• Bactericidal activity and ability to prevent *Pseudomonas aeruginosa* biofilm formation | [58]<br>[59] |

## 1.7 Antimicrobial peptides modification for medical application

The AMPs are potent against several multidrug-resistant bacteria and the mechanism of AMPs depends on their size, membrane permeability, and low anionic charge. There are four families of AMPs: cathelicidins, defensins, cercopins, and magainins. Peptides derived from eukaryotes are found to have more effective antimicrobial properties than those of commercial antibiotics [62]. Some of the AMPs from eukaryotes such as cathelicidins, bacitracin, telaprevir, and oritavancin are used for therapeutic applications.

Cathelicidins exhibit direct antimicrobial and excellent wound-healing activities by inducing chemotaxis and phagocytosis activities. Human cathelicidin (LL37), chicken cathelicidin (CATH), and porcine cathelicidin (protegrin, prophenin, and porcine myeloid antibacterial peptide) have a C-terminal domain rich in proline/arginine residues, which penetrates bacterial membranes and activates macrophages [63]. In addition to its antimicrobial and antibiofilm properties, they exhibit tissue repair and wound-healing properties. The origin of secretion and C-terminal domain of cathelicidins play a major role in determining its activity. For example, human cathelicidin (LL 37) has been linked with lung cancer due to its formyl-peptide receptors, which stimulates chemotaxis and suppresses tumorigenesis and plays a prominent role in metastasis of several types of cancer. On contrary, mouse cathelicidin-based AMPs are essential in downregulating azoxymethane-induced colon carcinogenesis [64]. Such drastic deviation in properties usually occurs in synthetic peptides than in natural ones. In some cases, immunotolerance of organism might chemically modify the amino acid terminal regions of peptides, resulting in downregulation of peptides. Halogenation is a common chemical modification occurring in cathelicidin from hagfish and centrocins, which exhibit lesser antimicrobial activity than their natural form [65].

Defensin peptides are another well-studied class of peptides, which is found excessively in plants, insects, and humans. The origin of human defensin occurs from neutrophils, as a part of innate immune response. First isolated human defensins (HD 5 and HD 6) were secreted from intestinal crypts and are found to exhibit antimicrobial and anti-tumor properties. There are two types of defensins existing in all organisms: α- and β-defensins, with different terminal amino acid residues contributing to their clinical applications. The change in alanine to aspartate in N-terminal domain of HDs results in reduced antimicrobial activity. Such single nucleotide changes reduce the net charge of peptide, limiting the antimicrobial activity. But, such modified natural defensins were found to be efficient in balancing the composition of intestinal microbiota. β-defensins, in addition to their bactericidal properties, were reported to be a protective agent against respiratory disorders, diarrhea, and sexual diseases. Human β-defensins (hBD), especially hBD-2,5, and 6, use lactoferrin and NaCl as co-factors to block viral replication. hBD-19,23 and 27 are modified forms of HD 6 that help in protecting male reproductive tract against colon infections. θ-defensin, identified in *Macaca mulatta* leukocytes exhibit stronger antiviral action against HIV and influenza virus. Similarly, rhesus θ-defensins exhibit bactericidal action against methicillin-resistant *Staphylococcus aureus* ((((MRSA) and *Pseudomonas aeruginosa*, which are resistant to several commercial antibiotics [66].

Many skin infections such as dermatitis, psoriasis, acne vulgaris, folliculitis, and rosacea can be treated/regulated with the help of AMPs. The wound healing and tissue repair ability of AMPs helps in maintaining the homeostasis of epithelial layer and increases the levels of cytokinins and interferons in the body, thereby stimulating innate immunity. Natural or synthetic modifications in AMPs resulted in more specific and targeted modes of action, thereby reduces the need for antibiotics. Some examples of therapeutic AMPs to treat skin infections are LL37, HD-6, Psoriasin, Pexiganan, Omiganan, Protegren-1, HBD-1,2, and P-novospirin G10 [66].

## 1.8 Antimicrobial peptides modification for industrial applications

Because of their antimicrobial nature, AMPs serve a direct linkage to crop protection. Cecropin and defensins isolated from several plant sources offer resistance to several crop infections. Alfalfa antifungal peptide isolated from the seeds of Alfalfa shows resistance toward *Vetricillium dahliae*, a harmful fungal pathogen of potatoes. Tachyplesin, an AMP isolated from crab, has been evaluated for its antifungal ability against *Sclerotinia sclerotiorum*. Inoculation of a specific AMP melittin to tobacco leaves helps in preventing the Tobacco mosaic virus entry into cells. The similarity of viral coat proteins with peptide sequence strengthens the immunity, thereby heightening the immune response. Expression of Attacin E in transgenic apple and pear resulted in bactericidal activity against fire blight disease. The introduction of magainin in grapevine resists the crown gall disease at a very early stage. AMPs also exhibit insecticidal activity and postharvest crop management. HP peptide derived from ribosomal protein of *Helicobacter pylori* has nematocidal activity against roundworm and insecticidal activity. To prevent postharvest crop decay, defensin genes in plants have to be activated, which gives long-term protection to crops. The use of AMPs such as tachyplesin 1, cecropin B, defensin A, and some antifungal peptides can be an environmental-friendly alternative to synthetic pesticides [67].

Despite having several advantages, peptides have certain undesirable characteristics which make it difficult to work on the targeted applications. Those reasons are low yield, toxicity index, and pharmacokinetic stability. Such issues can be dealt with the help of nanotechnology. Nanostructured peptides have lesser cytotoxicity, higher stability, and desired properties toward targeted applications. Some examples of nanostructured peptides are Cyclosporin A encapsulated with poly lactic acid-co-glycolide co-caprolactone, liposome-encapsulated nisin, vancomycin encapsulated with polycaprolactone or polylactic acid, and phospholipid encapsulated Polymyxin

B. All these nanostructured peptides have higher cellular uptake, lesser toxicity, higher stability, and greater bioavailability than their free form [68].

Because of its greater antimicrobial potential, AMPs possess higher scope in food industry as well. Milk-derived AMPs were more promising food preservatives than other peptides. Lactoferrin and Lactoferricin B control spoilage of mozzarella cheese by limiting the growth of mesophilic bacteria. Lactoferrin-based spray has been utilized in carcass processing to prevent spoilage due to bacterial contamination. Several reports of using milk-based peptides in fruits and vegetables help in preventing food spoilage [69].

Several other AMPs have been discovered from marine microbiota, which controls pathogens in aquaculture as well as helps in maintaining immunity. Recent advancements of AMPs in aquaculture are in grafting process for the pearl industry. Tachyplesin when combined with exopolysaccharides can be used as a filming agent to reduce oyster postoperative mortality, thereby increasing pearl quantity. In case of seafood sectors, shrimps get contaminated with *Vibrio penaeicida* which makes it unsuitable for consumption, as it can cause several infections while consumed. This issue can be sorted by using Type II crustin, making it naturally immune to *Vibrio* infections. Other AMPs identified from Stylicins, Crustins, and Penaeidins have antimicrobial and anticancer applications. Interestingly, these marine classes of AMPs have reported wound healing and antiprotease activities. A special class of AMPs called arminins defines the host−pathogen interaction in marine microbiota and maintains the homeostasis of beneficial microbiota [70].

## 1.9 An interdisciplinary upgrade to AMPs

Although AMPs have contributed much to drug discovery, the structure−activity relationship was not completely explained. The structural similarity of peptides within species might not work toward common targets due to its unpredictable mechanism of action. Experimental and computational studies have to be carried out simultaneously to ensure the immunomodulatory action of peptides. For solving such complex issues, machine learning strategies have been adapted to understand the relationship between biochemical attributes and their antibacterial activity. An adapted algorithm-based model of machine learning was developed to understand and classify shrimp AMPs reported in the literature. The MAMPs-Preds method was designed into a multilayer classifier model to identify and classify peptides accordingly. The first layer classifier uses the Random Forest algorithm to sort AMPs over other peptides. The second layer classifier involves Java integrated Mulan framework to identify the functional part of AMPs with a set of preloaded data. For this study, shrimp-based AMPs were preloaded and information regarding amino acid sequences and their physical properties were collected from SVM-prot 188 D web server. The processed sequences from the first layer will be subsequently fed to the second layer and a prediction of its activity was performed. An accuracy of 87.14% and 89.1% was achieved from the first and second layers, respectively [71].

Several new gene editing technologies have an immense potential in manipulating genes that code for AMPs paving new ways to synthesize functional AMPs. Gene editing tools like Zinc-Finger nucleases, Transcriptor activator like effector proteins, and CRISPR-Cas proteins helps in gene repair, insertion, and deletion of genes, thereby regulating peptide synthesis. Posttranslational modification of peptides can give rise to naturally modified AMPs with efficient immune memory. Such modified peptides are highly useful in the field of agriculture (pest-resistant crops, postharvest crop protection, etc.), food packaging, health care, and nanoapplications [72].

Antimicrobial nature of AMPs makes it a suitable drug candidate for several plant, animal, and human diseases. Reports also prove that the AMPs play a major role in neutrophile activation in all higher-level organisms. Activation of neutrophil by peptides, especially defensins, helps in improving the immune memory and also it is believed to play a major role in microbiota content of animal and human body. One such study was conducted in Juvenile goats with AMPs as feed additives to evaluate the growth, enzyme activity, and rumen microbial community of goats. Control and AMP-treated (recombinant swine defensin and fly antibacterial peptides at a ratio of 1:1) groups were tested for its average weight gain and enzymatic activity. The results showed that goats which consumed AMPs gained 2 kg weight more than that of control ones, with an average weight gain of 88.12 g/day. Enzyme activity of peptide treated goats exhibited higher lipase activity than untreated ones. Cellulase and protease activity was found to be equal, whereas decrease in amylase activity was observed in treated goats. *Fibrobacter* and *Anaerovibrio microbiota* were abundant in supplemented ones, indicating the increase in fermenting ability of goats. Higher abundancy of these microbes is necessary to degrade carbohydrates and is required to increase acetate, ammonia, and volatile fatty acid production which ultimately enhances the immunity and performance of goats [73].

## 1.10 Conclusion

The exploration of AMPs is much researched by academicians, and in recent years there has been an increase in the rate of identification and characterization. AMPs have managed to clear the FDA approval indicates its potential in clinical therapeutics. The availability of various databases has also provided much clarity in understanding the lipid−peptide interaction within the system. Such database also helps in developing techniques by learning about evolutionary and structural information of the AMPs. This chapter walkthrough historical developments of AMPs that facilitated in acceptance and prominence. It also remarks on the modifications adopted in the course of development, which highlights its versatile nature for future endeavors.

# References

[1] I. Levin-Reisman, I. Ronin, O. Gefen, I. Braniss, N. Shoresh, N.Q. Balaban, Antibiotic tolerance facilitates the evolution of resistance, Science 355 (2017) 826–830.

[2] R.J. Fair, Y. Tor, Antibiotics and bacterial resistance in the 21st century, Perspectives in Medicinal Chemistry 6 (2014) 25–64.

[3] F. Findlay, L. Proudfoot, C. Stevens, P.G. Barlow, Cationic host defense peptides; novel antimicrobial therapeutics against Category A pathogens and emerging infections, Pathogens and Global Health 110 (2016) 137–147.

[4] A.A. Bahar, D. Ren, Antimicrobial peptides, Pharmaceuticals 6 (2013) 1543–1575. Basel.

[5] C.J. Starr, J.L. Maderdrut, J. He, D.H. Coy, W.C. Wimley, Pituitary adenylate cyclase-activating polypeptide is a potent broad-spectrum antimicrobial peptide: structure-activity relationships, Peptides 104 (2018) 35–40.

[6] H.P. Jenssen, Hamill, R.E. Hancock, Peptide antimicrobial agents, Clinical Microbiology Reviews 19 (2006) 491–511.

[7] R.E. Hancock, A. Patrzykat, Clinical development of cationic antimicrobial peptides: from natural to novel antibiotics, Current Drug Targets. Infectious Disorders 2 (2002) 79–83.

[8] R.E. Hancock, Cationic peptides: effectors in innate immunity and novel antimicrobials, Lancet Infectious Disease 1 (2001) 156–164.

[9] M. Pushpanathan, P. Gunasekaran, J. Rajendhran, Antimicrobial peptides: versatile biological properties, International Journal of Peptides 2013 (2013) 675391.

[10] W. van 't Hof, E.C. Veerman, E.J. Helmerhorst, A.V. Amerongen, Antimicrobial peptides: properties and applicability, Biological Chemistry 382 (2001) 597–619.

[11] Z. Jiang, A.I. Vasil, J.D. Hale, R.E. Hancock, M.L. Vasil, R.S. Hodges, Effects of net charge and the number of positively charged residues on the biological activity of amphipathic alpha-helical cationic antimicrobial peptides, Biopolymers 90 (2008) 369–383.

[12] B. Stec, Plant thionins–the structural perspective, Cellular and Molecular Life Sciences 63 (2006) 1370–1385.

[13] A. Flemming, On a remarkable bacteriolytic element found in tissues and secretions, Proceedings of the Royal Society of London. Series B, Containing Papers of a Biological Character 93 (1922) 306–317.

[14] N. Benkerroum, Antimicrobial activity of lysozyme with special relevance to milk, African Journal of Biotechnology 7 (2008).

[15] A. Flemming, On the antibacterial action of cultures of a penicillium, with special reference to their use in the isolation of B. influenzae, British Journal of Experimental Pathology 10 (1929).

[16] L. Zaffiri, J. Gardner, L.H. Toledo-Pereyra, History of antibiotics. From salvarsan to cephalosporins, Journal of Investigative Surgery 25 (2012) 67–77.

[17] G.F. Gause, M.G. Brazhnikova, Gramicidin S and its use in the treatment of infected wounds, Nature 154 (1944). 703-703.

[18] H.L. Van Epps, Rene Dubos: unearthing antibiotics, Journal of Experimental Medicine 203 (2006) 259.

[19] R. Fernandez de Caleya, B. Gonzalez-Pascual, F. Garcia-Olmedo, P. Carbonero, Susceptibility of phytopathogenic bacteria to wheat purothionins in vitro, Applied Microbiology 23 (1972) 998–1000.

[20] S. Ohtani, T. Okada, H. Yoshizumi, H. Kagamiyama, Complete primary structures of two subunits of purothionin A, a lethal protein for brewer's yeast from wheat flour, Journal of Biochemistry 82 (1977) 753–767.
[21] M.L. Katz, L.V. Mueller, M. Polyakov, S.F. Weinstock, Where have all the antibiotic patents gone? Nature Biotechnology 24 (2006) 1529–1531.
[22] T. Nakatsuji, R.L. Gallo, Antimicrobial peptides: old molecules with new ideas, Journal of Investigative Dermatology 132 (2012) 887–895.
[23] H.I. Zeya, J.K. Spitznagel, Cationic proteins of polymorphonuclear leukocyte lysosomes. II. Composition, properties, and mechanism of antibacterial action, Journal of Bacteriology 91 (1966) 755–762.
[24] J.G. Hirsch, Phagocytin: a bactericidal substance from polymorphonuclear leucocytes, Journal of Experimental Medicine 103 (1956) 589–611.
[25] A. Tossi, C. Tarantino, D. Romeo, Design of synthetic antimicrobial peptides based on sequence analogy and amphipathicity, European Journal of Biochemistry 250 (1997) 549–558.
[26] Z. Oren, J. Hong, Y. Shai, A comparative study on the structure and function of a cytolytic alpha-helical peptide and its antimicrobial beta-sheet diastereomer, European Journal of Biochemistry 259 (1999) 360–369.
[27] M. Stawikowski, G.B. Fields, Introduction to peptide synthesis, Current Protocol in Protein Science Chapter 18 (2012). Unit 18 11.
[28] K. Andreev, M.W. Martynowycz, A. Ivankin, M.L. Huang, I. Kuzmenko, M. Meron, et al., Cyclization improves membrane permeation by antimicrobial peptoids, Langmuir 32 (2016) 12905–12913.
[29] A.D. de Bruijn, G. Roelfes, Chemical modification of dehydrated amino acids in natural antimicrobial peptides by photoredox catalysis, Chemistry 24 (2018) 11314–11318.
[30] S. Gunasekera, T. Muhammad, A.A. Stromstedt, K.J. Rosengren, U. Goransson, Backbone cyclization and dimerization of LL-37-derived peptides enhance antimicrobial activity and proteolytic stability, Frontiers in Microbiology 11 (2020) 168.
[31] C.H. Chen, T.K. Lu, Development and challenges of antimicrobial peptides for therapeutic applications, Antibiotics 9 (2020). Basel.
[32] R.E. Hancock, Cationic antimicrobial peptides: towards clinical applications, Expert Opinion in Investigative Drugs 9 (2000) 1723–1729.
[33] M. Hassan, M. Kjos, I.F. Nes, D.B. Diep, F. Lotfipour, Natural antimicrobial peptides from bacteria: characteristics and potential applications to fight against antibiotic resistance, Journal of Applied Microbiology 113 (2012) 723–736.
[34] R.E. Hancock, G. Diamond, The role of cationic antimicrobial peptides in innate host defences, Trends in Microbiology 8 (2000) 402–410.
[35] K.V. Reddy, R.D. Yedery, C. Aranha, Antimicrobial peptides: premises and promises, International Journal of Antimicrobial Agents 24 (2004) 536–547.
[36] A.T. Yeung, S.L. Gellatly, R.E. Hancock, Multifunctional cationic host defence peptides and their clinical applications, Cell and Molecular Life Sciences 68 (2011) 2161–2176.
[37] R.E. Hancock, Peptide antibiotics, Lancet 349 (1997) 418–422.
[38] R.E. Hancock, D.S. Chapple, Peptide antibiotics, Antimicrobial Agents and Chemotherapy 43 (1999) 1317–1323.
[39] H. Kunishima, Diagnosis, treatment and prevention of infectious diseases. Topics: I. Countermeasures against epidemic infectious diseases. 6. Multidrug resistant gram-negative bacterial infections, Nihon Naika Gakkai Zasshi 102 (2013) 2839–2845.
[40] H.G. Boman, Peptide antibiotics and their role in innate immunity, Annual Review in Immunology 13 (1995) 61–92.

[41] Y. Lai, R.L. Gallo, AMPed up immunity: how antimicrobial peptides have multiple roles in immune defense, Trends in Immunology 30 (2009) 131−141.
[42] R. Bals, Epithelial antimicrobial peptides in host defense against infection, Respiratory Research 1 (2000) 141−150.
[43] G. Ye, H. Wu, J. Huang, W. Wang, K. Ge, G. Li, et al., LAMP2: a major update of the database linking antimicrobial peptides, Database (Oxford) (2020).
[44] R. Hammami, J. Ben Hamida, G. Vergoten, I. Fliss, PhytAMP: a database dedicated to antimicrobial plant peptides, Nucleic Acids Research 37(Database issue) (2009) D963−D968.
[45] N.Y. Yount, A.S. Bayer, Y.Q. Xiong, M.R. Yeaman, Advances in antimicrobial peptide immunobiology, Biopolymers 84 (2006) 435−458.
[46] S.S. Usmani, G. Bedi, J.S. Samuel, S. Singh, S. Kalra, P. Kumar, et al., THPdb: Database of FDA-approved peptide and protein therapeutics, PLoS One 12 (2017) e0181748.
[47] F. Albericio, H.G. Kruger, Therapeutic peptides, Future Medicinal Chemistry 4 (2012) 1527−1531.
[48] A. Henninot, J.C. Collins, J.M. Nuss, The current state of peptide drug discovery: back to the future? Journal of Medicinal Chemistry 61 (2018) 1382−1414.
[49] B.G. de la Torre, F. Albericio, Peptide therapeutics 2.0, Molecules 25 (2020).
[50] B.G. de la Tore, F. Albericio, The pharmaceutical industry in 2018. An analysis of FDA drug approvals from the perspective of molecules, Molecules 24 (2019).
[51] P. Kumar, R.A. Shenoi, B.F. Lai, M. Nguyen, J.N. Kizhakkedathu, S.K. Straus, Conjugation of aurein 2.2 to HPG yields an antimicrobial with better properties, Biomacromolecules 16 (2015) 913−923.
[52] S. Singh, P. Papareddy, M. Morgelin, A. Schmidtchen, M. Malmsten, Effects of PEGylation on membrane and lipopolysaccharide interactions of host defense peptides, Biomacromolecules 15 (2014) 1337−1345.
[53] E. Maurício, C. Rosado, M. Duarte, J. Verissimo, S. Bom, L. Vasconcelos, Efficiency of Nisin as preservative in cosmetics and topical products, Cosmetics 4 (2017).
[54] Q. Zhang, J. Tang, R. Ran, Y. Liu, Z. Zhang, H. Gao, et al., Development of an antimicrobial peptide-mediated liposomal delivery system: a novel approach towards pH-responsive anti-microbial peptides, Drug Delivery 23 (2016) 1163−1170.
[55] P. Kumar, J.N. Kizhakkedathu, S.K. Straus, Antimicrobial peptides: diversity, mechanism of action and strategies to improve the activity and biocompatibility in vivo, Biomolecules 8 (2018).
[56] M. Vignoni, H. de Alwis Weerasekera, M.J. Simpson, J. Phopase, T.F. Mah, M. Griffith, et al., LL37 peptide@silver nanoparticles: combining the best of the two worlds for skin infection control, Nanoscale. 6 (2014) 5725−5728.
[57] H. Rahimi, S. Roudbarmohammadi, H.H. Delavari, M. Roudbary, Antifungal effects of indolicidin-conjugated gold nanoparticles against fluconazole-resistant strains of Candida albicans isolated from patients with burn infection, International Journal of Nanomedicine 14 (2019) 5323−5338.
[58] J.A. Bruce, J.C. Clapper, Conjugation of carboxylated graphene quantum dots with cecropin P1 for bacterial biosensing applications, ACS Omega 5 (2020) 26583−26591.
[59] K. Niemirowicz, U. Surel, A.Z. Wilczewska, J. Mystkowska, E. Piktel, X. Gu, et al., Bactericidal activity and biocompatibility of ceragenin-coated magnetic nanoparticles, Journal of Nanobiotechnology 13 (2015) 32.
[60] P. Kumar, A. Takayesu, U. Abbasi, M.T. Kalathottukaren, S. Abbina, J.N. Kizhakkedathu, et al., Antimicrobial Peptide-polymer conjugates with high activity: influence of polymer molecular weight and peptide sequence on antimicrobial activity, proteolysis, and biocompatibility, ACS Applied Materials and Interfaces 9 (2017) 37575−37586.

[61] A. Mahajan, A.S. Rawat, N. Bhatt, M.K. Chauhan, Structural modification of proteins and peptides, Indian Journal of Pharmaceutical Education and Research 48 (2014) 34−47.
[62] K. Midura-Nowaczek, A. Markowska, Antimicrobial peptides and their analogs: searching for new potential therapeutics, Perspectives in Medicinial Chemistry 6 (2014) 73−80.
[63] M.R. Scheenstra, M. van den Belt, J.L.M. Tjeerdsm-van Bokhoven, V.A.F. Schneider, S.R. Ordonez, A. van Dijk, et al., Cathelicidins PMAP-36, LL-37 and CATH-2 are similar peptides with different modes of action, Scientific Reports 9 (2019) 4780. a.
[64] K. Kuroda, K. Okumura, H. Isogai, E. Isogai, The human cathelicidin antimicrobial peptide LL-37 and mimics are potential anticancer drugs, Frontiers in Oncology 5 (2015) 144.
[65] G. Wang, Post-translational modifications of natural antimicrobial peptides and strategies for peptide engineering, Current Biotechnology 1 (2011) 72−79.
[66] M. Amerikova, I. Pencheva El-Tibi, V. Maslarska, S. Bozhanov, K. Tachkov, Antimicrobial activity, mechanism of action, and methods for stabilisation of defensins as new therapeutic agents, Biotechnology & Biotechnological Equipment 33 (2019) 671−682.
[67] K. Keymanesh, S. Soltani, S. Sardari, Application of antimicrobial peptides in agriculture and food industry, World Journal of Microbiology and Biotechnology 25 (2009) 933−944.
[68] L.S. Biswaro, M.G. da Costa Sousa, T.M.B. Rezende, S.C. Dias, O.L. Franco, Antimicrobial peptides and nanotechnology, recent advances and challenges, Frontiers in Microbiology 9 (2018) 855.
[69] J. Théolier, I. Fliss, J. Jean, R. Hammami, Antimicrobial peptides of dairy proteins: from fundamental to applications, Food Reviews International 30 (2014) 134−154.
[70] D. Destoumieux-Garzon, R.D. Rosa, P. Schmitt, C. Barreto, J. Vidal-Dupiol, G. Mitta, et al., Antimicrobial peptides in marine invertebrate health and disease, Philosophical Transaction of the Royal Society B: Biological Sciences 371 (2016).
[71] Y. Lin, Y. Cai, J. Liu, C. Lin, X. Liu, An advanced approach to identify antimicrobial peptides and their function types for penaeus through machine learning strategies, BMC Bioinformatics 20 (2019) 291.
[72] R. Sinha, P. Shukla, Antimicrobial peptides: recent insights on biotechnological interventions and future perspectives, Protein and Peptide Letters 26 (2019) 79−87.
[73] Q. Liu, S. Yao, Y. Chen, S. Gao, Y. Yang, J. Deng, et al., Use of antimicrobial peptides as a feed additive for juvenile goats, Scientific Reports 7 (2017) 12254.

# Biosynthesis of peptide antibiotics and innate immunity

**2**

K. Ajesh and K. Sreejith
Department of Biotechnology and Microbiology, Kannur University, Kannur, Kerala, India

## 2.1 Introduction

A century back noted zoologist, Elie Metchnikoff discovered the antimicrobial characteristics of epithelial surfaces, emphasizing the protective role of phagocytes in the presence of invading microbes. According to him, phagocytic cells must possess bactericidal substances that aid in the killing of invading bacteria [1]. Alexander Fleming discovered lysozyme in 1921, an antimicrobial peptide (AMP)-like protein (1,4-$N$-acetyl muramidase) while demonstrating his own nasal mucus to inhibit certain bacterial strains in culture. Later, the mechanism was identified as the cleaving of the peptidoglycan cell wall in the gram-positive bacteria and he reported its activity in tears, saliva, nasal secretions, sputum, and egg white [1,2]. The peptide, bombinin was discovered in the orange-speckled frog *Bombina variegata* in 1962, and some regard it to be the first description of an animal AMP [3]. In 1981 Boman and colleagues reported on the purification of cecropins, the first inducible AMPs and first major α-helical AMPs from the hemolymph of the giant moth *Hyalaphora cecropia* [4]. In the 1990s another landmark study occurred when Brodgen et al. purified the anionic AMPs from *Xenopus laevis* (African-clawed frog) and described numerous peptides of that kind in the ruminants such as cattle and sheep [5]. Over the last few decades, many numbers of AMPs expressed by the circulating phagocytic cells and epithelial cells in the respiratory, gastrointestinal, and urinary tracts have been identified and these molecules serve as a chemical barrier against invading microorganisms [6,7]. In primitive insect species, AMPs have been suggested to be a substitute for immune system processes, such as cytokine release, that characterize the bactericidal response of higher organisms [8]. In invertebrates, the hemolymph, as well as the hemocytes contains AMPs as the key components of innate immunity. Multicellular organisms contain a huge number of host defense peptides and proteins that play a defensive role in restricting diverse pathogenic microbes [9]. These peptide compounds were discovered to have a wide variety of actions such as bactericidal [10], fungicidal [11], viricidal [12], and immunomodulatory characteristics [13]. These effector molecules are polypeptides with less than 100 amino acid residues that vary greatly in sequence and secondary structure and are gene-encoded and synthesized by ribosomes. In humans and other mammalian species, defensin and cathelicidin are the two most common families of AMPs; however numerous peptides derived from larger proteins have also been identified. β-Defensins form a large family of AMPs that occurs widely in mammalian epithelial and phagocytic cells at higher levels. Cathelicidins, such as defensins are abundant and widely distributed and have a conserved N-terminal region and an altered region that encodes the mature C-terminal peptide [6,14,15].

Ribosomally synthesized AMPs are a part of eukaryotic and prokaryotic domains and represent a major component of defense against pathogens. Although they vary significantly in primary structure, they are more cationic and frequently amphiphilic [13,16]. In the past few decades, researchers have also established the biosynthesis of nonribosomal peptides (NRPs) that is mediated by large multienzyme machinery called the nonribosomal peptide synthetase (NRPS) as in bacteria, producing a wide range of structurally and biologically active peptides. The NRPS enzymes are positively and negatively regulated at the transcriptional and posttranscriptional levels and are encoded in operons [17].

## 2.2 Antimicrobial peptides in innate immunity

AMPs are an essential constituent of innate immunity in many species, including mammals, insects, and plants, and they serve in the front lines of the host's immunological defenses against infections (Table 2.1). Even though mammals produce a plethora of AMPs that play significant roles, the most thoroughly characterized families are the defensins and the cathlecidins [5,18]. During early life, when the adaptive immunity is underdeveloped, infants may gain protection from various mechanisms of innate immunity to counter the increased susceptibility to infections. Ingestion of maternal breast milk adds up to a remarkable repertoire of antimicrobial peptides/proteins for the infant's intestinal mucosal surface. Several antimicrobial peptides and proteins, including lactoferrin (LF), lysozyme, LL-37, α-defensins, and β-defensins, are found in breast milk. These peptides and proteins may help to protect newborns from inflammatory and infectious diseases [19,20]. Lactoferricin (25-amino acid peptide) is a breakdown product that is liberated from LF by pepsin digestion and has specific direct bacterial and antifungal effects [21]. Lactoferricin B is active against a physiologically diverse range of bacteria such as *Escherichia coli, Proteus vulgaris, Pseudomonas aeruginosa, Klebsiella pneumoniae, Salmonella enteritidis, Campylobacter jejuni, Yersinia enterocolitica, Streptococcus mutans, Staphylococcus aureus, Listeria monocytogenes, Corynebacterium diphtheriae,* and *Clostridium perfringens* [22] and is also effective against biofilms of *Aspergillus fumigatus, Fusarium solani,* and *Candida albicans* [23]. Ileal tissue from newborns with necrotizing enterocolitis had higher defensin levels than age-matched controls, showing that Paneth cells are stimulated to produce more defensins at some point during the pathogenesis of the diseases [24].

By constitutively and actively manufacturing diverse AMPs, our skin supports the first line of immunological protection against invading microbes from the external environment [40]. Many skin cell types, including epidermal keratinocytes, primary sebocytes, mastocytes, and eccrine glands release AMPs.Circulating cells that are engaged in the skin (neutrophils and natural killer cells), also contribute significantly to the total amount of AMPs present [7]. The most well characterized and produced by epithelial cells are β-defensin (HBD1), proinflammatory cytokine-inducible β-defensin (HBD2 and HBD 3), and cathelicidin [41]. Six human α-defensins have

Table 2.1 Overview of antimicrobial peptides in innate immune responses.

| Peptide | Source | Sequence | References |
|---|---|---|---|
| Human β-defensin 3 | Skin, tonsils, saliva, of *Homo sapiens* | GIINTLQKYYCRVRGGRCAVLSCLPKEEQIGKCSTRGRKCCRRKK | [25] |
| Human neutrophil peptide-1 | Neutrophils; NK cells, monocytes from *Homo sapiens* | ACYCRIPACIAGERRYGTCIYQGRLWAFCC | [26] |
| Bovine β-defensin 1 | *Bos Taurus* | DFASCHTNGGICLPNRCPGHMIQIGICFRPRVKCCRSW | [27] |
| Indolicidin | Bovine neutrophils, *Bos taurus* | ILPWKWPWWPWRR | [28] |
| Bactenecin | Bovine neutrophils, *Bos taurus* | RLCRIVVIRVCR | [29] |
| Lactoferricin B | 25-Residue peptide released by pepsin digest of lactoferrin | FKCRRWQWRMKKLGAPSITCVRRAF | [21] |
| Magainin 2 | African-clawed frog, *Xenopus laevis*, Africa | GIGKFLHSAKKFGKAFVGEIMNS | [11] |
| Brevinin-1E | Edible frog, *Rana boylii* | FLPLLAGLAANFLPKIFCKITRKC | [30] |
| Bombinin | Yellow-bellied toad, *Bombina variegate* | GIGALSAKGALKGLAEHFAN | [31] |
| Dermaseptin-S1 | Sauvage's leaf frog, *Phyllomedusa sauvagii* | ALWKTMLKKLGTMALHAGKAALGAAADTISQGTQ | [32] |
| Ranatuerin-2Ca | North American green frog *Rana clamitans* | GLFLDTLKGAAKDVAGKLLEGLKCKIAGCKP | [33] |
| Polyphemusin I | *Limulus polyphemus* | RRWCFRVCYRGFCYRKCR | [34] |
| Drosocin | Fruit fly, *Drosophila melanogaster* | GKPRPYSPRPTSHPRPIRV | [8] |
| Melittin | Honeybee venom, *Apis mellifera* | GIGAVLKVLTTGLPALISWIKRKRQQ | [35] |
| Heliomicin | Tobacco budworm, *Heliothis virescens* | DKLIGSCVWGAVNYTSDCNGECKRRGYKGGHCGSFANVNCWCET | [36] |
| Abaecin | Honeybee, *Apis mellifera* | YVPLPNVPQPGRRPFPTFPGQGPFNPKIKWPQGY | [37] |
| Lebocin 1/2 | Silk moth, *Bombyx mori* | DLRFLYPRGKLPVPTPPFNPKPIYIDMGNRY | [38,39] |

been identified to date, and they are further grouped into two important classes based on their gene structures and pattern of expression: Human enteric defensins (HDs) 5 and 6 and myeloid defensins (HNPs) 1−4 [42,43]. Several forms of HBD1 sized from 36 to 47 amino acids occupy the blood plasma, where they are attached to carrier molecules, which release the peptide in acidic conditions, as observed on vaginal mucous membranes [44]. Garcia-Lopez et al. [45], in their study, discovered that, extraplacental membranes exhibit different reactions to the presence of E. coli, by releasing HBD2 and HBD3 primarily along the choriodecidual region. Like HBD-3, HBD-1 and HBD-2, is antibacterial (P. aeruginosa and E. coli), as well as fungicidal (C. albicans and Malassezia furfur) [46].

Although the defensin classes are widely distributed in nature, the α-defensins are only found in vertebrates. Mammalian α-defensin is primarily present in small intestinal Paneth cells and neutrophils, whereas mammalian β-defensin can be isolated from both leukocytes and epithelial cells [46,47]. Cathelicidin in the skin provides enhanced protection from bacterial and viral infections [48]. Antibacterial peptides produced by neutrophils are released from their cytoplasmic granules on demand. Cathepsin G and chymase enzymes, azurocidin, defensins, and cathelicidin are examples [49]. Neutrophils generate β-defensins and cathelicidin LL-37/hCAP-18 in the airways. These molecules are also secreted into airway surface fluid by the respiratory epithelial cells and alveolar macrophages.

A large number of peptides have been found in bovines that show a broad spectrum of activity against infectious pathogens. One such peptide is a 38-amino acid lengthier, tracheal antimicrobial peptide (TAP). This molecule has shown its presence in the bovine tracheal mucosa and has bactericidal activity against those causing bovine respiratory diseases. The mRNA encoding this peptide has been found in abundance in the respiratory mucosa. The purified peptide was strongly inhibitory against various bacteria including E. coli, S. aureus, P. aeruginosa, and K. pneumoniae [50]. Upregulation of the TAP gene is noticed in tracheal epithelial cells in response to the stimulation by bacterial products such as lipopolysaccharide, Pam3CSK4, and interleukin (IL)-17A [51].

Pathogens are restricted in cold-blooded vertebrates with the aid of cationic AMPs that are synthesized from specialized structures observed in epithelial layers, as demonstrated in certain amphibians [52]. The skin of amphibians has been shown to be a major resource for novel APPs [53]. This peptide family comprises magainin I/ II, caerulein-precursor fragment (CPF) peptides, and at last, the PGLa/ peptide glycine-leucine-amide [54]. Magainins are significant polycationic peptides generated by mucous glands from frogs [55]. Maginin I and II are amphipathic peptides purified from dermal secretions of Xenopus laevis (African-clawed frog) and exhibit strong antimicrobial activity toward many bacteria, and fungi and induce osmotic destruction of protozoa [11,56]. Brevinins are natural AMPs that are seen in various frog species [30]. Over 350 types of Brevenins have been found in the database of anuran defense peptides (DADP), which are categorized into two groups: Brevinin-1 (around 24 residues) and Brevinin-2 (carry 33−34 residues) [57]. Dermaseptins are α-helical-shaped polycationic peptides (20−30 amino acids in length) [32] released by amphibian skin. Dermaseptins also take part a significant role in defensive mechanisms against human bacterial pathogens such as *S. aureus*, *P. aeruginosa*, and viruses including HIV, HPV, and HSV [58].

Polyphemusin I, an 18-amino acid long amphiphilic AMP from the horseshoe crab, *Limulus polyphemus* are strongly bactericidal against many pathogenic bacteria [34]. Polyphemusin I accumulate in the cytoplasm of the target organism without producing considerable membrane disruption, as observed by Powers et al. [59].

The insect species also have a robust line of defense that is solely dependent on a potent innate immune system, carrying the humoral and cellular factors that eliminate the invading pathogens instantly. A battery of AMPs is rapidly synthesized as a result of the humoral immune response [60]. These peptides are released from the fat body (vertebrates analogous to the liver and adipose tissue) and are generated during injury or in response to numerous inducers such as cell wall components, live or killed microbes, etc. [61,62]. Like vertebrates, epithelial cells of the insect gut, also secrete tissue-specific AMPs, in a response that is thought to be crucial in insect resistance against various intestinal parasitic species [61,63,64]. More than 100 AMPs from diverse insect species have now been identified. Defensins, cecropins, attacins, and proline-rich peptides are a common class of insect AMPs although gloverins and moricins are exclusively grouped under Lepidoptera [65,66]. These cationic peptides have extensive antibacterial, antifungal, antiparasitic, and antiviral action. Based on the structures and unique sequences, insect AMPs are grouped into four families: α-helical, cysteine-rich peptides, proline-rich peptides, and glycine-rich peptides/ proteins [66].

Cecropins, the first inducible antibacterial peptides identified, have major roles in insect innate immune system. Apart from membrane damaging property, it also inhibits proline absorption [67]. Cecropins are 31−37 amino acid long, linear 4-kDa peptides that are proven for their inhibitory activity against both groups of bacterial pathogens [68]. Insect defensins comprise 29−34 residues, and are active against both gram-negative and gram-positive groups of bacteria, with gram-positive bacteria being more susceptible [69]. Insect orders viz. Diptera, Hymenoptera, Coleoptera, Trichoptera, Hemiptera, and Odonata contribute to the Defensins group [63].

AMPs are a key component of the host's innate immune defense against infections in a wide group of species, and they serve as the foundation for discovering novel therapeutic medicines. In the literature, one may find numerous articles and reviews on the discovery, diversity, structural features, and activities (*in vitro* and *in vivo*) of these innate defense peptides. Though these compounds were first identified for their action against distinct kinds of bacteria, fungi, and viruses, their significance as multifunctional molecules are generally recognized. At the same time, their vital roles in the human immune system are yet to be described fully.

## 2.3 Biosynthesis of nonribosomal and ribosomal peptides

AMPs are classified into two types: nonribosomally generated peptides (NRPs) and ribosomally synthesized peptides (RPs). Both of these share a few common characteristics, that is, they are formed from relatively smaller precursor peptide sequences that are subsequently translationally modified [70]. When compared with the RPs, NRPs are "posttranslationally" altered to a great extent; in this instance, translation occurs on the enzyme complex instead of ribosomal structures. Also, NRPs often encode nonproteinogenic amino acids which are basically not found in RPs. Furthermore, while biosynthesis, polyketide and fatty acid motifs are considerably easier to integrate into NRP peptide chains [71].

Ribosomal synthesis of proteins and peptides is a fundamental process. In mammals, several gene families code for the AMPs that are expressed on epithelial surfaces as well as in neutrophils, which play a significant role in the frontline defense against pathogenic invasion [48]. NRPs are produced and assembled by large NRPSs complexes rather than ribosomal machinery. The NRPS enzymes have been found to exist in all three domains of life, with bacteria being the most predominant. Bacitracin, vancomycin, gramicidin, and polymyxin B are a few examples of nonribosomally produced AMPs [72].

## 2.3.1 Nonribosomal peptides

Peptide biosynthetic pathways received a great deal of attention in recent decades because of the broad-spectrum activity of these molecules [72]. The mode of synthesis of NRPs is highly conserved and is characterized as linear (type A), iterative (type B), nonlinear (type C), and stand-alone NRPSs [73]. The module numbers in linearly operating NRPS (type A) match the number and amino acid sequences in their product. Important examples of linear NRPSs comprise sivadicin from *Paenibacillus* to decameric tyrocidine from genus *Bacillus*. In the type B NRPSs, dedicated modules are reused numerous times by some NRPS machinery over the course of a biosynthetic cycle. In many circumstances, this results in the product's molecular symmetry as seen in gramicidin S. Other peptides of this class are echinomycin, valinomycin, and beauvericin. Finally, in the nonlinear NRPs, peptides are generated for which the amino acid doesn't correlate to the module arrangement on the synthetases template [74]. The aforementioned mechanism applies to the synthesis of capreomycin and viomycin. Although NRPSs follow a modular logic, exceptional cases are dissociated NRPSs in which the complete modules or even single domain function as stand-alone enzymes, which are very much seen in bacteria [73,75].

The NRPS is a series of enzyme modules (Fig. 2.1) that work together to extend the peptide through the coordinated action of several domains named adenylation (A), peptidyl carrier proteins (PCPs) and condensation (C). NRPS uses a PCP or PCP domain, of around 70−90 amino acids lengthier for transporting substrates and peptide intermediates among various catalytic domains. The PCP domains contain a serine residue which is conserved and denotes the position for covalent modification with a prosthetic moiety 4′-phosphopantetheinyl cofactor (Ppant) that is derived from coenzyme A. The first stage in the peptide assembly process is analogous to protein synthesis mechanisms in that amino acids are activated by transesterification with ATP producing appropriate aminoacyl-adenylate. Within its

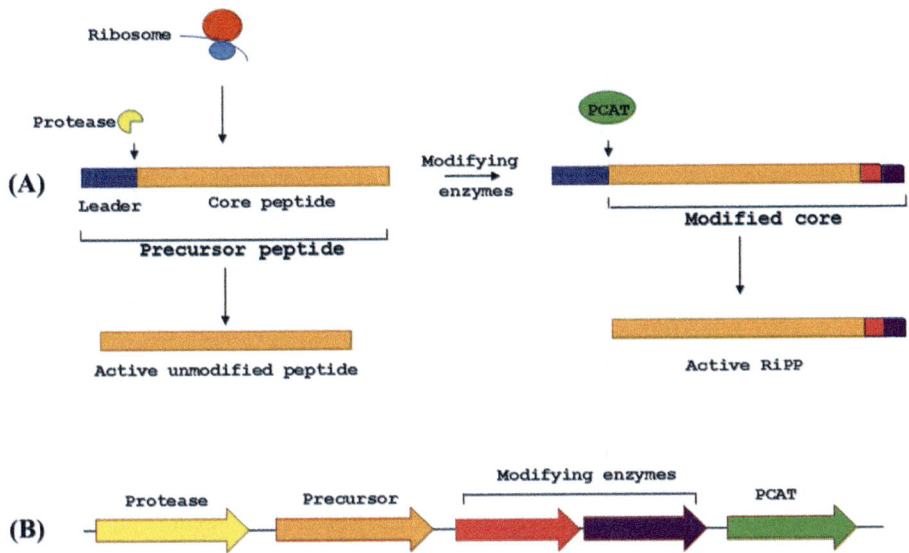

**Figure 2.1** (A) General biosynthetic pathway for ribosomally Synthesized and post-translationally modified peptides (RiPPs) (B) General Gene clusterribosomally.

module, domain A additionally catalyzes aminoacyl-adenylate movement onto the free thiol group of PCP-Ppant, which covalently links the enzyme and substrate. The substrate can then be modified through neighboring editing or "E" domains, by a mechanism such as epimerization or N-methylation. The product is further assembled in the elongation process through a sequential peptide bond formation between the downstream building block with its free amine and the upstream substrate's carboxy-thioester. Condensation or "C" domain, which is a large 450-amino acid enzyme, activates the peptide bond formation of two aminoacyl substrates through nucleophilic attack of the downstream PCP-bound acceptor amino acid with its free amino group on the upstream donor's activated thioester group. Through this mechanism, NRPSs utilize more reactive PCP thioesters than ribosomal protein synthesis. Ultimately, the full-length peptide chain is carved out by the C-terminal thioesterase (TE) domain activity [73,76,77]. Domains such as the cyclisation (Cy), methyltransferase (MT), epimerase (E), formylation (F), and oxidation (Ox) bring out chemical alterations to the structure. The formation of D-amino acids through epimerization is a striking characteristic of NRPs. Bioactivity and the structural conformation of the peptide are greatly influenced by the C and N methylation of amino acids produced in the methyl transferase domain. The heterocyclic rings such as oxazolines and thiazolines are integrated into the peptide by cyclization domain (Cy) and modified by the condensation domain (C). The formylation (F) domain catalyzes N-formylation as demonstrated in gramicidins from *Bacillus brevis*. Following this step, the intermediate can be released from PCP of the termination module through hydrolysis as a linear acid via macrocyclization.

Macrocyclization can be considered a remarkable feature for many peptides' natural products. Macrocyclization can occur via head-to-tail, backbone—backbone, or side chain—side chain linkages. Once released, NRP products undergo additional interactions with external enzymes viz halogenase, oxygenases, and glycotransferases. Halogenation of amino acids is often necessary to achieve complete bioactivity. Flavin-dependent and nonheme iron-dependent halogenases are the two types of halogenating enzymes responsible for halogen incorporation into NRPs. Flavin-dependent halogenases modify NRPs inside their aromatic or heteroaromatic ring structures (e.g., balhimycin) whereas nonheme iron-dependent halogenases, halogenate NRPs at unactivated aliphatic carbon centers (e.g., syringomycin E). The glycopeptide antibiotics (GPAs) such as vancomycin and teicoplanin are significantly altered to attain their final, active shape by undergoing cytochrome P450-catalyzed oxidative crosslinking of aromatic side chains. Hydroxylations modify a significant number of amino acids. Ramnoplanin, echinocandins (antifungal) etc. are the hydroxylated peptide antibiotics known [73,78−84]. The current known NRP structure denotes its complexity. The largest groups are head-to-tail cyclized peptides (e.g., cyclosporine and gramicidin S) and lipopeptides (e.g., fengycin and kannurin) [85,86]. The linear structures are also found in large numbers and range from sevadicin to peptaibols [87,88]. NRPSs are also the main mode of fungal peptide natural product assembly that produces some of the most well-known and important fungal-derived drugs such as cyclosporine and the echinocandins [89].

## 2.3.2 Ribosomally synthesized and posttranslationally modified peptides

Gene-encoded ribosomally produced antimicrobial peptides (RPs) are distributed among various cellular life forms—microorganisms, animals, and plants. Peptides are generally smaller in size, with 20−60 amino acid residues that are amphiphilic, and/or hydrophobic and cationic [90].

Nearly, all the RPs share some "universal characteristics." First of all, they are generated as a larger precursor peptide of ∼20−110 amino acid residues in length. It is encoded in a structural gene and is subsequently translated and modified through different enzymes that activate the production of an extensive range of chemical motifs [71,84]. Second, at least as seen in bacteria, the genes coding for precursor peptides are found in close proximity with those encoding modifying enzymes, secretion, and resistance. Third, several precursor peptides keep guiding sequences that direct the recruitment of modifying enzymes and identify proteolytic sites. All known RPs are formed from gene-encoded precursors (prepropeptides), from which mature peptides are derived by sequentially removing the signal peptide [91].

Ribosomally synthesized and posttranslationally modified peptides (RiPPs) are an added advantage along with the antibiotics synthesized through polyketide or nonribosomal pathways to combat the problems of multidrug resistance in bacterial and fungal pathogens. Despite the vivid structural classes covered by RiPPs, they all adhere to the same basic biosynthetic principles. First, a precursor peptide encoded by a single gene with an N-terminal leader peptide sequence and a C-terminal core sequence is translated. This step is followed by the deletion of the leader sequence through a succession of transporters, peptidases, or both (Fig. 2.2). The resulting active moiety is then treated further by other enzymes encoded by genes adjacent to the precursor gene [92,93].

Main chain modifications occur via processes such as proteolysis and macrocyclization. To achieve their active form, most RPs are excised from precursor peptides that have been modified or unmodified. In most circumstances, the pro peptide's N-terminus is sliced from the leader sequence, while the C-terminus is seldom altered. Peptide cyclization is seen in both eukaryotic and prokaryotic microorganisms. In eukaryotes, these peptides are normally N-to-C terminally cyclic; however, in bacterial RPs, amide linkages or cyclization with nucleophilic side chain residues can occur [91]. Many different RPs contain heterocyclized Cys, Ser, and Thr residues (e.g., microcin B17). Heterocyclization is anticipated to provide several advantages to RPs; mainly the presence of heterocycles conformationally restrains the peptide backbone structure. MicrocinB17, which comprises four oxazoles and four thiazoles, is an example of a heterocycle-containing RP [71]. Lanthionine-containing peptides are ribosomally synthesized and posttranslationally modified with thioether cross-links generated by adding a cysteine to a dehydroalanine (lanthionine) or a dehydrobutyrine (3-methyllanthionine) [94]. Prenylation, that is, the incorporation of isoprene-derived subunits into a core molecule is also reported from the RPs. Examples include ComX signaling peptides purified from *B. subtilis* and cyanobactin peptides isolated from *Prochloron* spp. [95,96]. Several bacterial RPs including microcin and lacticin, as well as the ribosomal synthesis of proteins in mitochondria and chloroplasts, have fMet at their N-terminus and are a part of the regular ribosomal biosynthetic machinery. With the help of a formyl group donor (10-formyltetrahydrofolate), the enzyme formyl transferase formylates the amino group of the methionine present in the initiator tRNAiMet and Formyl-Met (fMet) thus formed become the foremost residue of the nascent peptide that materializes by a bacterial protein synthesis machinery. Peptide deformylase removes the formyl moiety of N-terminal fMet during cotranslation [97]. Disulfide bonds are relatively frequent in proteins and smaller peptides. In RPs, they keep smaller peptide sequences inactive or potent forms. In addition, other modifications in RPs include glycosylation, lanthionine synthesis, and lactone formation [71].

AMPs are encoded by single genes that form extremely homologous gene families that occur in clusters. As we discussed in the earlier part of this chapter the most widely distributed AMPs in humans and other mammals are defensins and cathelicidins. In humans, α- and β-defensins colocalize at the defensin gene on chromosome 8p21−23. Within each group, these genes maintain a high level of sequence identity, although they differ greatly from one another. Decoding the defensins mRNA envisages that these peptides are created as a precursor molecule, which includes a signal sequence followed by a pro-peptide region, which is then further processed to produce the mature peptide [98−100]. Yet, only one cathelicidin—LL-37/hCAP18 an amphiphilic helical AMP has been discovered in humans, that is released by both epithelial and immune cells. Cathelicidins are encoded by four exon genes, with the active mature peptide ranging in size from 12 to 79 amino acids situated in the fourth exon 15. Like defensin-encoding genes, cathelicidin-encoding genes, also have a highly similar 59 region that encodes the conserved cathelin domain 16 [100–102]. In insects, AMPs are transcriptionally regulated by the Toll and immune deficiency (IMD) signaling pathways, along with the Janus Kinase/signal transducer and activator of transcription (JAK/STAT) and c-Jun N-terminal kinase (JNK) pathways, which are prominent and conserved over time [103].

The genes encoding the biosynthetic pathways are often clustered together on the chromosomes in the biosynthetic gene clusters (BGCs). Genome mining has been a crucial tool for identifying novel molecules in recent years, leading to the new discovery of these secondary metabolites. NP searcher, lustScan, and SBSPKS are computer tools that address NRP and polyketide biosynthesis pathways, whereas BAGEL3 concentrates on RiPPs. SMURF focuses on fungal secondary metabolite producers [104].

## 2.4 Summary and conclusion

AMPs benefit invertebrate and vertebrate host defense by destroying invading microorganisms ranging from bacteria to certain viruses and play a vital role in innate immunity. Recent studies also substantiate their immunomodulatory role by activating the immune system of the host. Therefore, it has the potential to overcome the problems of drug resistance if the shortcomings such as poor stability and cytotoxicity are addressed. At the same time, NRPs are widely used in pharmaceutics and recent studies have focused on their biosynthetic, physiological, and ecological features. The genes coding for the biosynthetic machinery are organized in clusters and the peptide synthesis follows a general scheme, that is, precursor peptide modification and proteolytic activation. Despite their vast differences in the biosynthetic pathways, NRPs and RPs share several posttranslational modifications

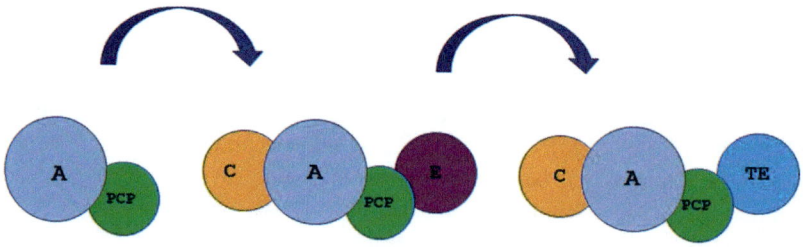

A: Adenylation domain; PCP: Peptidyl carrier protein; C: Condensation domain; E: Epimerisation domain; TE: Thioesterase domain

Figure 2.2 Schematic representation of NRPS modules.

in common. One of the most obvious similarities between these two groups is macrocyclization and differences such as the presence of D-amino acids in NRPs. With the advent of bioinformatics, many secondary metabolite BGCs have been characterized. In the era of drug resistance, the study of these peptides and its biosynthesis enzymes will pave the way for the construction of novel effective peptide-based drugs.

# References

[1] G. Tomas, Defensins and other antimicrobial peptides: a historical perspective and an update, Combinatorial Chemistry & High Throughput Screening 8 (2005) 209−217.
[2] R.I. Lehrer, Overview: antimicrobial peptides, as seen from a rear view mirror, Mammalian Host Defense Peptides (2004) 5−8.
[3] M. Simmaco, G. Kreil, D. Barra, Bombinins, antimicrobial peptides from Bombina species, Biochimica et Biophysica Acta (BBA) - Biomembranes 2009 (1788) 1551−1555.
[4] H. Steiner, D. Hultmark, A. Engstrom, H. Bennich, H.G. Boman, Sequence and specificity of two antibacterial proteins involved in insect immunity, Nature 292 (1981) 246−248.
[5] K.A. Brogden, M. Ackermann, P.B. Mc, B.F. Cray Jr, Tack, antimicrobial peptides in animals and their role in host defences, International Journal of Antimicrobial Agents 22 (2003) 465−478.
[6] D. Xu, W.Lu Xu, Defensins: a double-edged sword in host immunity, Frontiers in Immunology 11 (2020) 764.
[7] J. Schauber, R.L. Gallo, Antimicrobial peptides and the skin immune defense system, Journal of Allergy and Clinical Immunology 122 (2008) 261−266.
[8] M. Gobbo, L. Biondi, F. Filira, R. Gennaro, M. Benincasa, B. Scolaro, et al., Antimicrobial peptides: synthesis and antibacterial activity of linear and cyclic drosocin and apidaecin 1b analogues, Journal of Medicinal Chemistry 45 (2002) 4494−4504.
[9] E.E. Avila, Functions of antimicrobial peptides in vertebrates, Current Protein and Peptide Science 18 (2017) 1098−1119.
[10] K. Marr, W.J. Gooderham, R.E.W. Hancock, Antibacterial peptides for therapeutic use: obstacles and realistic outlook, Current Opinion in Pharmacology 6 (2006) 468−472.
[11] K. Ajesh, K. Sreejith, Peptide antibiotics: an alternative and effective antimicrobial strategy to circumvent fungal infections, Peptides 30 (2009) 999−1006.
[12] I.-N. Hsieh, K.L. Hartshorn, The role of antimicrobial peptides in influenza virus infection and their potential as antiviral and immunomodulatory therapy, Pharmaceuticals 9 (2016) 53.
[13] E.F. Haney, R.E.W. Hancock, Peptide design for antimicrobial and immunomodulatory applications, Peptide Science 100 (2013) 572−583.
[14] D.-Y. Lee, K. Yamasaki, J. Rudsil, C.C. Zouboulis, G.T. Park, J.-M. Yang, et al., Sebocytes express functional cathelicidin antimicrobial peptides and can act to kill *Propionibacterium acnes*, The Journal of Investigative Dermatology 128 (2008) 1863.
[15] M. Masamoto, T. Ohtake, R.A. Dorschner, R.L. Gallo, B. Schittek, C. Garbe, Cathelicidin anti-microbial peptide expression in sweat, an innate defense system for the skin, Journal of Investigative Dermatology 119 (2002) 1090−1095.
[16] N. Singh, J. Abraham, Ribosomally synthesized peptides from natural sources, The, Journal of Antibiotics 67 (2014) 277−289.

[17] M.A. Martinez-Nunez, Nonribosomal peptides synthetases and their applications in industry, Sustainable Chemical Processes 4 (2016) 1–8.
[18] E. Guaní-Guerra, T. Santos-Mendoza, S.O. Lugo-Reyes, L.M. Teran, Antimicrobial peptides: general overview and clinical implications in human health and disease, Clinical Immunology 135 (2010) 1–11.
[19] J. Battersby, J. Khara, V.J. Wright, O. Levy, B. Kampmann, Antimicrobial proteins and peptides in early life: ontogeny and translational opportunities, Frontiers in Immunology 7 (2016) 309.
[20] O. Levy, Antimicrobial proteins and peptides: anti-infective molecules of mammalian leukocytes, Journal of Leukocyte Biology 76 (2004) 909–925.
[21] E.M. Jones, A. Smart, G. Bloomberg, L. Burgess, M.R. Millar, Lactoferricin, a new antimicrobial peptide, Journal of Applied Bacteriology 77 (1994) 208–214.
[22] W. Bellamy, M. Takase, H. Wakabayashi, K. Kawase, M. Tomita, Antibacterial spectrum of lactoferricin B, a potent bactericidal peptide derived from the N-terminal region of bovine lactoferrin, Journal of Applied Bacteriology 73 (1992) 472–479.
[23] J. Sengupta, S. Saha, A. Khetan, S.K. Sarkar, S.M. Mandal, Effects of lactoferricin B against keratitis-associated fungal biofilms, Journal of Infection and Chemotherapy 18 (2012) 698–703.
[24] S.R. Lueschow, S.J. McElroy, The Paneth cell: the curator and defender of the immature small intestine, Frontiers in Immunology 11 (2020) 587.
[25] D. Vishnu, A. Krukemeyer, A. Ramamoorthy, The human beta-defensin-3, an antibacterial peptide with multiple biological functions, Biochimica et Biophysica Acta (BBA) - Biomembranes 2006 (1758) 1499–1512.
[26] T.W.L. Groeneveld, T.H. Ramwadhdoebé, L.A. Trouw, D.L. van den Ham, V. van der Borden, J.W. Drijfhout, et al., Human neutrophil peptide-1 inhibits both the classical and the lectin pathway of complement activation, Molecular Immunology 44 (2007) 3608–3614.
[27] R. Susanne, K. Exner, S. Paul, J.-M. Schröder, E. Kalm, C. Looft, Bovine β-defensins: identification and characterization of novel bovine β-defensin genes and their expression in mammary gland tissue, Mammalian Genome 15 (2004) 834–842.
[28] S. Narasimhaiah, C. Subbalakshmi, R. Nagaraj, Indolicidin, a 13-residue basic antimicrobial peptide rich in tryptophan and proline, interacts with $Ca^{2+}$-calmodulin, Biochemical and Biophysical Research Communications 309 (2003) 879–884.
[29] Y.-S. Morgan, D. Drouin, P.A. Cavalcante, H.W. Barkema, E.R. Cobo, Host defense-cathelicidins in cattle: types, production, bioactive functions and potential therapeutic and diagnostic applications, International Journal of Antimicrobial Agents 51 (2018) 813–821.
[30] J.M. Conlon, Á. Sonnevend, M. Patel, C. Davidson, P.F. Nielsen, T. Pál, et al., Isolation of peptides of the brevinin-1 family with potent candidacidal activity from the skin secretions of the frog *Rana boylii*, The Journal of Peptide Research 62 (2003) 207–213.
[31] M. Simmaco, G. Kreil, D. Barra, Bombinins, antimicrobial peptides from Bombina species, Biochimica et Biophysica Acta (BBA) - Biomembranes 1788 (2009) 1551–1555.
[32] A. Mor, N. Van Huong, A. Delfour, D. Migliore-Samour, P. Nicolas, Isolation, amino acid sequence and synthesis of dermaseptin, a novel antimicrobial peptide of amphibian skin, Biochemistry 30 (1991) 8824–8830.
[33] T. Halverson, Y.J. Basir, F.C. Knoop, J.M. Conlon, Purification and characterization of antimicrobial peptides from the skin of the North American green frog *Rana clamitans*, Peptides 21 (2000) 469–476.

[34] T. Miyata, F. Tokunaga, T. Yoneya, K. Yoshikawa, S. Iwanaga, M. Niwa, et al., Antimicrobial peptides, isolated from horseshoe crab hemocytes, tachyplesin II, and polyphemusins I and II: chemical structures and biological activity, The Journal of Biochemistry 106 (1989) 663−668.

[35] G. Kreil, Biosynthesis of Melittin, A toxic peptide from bee venom: amino-acid sequence of the precursor, European Journal of Biochemistry 33 (1973) 558−566.

[36] M. Lamberty, A. Caille, C. Landon, S. Tassin-Moindrot, C. Hetru, P. Bulet, et al., Solution structures of the antifungal heliomicin and a selected variant with both antibacterial and antifungal activities, Biochemistry 40 (2001) 11995−12003.

[37] P. Casteels, C. Ampe, L. Riviere, J.V. Damme, C. Elicone, M. Fleming, et al., Isolation and characterization of abaecin, a major antibacterial response peptide in the honeybee (*Apis mellifera*), European Journal of Biochemistry 187 (1990) 381−386.

[38] H. Seiichi, M. Yamakawa, A novel antibacterial peptide family isolated from the silkworm, *Bombyx mori*, Biochemical Journal 310 (1995) 651−656.

[39] J. Nesa, A. Sadat, D.F. Buccini, A. Kati, A.K. Mandal, O.L. Franco, Antimicrobial peptides from *Bombyx mori*: a splendid immune defense response in silkworms, RSC Advances 10 (2020) 512−523.

[40] D.G. Lee, H.K. Kim, S. Am. Kim, Y. Park, S.C. Park, S.H. Jang, et al., Fungicidal effect of indolicidin and its interaction with phospholipid membranes, Biochemical and Biophysical Research Communications 305 (2003) 305−310.

[41] K.D. Smet, R. Contreras, Human antimicrobial peptides: defensins, cathelicidins and histatins, Biotechnology Letters 27 (2005) 1337−1347.

[42] J.M. Schroder, J. Harder, Human beta-defensin-2, The International Journal of Biochemistry & Cell Biology 31 (1999) 645−651.

[43] D. Xu, W. Lu, Defensins: a double-edged sword in host immunity, Frontiers in Immunology 11 (2020) 764.

[44] E.V. Valore, C.H. Park, A.J. Quayle, K.R. Wiles, P.B. McCray Jr, T. Ganz, Human beta-defensin-1: an antimicrobial peptide of urogenital tissues, The Journal of Clinical Investigation 101 (1998) 1633−1642.

[45] G. Garcia-Lopez, P. Flores-Espinosa, V. Zaga-Clavellina, Tissue-specific human beta-defensins (HBD) 1, HBD2, and HBD3 secretion from human extra-placental membranes stimulated with *Escherichia coli*, Reproductive Biology and Endocrinology 8 (2010) 1−8.

[46] J.J. Schneider, A. Unholzer, M. Schaller, M. Schafer-Korting, H.C. Korting, Human defensins, Journal of Molecular Medicine 83 (2005) 587−595.

[47] R.N. Cunliffe, Y.R. Mahida, Antimicrobial peptides in innate intestinal host defence, Gut 47 (2000) 16−17.

[48] V. Nizet, T. Ohtake, X. Lauth, J. Trowbridge, J. Rudisill, R.A. Dorschner, et al., Innate antimicrobial peptide protects the skin from invasive bacterial infection, Nature 414 (2001) 454−457.

[49] D. Yang, A. Biragyn, L.W. Kwak, J.J. Oppenheim, Mammalian defensins in immunity: more than just microbicidal, Trends in Immunology 23 (2002) 291−296.

[50] D. Gill, D.E. Jones, C.L. Bevins, Airway epithelial cells are the site of expression of a mammalian antimicrobial peptide gene, Proceedings of the National Academy of Sciences 90 (1993) 4596−4600.

[51] T.-A. Khaled, L. Wyer, L. Berghuis, L.L. Bassel, M.E. Clark, J.L. Caswell, Regulation of tracheal antimicrobial peptide gene expression in airway epithelial cells of cattle, Veterinary Research 47 (2016) 1−7.

[52] J.W. Ashcroft, Z.B. Zalinger, C.R. Bevier, F.A. Fekete, Antimicrobial properties of two purified skin peptides from the mink frog (*Rana septentrionalis*) against bacteria isolated from the natural habitat, Comparative Biochemistry and Physiology Part C: Toxicology & Pharmacology 146 (2007) 325–330.

[53] C. Chianese, A. Zannella, A.D. Monti, N. Filippis, G. Doti, M.Galdiero Franci, The broad-spectrum antiviral potential of the amphibian peptide AR-23, International Journal of Molecular Sciences 23 (2022) 883.

[54] J.M. Conlon, M. Mechkarska, Host-defense peptides with therapeutic potential from skin secretions of frogs from the family Pipidae, Pharmaceuticals 7 (2014) 58–77. Basel.

[55] M.W. Lee, U.P. Kari, Structure–activity studies on magainins and other host defense peptides, Biopolymers: Original Research on Biomolecules 37 (1995) 105–122.

[56] R. Ramos, S. Moreira, A. Rodrigues, M. Gama, L. Domingues, Recombinant expression and purification of the antimicrobial peptide magainin-2, Biotechnology Progress 29 (2013) 17–22.

[57] M. Novkovic, J. Simunic, V. Bojovic, A. Tossi, D. Juretic, DADP: the database of anuran defense peptides, Bioinformatics 28 (2012) 1406–1407.

[58] E.J.H. Bartels, D. Dekker, M. Amiche, Dermaseptins, multifunctional antimicrobial peptides: a review of their pharmacology, effectivity, mechanism of action, and possible future directions, Frontiers in Pharmacology (2019) 1421.

[59] J.-P.S. Powers, M.M. Martin, D.L. Goosney, R.E.W. Hancock, The antimicrobial peptide polyphemusin localizes to the cytoplasm of *Escherichia coli* following treatment, Antimicrobial Agents and Chemotherapy 50 (2006) 1522–1524.

[60] M. Lamberty, D. Zachary, R. Lanot, C. Bordereau, A. Robert, J.A. Hoffmann, et al., Insect immunity: constitutive expression of a cysteine-rich antifungal and a linear antibacterial peptide in a termite insect, Journal of Biological Chemistry 276 (2001) 4085–4092.

[61] W. Eleftherianos, C. Zhang, A. Heryanto, G. Mohamed, G. Contreras, M. Tettamanti, et al., Bassal, diversity of insect antimicrobial peptides and proteins-a functional perspective: a review, International Journal of Biological Macromolecules 191 (2021) 277–287.

[62] D.G. Boucias, J.C. Pendland, Insect immune defense system, part I: innate defense reactions, Principles of Insect Pathology (1998) 439–468. Springer, Boston, MA.

[63] Q. Wu, J. Patocka, K. Kuca, Insect antimicrobial peptides, a mini review, Toxins 10 (2018) 461. Basel.

[64] H.G. Boman, I. Nilsson, B. Rasmuson, Inducible antibacterial defence system in drosophila, Nature 237 (1972) 232–235.

[65] M.D. Manniello, A. Moretta, R. Salvia, C. Scieuzo, D. Lucchetti, H. Vogel, et al., Insect antimicrobial peptides: potential weapons to counteract the antibiotic resistance, Cellular and Molecular Life Sciences 78 (2021) 1–3.

[66] H.-Y. Yi, M. Chowdhury, Y.-D. Huang, X.-Q. Yu, Insect antimicrobial peptides and their applications, Applied Microbiology and Biotechnology 98 (2014) 5807–5822.

[67] H. Steiner, Secondary structure of the cecropins: antibacterial peptides from the moth *Hyalophora cecropia*, FEBS Letters 137 (1982) 283–287.

[68] J. DeLucca, J.M. Bland, T.J. Jacks, C. Grimm, T.E. Cleveland, T.J. Walsh, Fungicidal activity of cecropin A, Antimicrobial Agents and Chemotherapy 41 (1997) 481–483.

[69] R. Maget-Dana, M. Ptak, Penetration of the insect defensin A into phospholipid monolayers and formation of defensin A-lipid complexes, Biophysics Journal 73 (1997) 2527–2533.

[70] Y. Huan, Q. Kong, H. Mou, H. Yi, Antimicrobial peptides: classification, design, application and research progress in multiple fields, Frontiers in Microbiology (2020) 2559.
[71] J.A. McIntosh, M.S. Donia, E.W. Schmidt, Ribosomal peptide natural products: bridging the ribosomal and nonribosomal worlds, Natural Product Reports 26 (2009) 537–559.
[72] D. Schwarzer, R. Finking, M.A. Marahiel, Nonribosomal peptides: from genes to products, Natural Product Reports 20 (2003) 275–287.
[73] R.D. Süssmuth, A. Mainz, Nonribosomal peptide synthesis—principles and prospects, Angewandte Chemie International Edition 56 (2017) 3770–3821.
[74] G.H. Hura, C.R. Vickerya, M.D. Burkarta, Explorations of catalytic domains in nonribosomal peptide synthetase enzymology, Natural Product Reports 29 (2012) 1074–1098.
[75] M.A. Martínez-Nunez, V.E.L. López, Nonribosomal peptides synthetases and their applications in industry, Sustainable Chemical Processes 4 (2016).
[76] C. Mercer, M.D. Burkart, The ubiquitous carrier protein—a window to metabolite biosynthesis, Natural Product Reports 24 (2007) 750–773.
[77] C.A. Mitchell, C. Shi, C.C. Aldrich, A.M. Gulick, Structure of PA1221, a nonribosomal peptide synthetase containing adenylation and peptidyl carrier protein domains, Biochemistry 51 (2012) 3252–3263.
[78] H.M. Patel, J. Tao, C.T. Walsh, Epimerization of an L-cysteinyl to a D-cysteinyl residue during thiazoline ring formation in siderophore chain elongation by pyochelin synthetase from *Pseudomonas aeruginosa*, Biochemistry 42 (2003) 10514–10527.
[79] C.T. Walsh Koglin, Structural insights into nonribosomal peptide enzymatic assembly lines, Natural Product Reports 26 (2009) 987–1000.
[80] H. Gan, J. Gaynord, S.M. Rowe, T. Deingruber, D.R. Spring, The multifaceted nature of antimicrobial peptides: current synthetic chemistry approaches and future directions, Chemical Society Review 50 (2021) 7820–7880.
[81] T. Stachelhaus, H.D. Mootz, V. Bergendahl, M.A. Marahiel, Peptide bond formation in nonribosomal peptide biosynthesis. Catalytic role of the condensation domain, Journal of Biological Chemistry 273 (1998) 22773–22781.
[82] S.A. Samel, B. Wagner, M.A. Marahiel, L.-O. Essen, The thioesterase domain of the fengycin biosynthesis cluster: a structural base for the macrocyclization of a nonribosomal lipopeptide, Journal of Molecular Biology 359 (2006) 876–889.
[83] C.T. Walsh, H. Chen, T.A. Keating, B.K. Hubbard, H.C. Losey, L. Luo, et al., Tailoring enzymes that modify nonribosomal peptides during and after chain elongation on NRPS assembly lines, Current Oinion in Chemical Biology 5 (2001) 525–534.
[84] E.W. Schmidt, The hidden diversity of ribosomal peptide natural products, BMC Biology 8 (2010).
[85] S.H. Joo, Cyclic Peptides as therapeutic agents and biochemical tools, Biomolecules and Therapeutics (Seoul) 20 (2021) 19–26.
[86] A.H. Shabeer, K. Ajesh, K.V. Dileep, P. Prajosh, K. Sreejith, Structural characterization of *Kannurin isoforms* and evaluation of the role of β-hydroxy fatty acid tail length in functional specificity, Scientific Reports 10 (2020) 1–12.
[87] E. Garcia-Gonzalez, S. Müller, P. Ensle, R.D. Süssmuth, E. Genersch, Elucidation of sevadicin, a novel non-ribosomal peptide secondary metabolite produced by the honey bee pathogenic bacterium *Paenibacillus* larvae, Environmental Microbiology 16 (2014) 1297–1309.

[88] E. Benedetti, A. Bavoso, B. Di Blasio, V. Pavone, C. Pedone, C. Toniolo, et al., Peptaibol antibiotics: a study on the helical structure of the 2–9 sequence of emerimicins III and IV, Proceedings of National Academy of Science U S A 79 (1982) 7951–7954.

[89] G. Bills, Y. Li, L. Chen, Q. Yue, X.-M. Niu, Z. An, New insights into the echinocandins and other fungal non-ribosomal peptides and peptaibiotics, Natural Product Reports 31 (2014) 1348–1375.

[90] T. Luders, G.A. Birkemo, G. Fimland, J. Nissen-Meyer, I.F. Nes, Strong synergy between a eukaryotic antimicrobial peptide and bacteriocins from lactic acid bacteria, Applied and Environmental Microbiology 69 (2003) 1797–1799.

[91] P.G. Arnison, M.J. Bibb, G. Bierbaum, A.A. Bowers, T.S. Bugni, G. Bulaj, et al., Ribosomally synthesized and post-translationally modified peptide natural products: overview and recommendations for a universal nomenclature, Natural Product Reports 30 (2013) 108–160.

[92] A.-C. Letzel, S.J. Pidot, C. Hertweck, Genome mining for ribosomally synthesized and post-translationally modified peptides (RiPPs) in anaerobic bacteria, BMC Genomics 15 (2014) 1–21.

[93] H.-M. Huang, H. Kries, Unleashing the potential of ribosomal and nonribosomal peptide biosynthesis, Biochemistry 58 (2018) 73–74.

[94] T. Denoel, C. Lemaire, A. Luxen, Progress in lanthionine and protected lanthionine synthesis, Chemistry- A European Journal 58 (2018) 15421–15441.

[95] R. Magnuson, J. Solomon, A.D. Grossman, Biochemical and genetic characterization of a competence pheromone from *B. subtilis*, Cell 77 (1994) 207–216.

[96] K. Sivonen, N. Leikoski, D.P. Fewer, J. Jokela, Cyanobactins—ribosomal cyclic peptides produced by cyanobacteria, Applied Microbiology and Biotechnology 86 (2010) 1213–1225.

[97] K.I. Piatkov, T.T.M. Vu, C.-S. Hwang, A. Varshavsky, Formyl-methionine as a degradation signal at the N-termini of bacterial proteins, Microbial Cell 2 (2015) 376–393.

[98] L. Liu, Lide, C. Zhao, H.H.Q. Heng, T. Ganz, The human β-defensin-1 and α-defensins are encoded by adjacent genes: two peptide families with differing disulfide topology share a common ancestry, Genomics 43 (1997) 316–320.

[99] X.-M. Fang, Q. Shu, Q.-X. Chen, M. Book, H.-G. Sahl, A. Hoeft, et al., Differential expression of alpha- and beta-defensins in human peripheral blood, European Journal of Clinical Investigation 33 (2003) 82–87.

[100] R.E.W. Hancock, G. Diamond, The role of cationic antimicrobial peptides in innate host defences, Trends in Microbiology 8 (2000) 402–410.

[101] K. Bandurska, A. Berdowska, R. Barczyńska-Felusiak, P. Krupa, Unique features of human cathelicidin LL-37, Biofactors 41 (2015) 289–300.

[102] U.H.N. Dürr, U.S. Sudheendra, A. Ramamoorthy, LL-37, the only human member of the cathelicidin family of antimicrobial peptides. Biochimica et biophysica acta (BBA), Biomembranes 2006 (1758) 1408–1425.

[103] Y.H. Jo, B.B. Patnaik, J. Hwang, K.B. Park, H.J. Ko, C.E. Kim, et al., Regulation of the expression of nine antimicrobial peptide genes by TmIMD confers resistance against gram-negative bacteria, Scientific Reports 9 (2019) 1–14.

[104] T. Weber, K. Blin, S. Duddela, D. Krug, H.U. Kim, R. Bruccoleri, et al., antiSMASH 3.0—a comprehensive resource for the genome mining of biosynthetic gene clusters, Nucleic Acids Research 43 (2015) W237–W243.

# Antimicrobial peptides: features and modes of action

Feba Francis and Nitin Chaudhary
Department of Biosciences and Bioengineering, Indian Institute of Technology Guwahati, Guwahati, Assam, India

## 3.1 Introduction

Humans and other multicellular organisms have been coevolving with pathogenic microbes. Evolution is intricately related to the mutations in the genome. Mutations that confer advantages get naturally selected. Mutations that impart a microbe better chances of survival in an antibiotic-containing medium get selected and the mutant microbe dominates the population in the medium. The indiscriminate use of antibiotics has led to the appearance of drug-resistant bacteria. Many clinical bacterial strains show resistance against most known antibiotics, thereby causing difficult-to-treat infections. The survival amid multipotent organisms is the elementary purpose that each organism strive for. Innate immunity provides rapid and nonspecific action against pathogenic microbes and protects the organisms from infection. Innate immunity, the first line of defense, includes physical barriers such as skin and cilia, as well as the chemical components. One of the various ways that living organisms employ to combat pathogens is the antimicrobial peptides (AMPs). AMPs have been identified in almost all the organisms that have been investigated for their presence. AMPs with antibacterial, antiviral, antifungal, antiparasitic, and antimycoplasmal properties have been reported in the literature [1]. Apart from displaying the direct microbicidal activity, AMPs may possess other biological activities that include, but are not limited to, immunomodulation [2], inhibition of lipopolysaccharide (LPS)-induced inflammation [3], wound healing [4], and antitumor activity [5]. The AMPs, therefore, are often termed as host defense peptides (HDPs) as well.

One of the big pluses of the AMPs over conventional antibiotics lies in their mechanism of action. As they kill the microbes primarily through their membrane perturbation, that is, without targeting any molecular receptors on the microbial surface, the chances of bacterial resistance are considerably less. According to the recent WHO report [6], antibiotic-resistant strains are rapidly increasing, and the strategies to combat them show hardly any success in their clinical trials. The increase in multidrug-resistant microorganisms and the decline in new antibiotics' discovery rate constitute a significant healthcare challenge today. As new infectious organisms emerge with high competence, discovering novel antimicrobial agents is the need of the hour. The very facts that all living forms use the AMPs and that

Antimicrobial Peptides. DOI: https://doi.org/10.1016/B978-0-323-85682-9.00016-7
© 2023 Elsevier Inc. All rights reserved.

they display broad-spectrum antimicrobial activity make them very promising candidates for developing the next-generation antibiotics. However, certain limitations and challenges need to be overcome before AMPs could be taken from the bench to the bed. For example, AMPs may have associated toxicity and hemolytic activity, and they usually have a short half-life [7,8]. It is, therefore, essential to develop synthetic and long-lasting AMP analogs that display high activity and little toxicity. Let us look into the pioneer studies that brought these fascinating molecules to the forefront.

## 3.2 Historical perspective

The history of AMPs dates back to the late 1930s when Dubos, in 1939, isolated an antibacterial agent from a sporulating soil Bacillus [9]. The agent displayed bactericidal activity against all the Gram-positive bacteria tested but failed to act against Gram-negative bacteria. It was shown that the agent protected the mice that were infected with virulent pneumococci [10]. Despite the antibacterial agent's success in the animal models, it could not be used for treating patients as it was toxic and worked poorly in immunosuppressed individuals. In the following year, Hotchkiss and Dubos isolated gramicidin and tyrocidine as the antibacterial agent's two active ingredients [11,12]. Gramicidin was found to be efficient in treating the mice infected with virulent pneumococci [11]. The bacterium that produced these antimicrobial agents was later identified as *Bacillus brevis* [13]. Gause and Brazhnikova applied Dubos' protocol on the *B. brevis* strains from Russian soil and isolated an antimicrobial agent that they named gramicidin S (Soviet gramicidin) [14]. Gramicidin and tyrocidine are cyclic decapeptides [15,16] and exhibit antibacterial activity by disrupting the cell membrane. However, both tyrocidine and gramicidin turned out to be poorly selective. They exhibited considerable hemolysis alongside the antimicrobial activity, which in turn reduced their effectiveness and limited their usage to the topical treatments of wounds and ulcers [16]. Purothionin, an AMP isolated from the plant *Triticum aestivum* by Stuart and Harris in 1942, was found to be effective against fungi and some pathogenic bacteria [17]. Later, the studies revealed different classes of purothionin, that is, α-, β-, and γ-purothionins [18,19]. They are sulfur-rich peptides belonging to the thionin family of plant AMPs [17,20]. HDPs from animals were first described in the 1960s by Kiss and Michl in Bombina species. They noted the presence of antimicrobial and hemolytic activity in the skin secretions of *Bombina variegata*, which led to the isolation of AMPs, bombinins [21,22]. Subsequently, many HDPs were discovered from various sources. In 1980, Hans Boman and coworkers injected live bacteria to pupae of *Hyalophora cecropia* and isolated AMPs named P9A and P9B that exhibit bactericidal activity against *Escherichia coli* and other Gram-negative bacteria [23]. The sequence and specificity were subsequently determined, and the two P9 peptides were termed cecropins A and B [24]. In 1983, Ganz and Lehrer isolated three microbicidal agents from granule-rich sediment of human neutrophils and called them human neutrophil peptides 1–3 (HNP-1–3), which were later coined as defensins [25]. The discovery of magainins from *Xenopus laevis* is another milestone in the history of AMPs. Zasloff noticed that these frogs seldom developed infection after experimental surgeries despite being in a nonsterile environment. He was curious about this "sterilizing activity" of frog skin, and ended up discovering the AMPs called magainins [26]. Since then, a large number of AMPs have been identified from very diverse sources and are recorded in various AMP databases [27–29].

## 3.3 Features of antimicrobial peptides

### 3.3.1 Diversity

The sequences, structures, and activities of several hundred AMPs have been characterized [27,30]. The analyses of the sequences and the structures show that the AMPs are highly diverse, both in sequence and structure. In terms of length, they range from very short (<10 amino acids) to fairly long (up to around 100 residues). They carry a net positive charge at neutral pH that could go up to +7 or more. Some AMPs are unusually rich in certain amino acids. A large majority of the naturally occurring AMPs are composed of exclusively L-amino acids [31]. However, AMPs with nonnatural amino acids are also quite common [32]. Owing to such diversity in the primary structure, the peptides are rarely classified based on their sequence. However, it is somewhat easier to classify them based on the structure. The diversity in the sequence and the structure attributes to the diverse activities that the AMPs display. These include, but are not limited to, the spectrum of killing, sensitivity to the salt or divalent cations, mechanism of action, wound healing, and immunomodulatory effects [33–35]. Every organism that has been investigated for the presence of AMPs is found to possess them. Even bacteria use peptides called bacteriocins to combat the other bacterial strains [36,37].

### 3.3.2 Cationicity and amphipathicity

Despite being exceedingly diverse in the sequences, amino acid compositions, and structures, AMPs harbor two features that are common to most AMPs: cationicity and amphipathicity. AMPs are unusually rich in basic and hydrophobic amino acids. The presence of basic amino acids, especially lysine and arginine, confers them a net positive charge at neutral pH. The cationicity imparts them the selectivity to bind the anionic microbial membranes [38]. Amphipathicity imparts membrane permeabilizing activity to the AMPs [39–42]. Unless already folded into an amphipathic structure, the peptides fold upon membrane binding to take up an amphipathic structure that facilitates their insertion into the membranes. Membrane perturbation, therefore, happens to be one of the most common modes of action of the cationic AMPs. Although mostly cationic, anionic AMPs (AAMPs) do exist in nature as a part of the innate defense system. For example, maximin H5 from

*Bombina maxima* has a net charge of −2 [43]. Some AAMPs seem to interact with membranes via divalent cations-mediated bridges [44].

### 3.3.3 Structure

As AMPs are highly diverse in their primary structure, they could be classified on other structural parameters such as (a) linear or cyclic (b) lacking or harboring disulfide bridges, and (c) secondary structure of the peptide. The cyclic structure is achieved through end-to-end ligation, as in θ-defensins, or via the disulfide bridge(s) between the Cys residues near the peptide termini, as in tigerinins and bactenecin [45,46]. Cyclic AMPs are less prone to proteolytic degradation and display enhanced structural stability [47]. As far as the secondary structure of AMPs is concerned, a large majority of them are unordered in solution and fold to take up an amphipathic structure only upon membrane binding. The peptides that have defined three-dimensional (3D) structures usually contain one or more disulfide bridges to impart stability. Even though several thousand AMPs have been discovered to date, the 3D structures have been characterized for a few (about 13%) of them, primarily by the solution nuclear magnetic resonance (NMR) spectroscopy [48–50]. Based on their secondary structure, AMPs are broadly classified into four major classes (Fig. 3.1) [51].

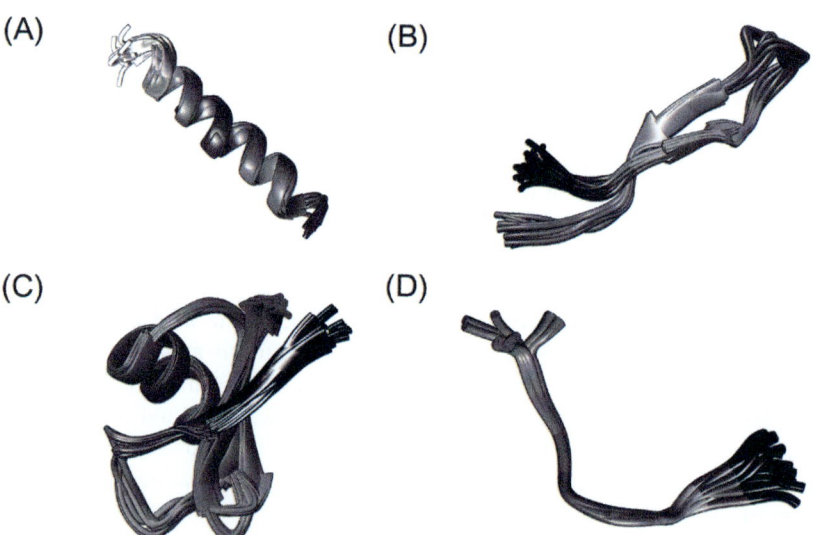

**Figure 3.1** Four major structural classes of AMPs. (A) α-helical structure, e.g., magainin 2 (PDB ID: 2MAG), (B) β-Sheet structure, e.g., β-lactoferricin B (PDB ID: 1LFC), (C) α/β structure, e.g., plant defensin Psd1 (PDB ID: 1JKZ), and (D) non-α/β peptide structure, e.g., bovine indolicidin (PDB ID: 1G89).

### 3.3.3.1 The α-helical antimicrobial peptides

The α-helical peptides constitute the most populous class of AMPs. They are the most widespread among all structural classes and are found in species ranging from invertebrates to higher organisms [26,39]. They usually adopt an unordered conformation in aqueous solutions, but quickly fold into an amphipathic α-helix upon membrane binding or in the membrane-mimetic environments such as trifluoroethanol, hexafluoroisopropanol, and sodium dodecyl sulfate micelles [52–54]. Cecropins from insects [55], magainins from anurans [53], and human cathelicidin LL-37 [56] are some of the well-studied α-helical AMPs. Some of the well-studied AMPs from this structural class are listed in Table 3.1, and the structures for the selected peptides are shown in Fig. 3.2.

### 3.3.3.2 β-Sheet antimicrobial peptides

The β-sheet AMPs are made up of two or more β-strands forming at least one β-sheet. The simplest architecture is that of a β-hairpin. The strands in the β-hairpin are often linked through disulfide bonds that impart conformational stability to the otherwise dynamic β-strands [57]. Tachyplesins isolated from *Tachypleus tridentatus* and protegrins isolated from porcine leukocytes are the examples of well-studied β-hairpin AMPs with membrane-disruptive action [58]. The HNP-1, the rabbit kidney defensin RK-1, and the bovine neutrophil beta-defensin-12 are some of the examples of β-sheet AMPs with more than two strands. The peptide LCI isolated from *Bacillus subtilis* is a Cys-lacking β-sheet AMP with a novel topology of four-stranded antiparallel β-sheet. Selected members of this structural class are presented in Table 3.2 and Fig. 3.3.

### 3.3.3.3 αβ-Antimicrobial peptides

Many AMPs consist of both α-helical and β-sheet structures, some of which are listed in Table 3.3 and Fig. 3.4. Human β-defensin 1, for example, possesses an N-terminal α-helical region along with a highly conserved β-sheet [59]. Drosomycin is an antifungal peptide derived from *Drosophila melanogaster* with potent activity against filamentous fungi. The 3D structure of drosomycin reveals α-helix and twisted three-stranded β-sheet structure stabilized by disulfide bridges [60]. Kalata B2, isolated from plant *Oldenlandia affinis*, is another classical example of the αβ peptides with unusually stable cyclic structure [61]. Kalata peptides are prototypic cyclotides with closely packed disulfide core conferring stability against thermal, chemical, and enzymatic degradation [62].

### 3.3.3.4 Non-αβ (extended structure)

The extended AMPs are predominantly rich in specific amino acids such as proline, arginine, glycine, histidine, tryptophan, and have no regular secondary structure (Table 3.4, Fig. 3.5). Extended AMPs do not take up stable secondary

**Table 3.1** List of selected α-helical AMPs.

| AMP | Sequence | Source |
|---|---|---|
| Brevinin 1BYa | FLPILASLAAKFGPKLFCLVTKKC | *Rana boylii* |
| Cathelicidin LL-37 | LLGDFFRKSKEKIGKEFKRIVQRIKDFLRNLVPRTES | *Homo sapiens* |
| Cecropin A | KWKLFKKIEKVGQNIRDGIIKAGPAVAVVGQATQIAK-NH$_2$ | *Hyalophora cecropia* |
| Clavanin A | VFQFLGKIIHHVGNFVHGFSHVF-NH$_2$ | *Styela clava* |
| Dermaseptin B2 | GLWSKIKEVGKEAAKAAAKAAGKAALGAVSEAV-NH$_2$ | *Phyllomedusa bicolor* |
| Dermcidin | SSLLEKGLDGAKKAVGGLGKLGKDAVEDLESVGKGAVHDVKDVLDSVL | *Homo sapiens* |
| Magainin II | GIGKFLHSAKKFGKAFVGEIMNS | *Xenopus laevis* |
| Melittin | GIGAVLKVLTTGLPALISWIKRKRQQ-NH$_2$ | *Apis mellifera* |
| Temporin L | FVQWFSKFLGRIL-NH$_2$ | *Rana temporaria* |
| Aurein 1.1 | GLFDIIKKIAESI-NH$_2$ | *Litoria aurea, Litoria raniformis* |
| Buforin 2 | TRSSRAGLQFPVGRVHRLLRK | *Bufo gargarizans* |
| Piscidin 1 | FFHHIFRGIVHVGKTIHRLVTG | *Morone saxatilis* |
| Pseudin 2 | GLNALKKVFQGIHEAIKLINNHVQ | *Pseudis paradoxa* |
| Phylloseptin 1 | FLSLIPHAINAVSAIAKHN-NH$_2$ | *Phyllomedusa hypochondrialis* |

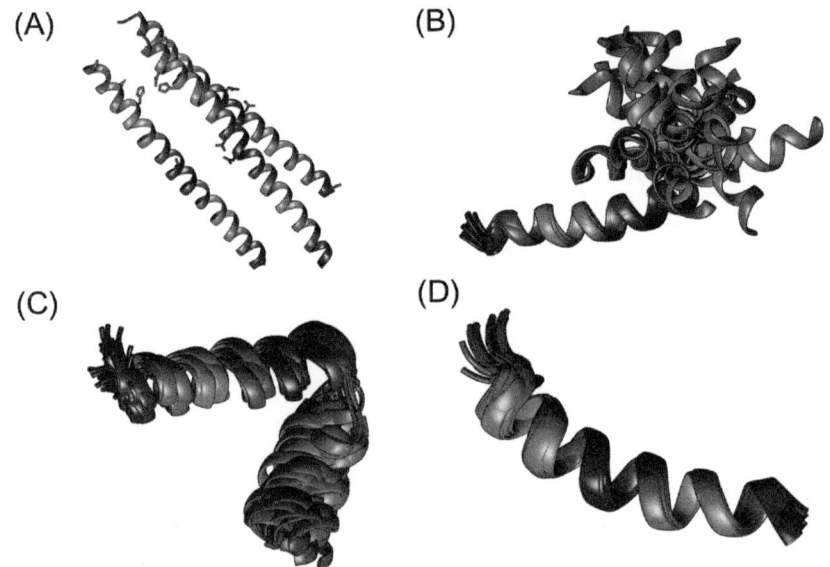

**Figure 3.2** The PDB structures of selected α-helical antimicrobial peptides. (A) Dermcidin (PDB ID: 2YMK), (B) CB1a (PDB ID: 2IGR), (C) Papiliocin (PDB ID: 2LA2), and (D) LL-37 (PDB ID: 2LMF).

structures through intramolecular H-bonding and interact with the membrane lipids via the H-bonding and Van der Waals interactions [57]. The Trp-rich peptide indolicidin, derived from bovine neutrophils, is the best characterized AMP of this class [57,63]. Pyrrhocoricin from *Pyrrhocoris apterus* [64], apidaecin from lymph fluid of *Apis mellifera* [65], and PR-39 from pig are examples of proline-rich AMPs of non-αβ structural class. Histatins are the histidine-rich antifungal peptides present in the salivary secretions of human submandibular and parotid glands [66]. Majorly, extended AMPs are not membranolytic peptides; instead, they translocate across the membrane and interact with intracellular targets [67,68]. However, some peptides could possess the membranolytic activity, for example, indolicidin [69]. Adepantins are anuran peptides with high glycine content and display high selectivity for Gram-negative bacteria [70]. Certain AMPs adopt a loop conformation with one intramolecular disulfide bridge, for example, bactenecin from bovine neutrophil granules [71] and lactoferricins derived from bovine lactoferrin [72].

The antimicrobial peptide database (APD3) contains data for more than 3200 AMPs. Among these, around 14% of peptides are α-helical, whereas the 3D structure for ∼75% of AMPs is unknown [27]. The α-helical AMPs, therefore, constitute the most populous structural class of AMPs.

**Table 3.2** List of selected β-sheet AMPs.

| AMP | Sequence | Source |
|---|---|---|
| Protegrin 1 | RGGRLCYCRRRFCVCVGR-NH$_2$ | Sus scrofa |
| Tachyplesin 1 | KWCFRVCYRGICYRRCR | Tachypleus tridentatus |
| HNP-1 | ACYCRIPACIAGERRYGTCIYQGRLWAFCC | Homo sapiens |
| Thanatin | GSKKPVPIIYCNRRTGKCQRM | Podisus maculiventris |
| LCI | AIKLVQSPNGNFAASFVLDGTKWIFKSKYYDSSKGYWVGIYEVWDRK | Bacillus subtilis |
| Bovine neutrophil β-defensin 12 | GPLSCGRNGGVCIPIRCPVPMRQIGTCFGRPVKCCRSW | Bos Taurus |
| Hepcidin-20 | ICIFCCGCCHRSKCGMCCKT | Homo sapiens |
| Gomesin | ZQCRRLCYKQRCVTYCRGR-NH$_2$ (Z = pyroglutamate) | Acanthoscurria gomesiana |
| Polyphemusin 1 | RRWCFRVCYRGFCYRKCR-NH$_2$ | Limulus polyphemus |
| Microcin J25 | GGAGHVPEYFVGIGTPISFYG | Escherichia coli AY25 |

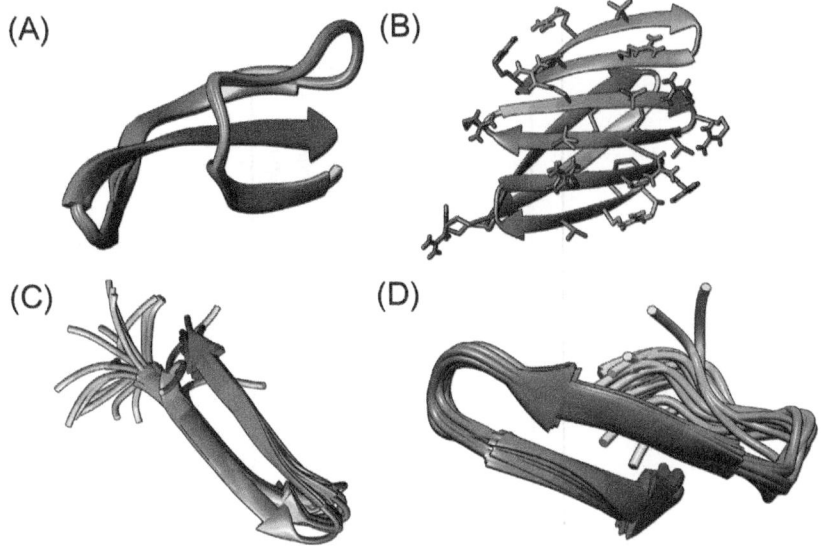

**Figure 3.3** The β-sheet antimicrobial peptides. (A) Human defensin 5 (PDB ID: 2LXZ), (B) antiparallel tetrameric β-hairpin structure of Baboon theta defensin 2 (PDB ID: 5INZ), (C) Protegrin 1 (PDB ID: 1PG1), and (D) Thanatin (PDB ID: 6AAB).

## 3.4 Biosynthesis and regulation

AMPs are synthesized in two different ways: ribosomally and nonribosomally. The ribosomally synthesized AMPs, as the name suggests, are genetically encoded. All species, including bacteria, produce such peptides. Nonribosomally synthesized peptides, on the other hand, are mainly produced by bacteria. The genes for the ribosomally synthesized peptides are mostly present as clusters in the genome, thereby producing multiple AMPs at a single site [3]. Ribosomally synthesized AMPs are generally produced as inactive precursors and require proteolytic cleavage to become active [73]. Their regulation depends not only on their expression but also on the abundance of appropriate proteases [3]. The AMPs can be expressed constitutively, or they can be inducible. In multicellular organisms, some AMPs are constitutively expressed, where they are stored at high concentrations in granules as inactive precursors and released locally at the site of infection. The expression of other AMPs is induced in response to pathogen-associated molecular patterns, lipopolysaccharides (LPSs), or cytokines [3,74]. In humans, histatins are expressed constitutively by submandibular, sublingual, and parotid glands [75]. On the other hand, drosomycin, a Drosophila AMP, gets induced when the fruitfly is exposed to the bacteria [60].

Nonribosomally synthesized peptides are synthesized on a large multimodular enzyme complex [76]. One of the classic examples of nonribosomally synthesized

**Table 3.3** List of selected αβ AMPs.

| AMP | Sequence | Source |
|---|---|---|
| Kalata B2 | GLPVCGETCFGGTCNTPGCSCTWPICTRD | *Oldenlandia affinis* |
| Plectasin | GFGCNGPWDEDDMQCHNHCKSIKGYKGGYCAKGGFVCKCY | *Pseudoplectania nigrella* |
| Phormicin | ATCDLLSGTGINHSACAAHCLLRGNRGGYCNGKGVCVCRN | *Phormia terranovae* |
| pBD-2 | DHYICAKKGGTCNFSPCPLFNRIEGTCYSGKAKCCIR | *Sus scofa* |
| Drosomycin | DCLSGRYKGPCAVWDNETCRRVCKEEGRSSGHCSPSLKCWCEGC | *Drosophila melanogaster* |
| Mytilin B | SCASRCKGHCRARRCGYYVSVLYRGRCYCKCLRC | *Mytilus edulis* |
| Human β-defensin 1 | DHYNCVSSGGQCLYSACPIFTKIQGTCYRGKAKCCK | *Homo sapiens* |

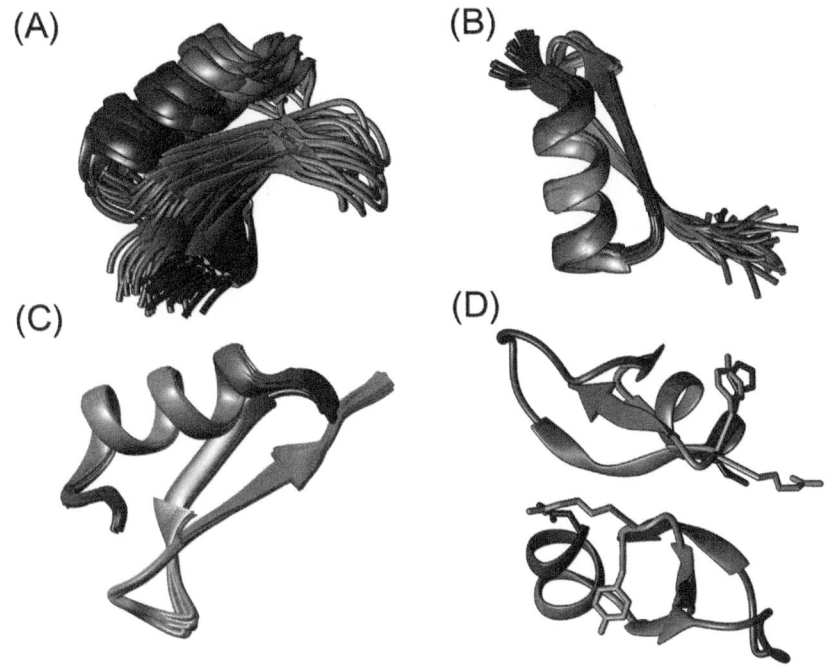

**Figure 3.4** The αβ-antimicrobial peptides: (A) NaD1 (PDB ID: 1MR4), (B) nasonin 1 (PDB ID: 2KOZ), (C) defensin MGD-1 (PDB ID: 1FJN), and (D) human β-defensin-1 (PDB ID: 1IJV).

AMPs is peptaibols. Peptaibols are a large class of membrane-active polypeptides derived from fungi. They show considerable sequence homology, but differ in biological properties and 3D structures [77,78]. Peptaibols are characterized by the presence of unusual amino acids principally, α-amino butyric acid (Aib), isovaleric acid (Iva), and imino hydroxyl proline. Their antibiotic activity is attributed to their ability to form ion channels [79]. Other nonribosomally synthesized peptides include polymyxins, actinomycin D, gramicidin S. Polymyxins are an old class of peptide antibiotics that are nonribosomally synthesized by *Bacillus polymyxa* [80]. Actinomycin D, the first antibiotic isolated from Streptomyces genus, and gramicidin S from *B. brevis* are also effective AMPs produced by nonribosomal peptide synthetase (NRPS) enzyme [81]. The NRPS genes are organized in a single operon in bacteria and as gene clusters in eukaryotes [82].

## 3.5 Some common families of antimicrobial peptides

A number of peptides can be grouped together into families. Some of these families are listed in this section.

**Table 3.4** List of selected non-αβ AMPs.

| AMP | Sequence | Source |
|---|---|---|
| PR-39 | RRRPRPPYLPRPRPPFFPPRLPRIPPGFPPRFPPRFP-NH$_2$ | *Sus scrofa* |
| Indolicidin | ILPWKWPWWPWRR-NH$_2$ | *Bos Taurus* |
| Drosocin | GKPRPYSPRPTSHPRPIRV | *Drosophila melanogaster* |
| Histatin 5 | DSHAKRHHGYKRKFHEKHHSHRGY | *Homo sapiens* |
| Lactoferricin B | FKCRRWQWRMKKLGAPSITCVRRAF | *Bos Taurus* |
| Tritrpticin | VRRFPWWWPFLRR | *Sus scrofa* |

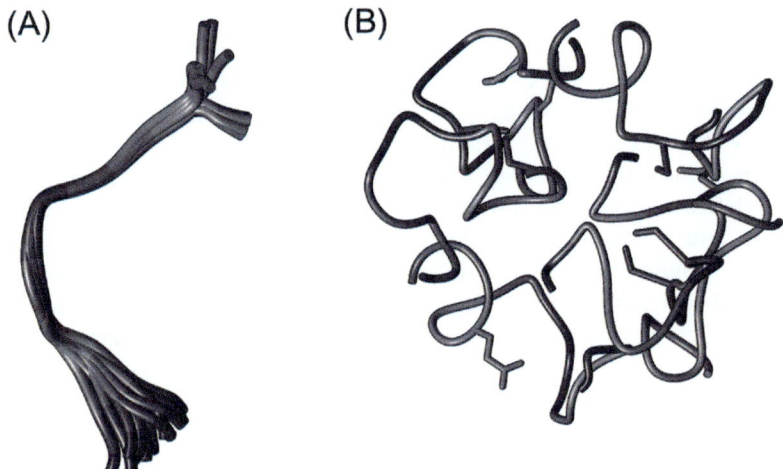

**Figure 3.5** The non-αβ antimicrobial peptides. (A) Indolicidin bound to dodecylphosphocholine micelles (PDB ID: 1G89). (B) Tetracyclic structure of mersacidin, a lantibiotic (PDB ID: 1QOW).

### 3.5.1 Cathelicidins

Cathelicidins constitute one of the major components of immune systems in the vertebrates. Cathelicidins are synthesized as inactive prepropeptides that are stored in the neutrophil granules. They are named cathelicidin due to the presence of a common domain cathelin in their pro-region [83]. Cecropin from the *Hyalophora cecropia* was the first cathelicidin to be discovered [84]. Magainins isolated by Zasloff from *Xenopus leavis* are also cathelicidins [84]. Bactenecins isolated from bovine neutrophils were among the first cathelicidins isolated from mammals [85]. Cathelicidins have now been identified in many vertebrates, including humans and farm animals. The well-studied porcine cecropin P1 is also a cathelicidin [85]. Human has only one cathelicidin, the 37-residue peptide named LL-37. LL-37 is an α-helical AMP that is highly charged; out of the 37 residues, 16 are charged (6 Lys, 5 Arg, 3 Glu, 2 Asp) [86]. The peptide, therefore, carries a net charge of +6 at neutral pH. Apart from displaying the antibacterial activity through membrane perturbation, LL-37 plays immunomodulatory roles such as regulating the inflammatory response [87], chemo-attracting the immune cells to the site of infection, neutralization of LPS, and wound healing. Moreover, recent studies suggest that it is also involved in the regulation of cancer [88–90].

### 3.5.2 Defensins

Defensins have a common β-sheet core stabilized with three or four disulfide bridges between six or eight conserved cysteine residues, respectively. Mammalian defensins contain three disulfide bridges and are subdivided into α-, β- and θ-defensins based on the linkage pattern of the cysteine residues. In α-defensins, the cysteine connectivity is $C^1$–$C^6$, $C^2$–$C^4$, $C^3$–$C^5$, while the connectivity in β-defensins is $C^1$–$C^5$, $C^2$–$C^4$, $C^3$–$C^6$ [91,92]. Several human α-defensins are highly expressed in neutrophils [25], and other α-defensins are produced and secreted by Paneth cells in the small intestine [93]. The β-defensins are ubiquitous and present in all vertebrates [94]. The human genome has more than 30 genes coding for β-defensins, and mice have even more genes coding for these peptides [95]. β-defensins are produced mainly in epithelial cells [96]. The cyclic θ-defensins arose from α-defensins after the divergence of primates and have been purified from the leukocytes of rhesus macaques and baboons [97]. These molecules are the only cyclic peptides found in mammals and exhibit antiviral activity as well [98,99].

### 3.5.3 Thionins

Plant AMPs are classified into various families: thionins, defensins, hevein- and knottin-type peptides, hairpinins, and macrocyclic peptides (cyclotides). Thionins are cysteine-rich AMPs with antimicrobial and toxic properties [100]. They are small in size (around 5 kDa) and exhibit selectivity, but happen to be toxic to a wide range of organisms and eukaryotic cell-lines though they show cell selectivity [101]. Thionins were first discovered by Ball and coworkers in 1942 when they observed an active ingredient inhibiting yeast growth [102]. They isolated and crystallized the peptide from the wheat endosperm (*T. aestivum*) and termed it as Purothionin due to high sulfur content. Purothionin was found to possess fungicidal and bactericidal properties [103]. Thionins are classified into five structural classes based on the amino acid sequences and disulfide bond arrangements. Thionins are usually extracted from natural sources by solvent extraction as proteolipid complexes indicating their association with lipids [102]. Although purothionin was first crystallized in 1942, the structure of a thionin was first reported in 1981 when Hendrickson and Teeter determined the structure of crampin [104]. Two distinct groups of thionins exist; they are α/β-thionins and γ-thionins. The γ-thionins, due to their similarity to defensins, are now classified as plant defensins [105].

### 3.5.4 Antimicrobial peptides rich in specific amino acids

Another special class of peptides, that cannot be considered a family though, is of peptides that have unusual amino acid compositions. These peptides are unusually rich in one or two amino acids. The selected members of this class are discussed here.

#### 3.5.4.1 Tryptophan-rich antimicrobial peptides

Indolicidin (named as it is an indole-rich peptide) is an AMP isolated from the cytoplasmic granules of bovine neutrophils [63]. It is a tridecapeptide amide with 5 tryptophan residues, and displays remarkable in vitro bactericidal activity [106].

It is the only indole-rich peptide in mammalian leukocytes that is amidated and stored in the granules in its mature form [106]. Due to the presence of 5 tryptophan and 3 proline residues, it takes up an extended conformation. Tritripticin is another tryptophan-rich AMP derived from a porcine precursor protein with sequence similarity to indolicidin [107,108]. Puroindolins are tryptophan-rich AMPs found in wheat that exhibit antibacterial and antifungal activities [109].

### 3.5.4.2 Proline-rich antimicrobial peptides

Several peptides are unusually rich in proline residues. PR-39, a 39-residue proline and arginine-rich peptide from porcine is one such example [110]. Bactenins Bac5 and Bac7 from *Bos taurus*, apidaecin from *Apis mellifera*, and drosocin from *D. melanogaster* are some other examples. PR-39, isolated from the porcine small intestine, is a particularly well-studied AMP of this class. Apart from the antibacterial activity, it plays various other roles. It is reported to be a neutrophil chemoattractant indicating its role in inflammation [111]. It induces syndecan expression in mesenchymal cells [112], inhibits neutrophil oxidase [111], and induces angiogenesis [113]. PR-39 does not kill the bacterium by membranolytic action; instead, it inhibits the protein synthesis, causing cell lysis [114].

### 3.5.4.3 Histatins

Histatins are histidine-rich peptides found in the saliva of humans and higher primates with potent antifungal activity [66]. They are secreted from the parotid and submandibular glands in humans [115]. Around 12 fragments of histatins have been identified that are derived from two genes—histatins 1 and 3 by proteolytic degradation [115,116]. The histatins do not cause membrane lysis, and the exact mechanism of action is not yet clear. Histatin 5, the most potent among all histatins with high activity against *Candida albicans*, is known to affect mitochondrial membrane integrity [117].

### 3.5.4.4 Unusual amino acid containing ribosomally synthesized antimicrobial peptides

AMPs produced by bacteria to combat and control the invasive microorganisms are termed bacteriocins [118]. They are a small, heat-stable, heterogeneous group of peptides with narrow or broad-spectrum activity toward other bacterial species. Certain bacteriocins contain an unusual amino acid, lanthionine, and exist with thioether rings; they are known as lantibiotics [119]. Lantibiotics are a class of ribosomally synthesized AMPs that are posttranslationally modified to have unusual amino acids, such as dehydrated and lanthionine residues. Lanthionine, a nonproteinogenic amino acid, consists of two alanine residues that are joined through their β-carbons via a thioether link. Nisin is a well-characterized lantibiotic that is commonly used as food preservative [120].

## 3.6 Relationship of structure with function

As AMPs are distributed throughout the taxonomical hierarchy, they are also diversified in their functions. Higher hydrophobicity is correlated with stronger hemolytic activity [121]. Besides, the high hydrophobicity adversely affects the peptide's antimicrobial activity. Hence, an optimum hydrophobicity is required for good antimicrobial activity [121]. The decreased antimicrobial activity of peptides with high hydrophobicity can be explained by the strong peptide self-association, which prevents the peptide from passing through the cell wall in prokaryotic cells. In contrast, increased peptide self-association does not affect the peptide's access to the eukaryotic membranes [122].

Structure—activity relationship studies of AMPs have revealed that there are many factors such as charge, secondary structure, hydrophobicity, and hydrophobic moment that affect these peptides' specificity and biological activity [123]. Each of these factors is intimately related such that altering one can result in the change in others. In general, AMPs interact with their membrane target, change the conformation by self-association, and perturb the membrane resulting in cell death [124].

Analyzing the 3D structure in solution and the membrane-bound state gives a clear understanding of the action mechanism. NMR spectroscopy happens to be the most useful tool in studying structural details. Due to the short sequences, conventional two-dimensional (2D) NMR is sufficient for solving the structure of many peptides [125]. NOESY spectra facilitate high-resolution solution structure determination for these peptides. By studying amide deuterium exchange kinetics or temperature dependence of amide chemical shifts, additional information such as hydrogen-bonding pattern can be known [126]. Even though many studies have been carried out to understand the mechanisms of AMP action, the exact sequence of actions remains a mystery. The diversity in the AMP characteristics, in conjunction with the complexity that they encounter on the microbial surface that varies from microbe to microbe, makes it challenging to understand the detailed mechanism of their action.

## 3.7 Modes of action

Microbial surfaces are decorated with anionic moieties that include negatively charged lipids such as phosphatidylglycerol and cardiolipin, lipopolysaccharides (in the case of Gram-negative bacteria), and lipoteichoic acids (in the case of Gram-positive bacteria) [127–129]. As most AMPs are cationic in nature, they selectively bind to microbial surfaces. Upon binding, the AMPs have to traverse through the thick peptidoglycan layer in the Gram-positive bacteria and the outer membrane of the Gram-negative bacteria. Once they gain access to the cytoplasmic membrane, they can either disrupt the membrane integrity or translocate across the membrane to act on the intracellular targets [41]. Several models have been proposed as far as membranolytic action is concerned.

## 3.7.1 Membrane-mediated action

The binding of the AMPs to the cytoplasmic membrane is accompanied by its folding into an amphipathic structure unless they are already folded. The amphipathicity facilitates their insertion into the membrane. This leads to the expansion of the outer leaflet accompanied by local membrane thinning. Besides, the high positive charge causes the crowding of negatively charged lipids around the peptides and their segregation from zwitterionic lipids. This phenomenon is known as "phase separation" or "phase segregation" [127,129,130]. As the concentration of AMPs in the outer leaflet of bacterial membrane increases, several molecular events are triggered, which ultimately lead to cell death; this latter set of events has been proposed to be mediated via different models [128,129,131,132]. The AAMPs can also display antimicrobial action through membrane perturbation. The binding to membranes is aided by metal ions to form cationic salt bridges with negatively charged components of the microbial membrane [44]. Interaction of dermcidin with negatively charged bacterial phospholipids is facilitated via $Zn^{2+}$-dependent oligomeric complex, leading to the ion channel formation, which results in membrane depolarization and cell death [133].

The mechanism of AMP action leading to cell lysis occurs in multiple ways. Though the exact mechanism is not clear, various models have been proposed to explain the idea (Fig. 3.6). While the mechanism of β-sheet forming peptides has

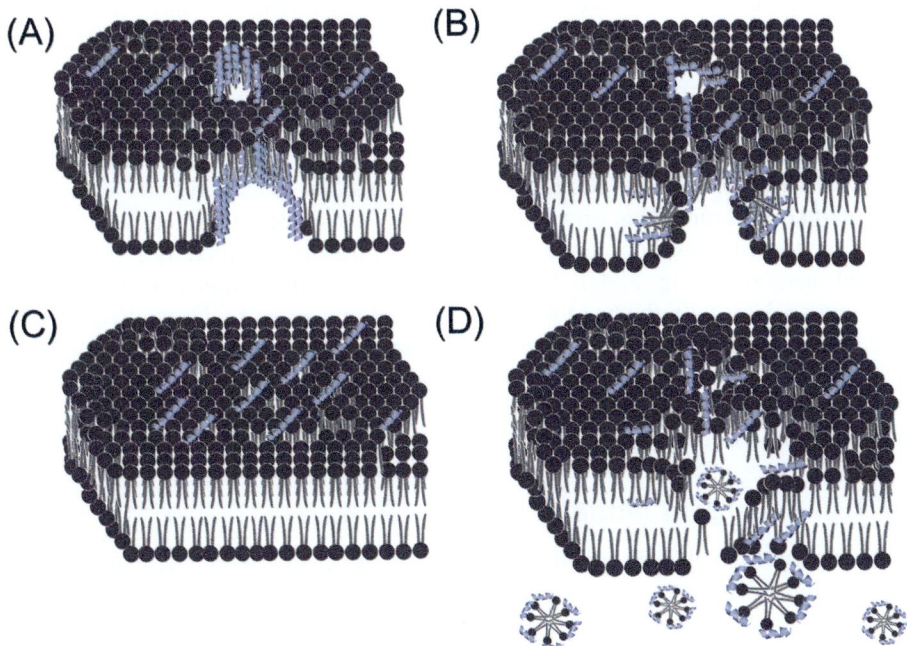

**Figure 3.6** Different models proposed for the membranolytic action of AMPs. (A) Barrel-stave model, (B) toroidal pore model, (C) carpet model, and (D) detergent model.

not been clearly elucidated, various models are proposed to demonstrate the pore-forming ability of α-helical AMPs.

### 3.7.1.1 Barrel-stave model

The first model was proposed by Baumann and Mueller in 1974 to explain the single-channel induced by alamethicin in lipid membrane [134]. The model is known as the barrel-stave model (Fig. 3.6A). In this model, the amphipathic peptides orient themselves perpendicular to the lipid membrane and form a barrel-like structure. The interior part of the channel is composed of hydrophilic regions of the peptide, while the outer hydrophobic part is surrounded by the fatty-acyl region of the phospholipid bilayer. Thus the channel acts as a conductance channel that disrupts transmembrane potential and ion gradients, leading to leakage of cell components and cell death. The fungal AMP, alamethicin, is one such example of AMPs that penetrates and aggregates to form the transmembrane pores [42].

### 3.7.1.2 Toroidal pore model

The toroidal pore model describes peptides' action by inducing a bend in the lipid bilayer where the lipid head groups orient themselves to form a pore. The pore is composed of both the peptide and phospholipid [135,136]. The toroidal pore differs from the barrel-stave model in that the AMPs are always associated with the lipid head group even when they are perpendicularly inserted into the bilayer (Fig. 3.6B). Magainin, protegrins, and melittin are proposed to act through this mode [137].

### 3.7.1.3 Carpet model

In the carpet model, the peptide molecules align parallel to the lipid bilayer forming a carpet-like structure. This orientation disrupts the membrane potential by destabilizing the packed phospholipids forming cracks in the bilayer, thereby disintegrating the cell membrane (Fig. 3.6C). The general feature of the carpet model is that the peptide associates with the head group of the lipid bilayer, not with the hydrophobic core of the membrane. Hence, without the insertion of peptide molecules into the membrane core, the membrane gets disrupted/dissolved in a dispersion-like manner and does not involve the formation of any pore/channels [57]. Aurein 1.2 is the exemplary peptide of the carpet model [138].

### 3.7.1.4 Detergent model

In the detergent model, the peptide disrupts the membrane integrity, causing leakage of cellular content, thereby causing cell death, much like a detergent [139]. It is a nonpore forming model where the catastrophic collapse of membrane integrity and probe-size-independent leakage of cytosolic components takes place (Fig. 3.6D). This mechanism of AMP action is observed mainly at the high concentrations of the peptides [139].

## 3.7.2 Membrane-independent/nonmembrane-disruptive mechanism

Though membrane-permeabilization is often associated with the antimicrobial activity of the AMPs, nonmembrane-disruptive mechanisms also exist wherein the peptides target intracellular molecules and processes [140,141]. Unlike membranolytic AMPs, these intracellularly active AMPs target DNA/protein synthesis, causing cell death [142]. Buforin 2 peptide derived from buforin 1 (from toad *Bufo gargarizans*), for example, kills the cell by binding to DNA and RNA after penetrating the cell membrane, thus inhibiting cellular functions [143,144]. Certain AMPs have chemotactic properties and induce inflammatory cytokine expression by innate immune cells, e.g., Bac5 exhibits bactericidal activity by inhibiting protein synthesis through cell-mediated immune response [145]. Translocated peptides can activate autolysins and phospholipases, cause flocculation, and inhibit septum formation (alters cytoplasmic membrane formation), e.g., PR-39, PR-26, indolicidin [146], microcin 25; inhibit cell wall formation, e.g., lantibiotic mersacidin inhibits peptidoglycan synthesis by interfering with transglycosylation [147]; bind to DNA or RNA and alter their functions, or inhibit the synthesis of nucleic acid and protein, e.g., buforin 2 [143] and tacyplesin [148]; inhibit the enzymatic activity, e.g., salivary histatins inhibit proteolytic activity of matrix metalloproteases enzymes (MMP 2 and MMP 9) that are elevated in periodontal disease [149].

## 3.8 Multifaceted roles of antimicrobial peptides

Functional diversity is another notable feature of AMPs. Besides having antimicrobial activity, they might possess other biological activities such as anticarcinogenic, antidiabetogenic, and immunomodulatory. Some of the nonantibiotic activities are discussed here.

### 3.8.1 Anticancer antimicrobial peptides

Certain AMPs can target the plasma membrane of cancer cells and are therefore termed anticancer peptides (ACPs) [150]. Certain mechanisms, both membranolytic and nonmembranolytic have been proposed, but the exact mechanisms by which these peptides kill cancer cells remain unclear [151]. Cancer cells possess a higher amount of phosphatidylserine on their cell surface compared to the normal cells. This imparts some selectivity to the ACPs toward cancerous cells [152]. The AMP dermaseptin PS-1 exhibits anticancer activity by inducing the intrinsic apoptotic pathway [153]. Tumor specificity also plays a key role in the efficacy of the ACPs. Though most ACPs are more cytotoxic to tumor cells than erythrocytes, tumor-selective peptides are not lacking [150]. Certain peptides target the cancer cells indirectly via an angiostatic mechanism. They inhibit the growth of endothelial cells, thereby inhibiting angiogenesis and the proliferation of cancer cells. Dermaseptins B2 and B3 are two such peptides [154]. The peptide pentadactylin induces DNA fragmentation and arrests the cell cycle at S phase that provokes apoptotic cell death in tumor cells [155]. Temporin 1CEa elevates the tumor cell membrane's permeabilization, thereby altering its membrane potential. It also induces intracellular calcium release and disrupts the mitochondrial membrane potential in tumor cells, causing cytotoxicity [156]. Unlike conventional drugs, these peptides trigger diverse mechanisms to kill the tumor cells; hence the chances of developing resistance against them is relatively slow.

### 3.8.2 Wound-healing antimicrobial peptides

The skin, being a highly protective barrier, prevents the organisms from infections. Unrepaired skin damage is prone to infection by various opportunistic pathogens. Wound healing is a slow and complicated process that comprises four stages: hemostasis, inflammatory response, cell proliferation, and tissue reconstruction [157]. Delayed wound healing, as in the case of diabetes patients, is also life-threatening. Accelerated wound healing, therefore, happens to be a defense mechanism against infections. A number of reports suggest that human AMPs are involved in wound healing [158]. The wounds in human skin, for example, induce expression of human β-defensins 2 and 3. Human cathelicidin LL-37 is another wound-healing peptide. Wound-inflicted human skin shows elevated levels of LL-37 [159]. The levels return to normal once the wound is closed. The mechanism behind wound-healing activity has been reviewed elsewhere [160]. Porcine cathelicidin PR-39 is another example of a wound-healing peptide [161].

### 3.8.3 Antidiabetogenic peptides

Brevinin 1ITa and brevinin 1ITb, the AMPs isolated from the skin secretions of *Rana italica*, induce insulin release from rat pancreatic β-cells [162]. Amolopin, a peptide from *Amolops loloensis* skin secretions, stimulates insulin secretion in INS-1 cells [163]. AMPs temporin A, temporin F, and temporin G from *Rana temporaria* cause concentration-dependent stimulation of insulin release without cytotoxicity (up to 3 μM concentration), and the combinational therapy involving temporin A and temporin G is proposed as a promising therapy for type II diabetes treatment [164]. Peptides from skin secretions of families Pipidae, Ranidae, and Dicroglossidae show potent insulinotropic activity by stimulating insulin release and lowering blood sugar, thus displaying antidiabetic activity [165].

### 3.8.4 Antiinflammatory and immunomodulatory peptides

HDPs have an imperative role in both innate and adaptive immune responses [2]. Nearly all mammalian AMPs have other roles alongside microbicidal activity [166]. They can induce various immunomodulatory functions such as chemotactic activity, attracting leukocytes, stimulation of angiogenesis, enhancement of leukocyte/monocyte activation and differentiation, modulation of the expression of proinflammatory cytokines/chemokines. AMPs AC12, DK16, and RC11, isolated from

*Hypsiboas raniceps* skin secretions, show antiinflammatory activity [167]. Esculentin 1GN exhibits inhibition of inflammatory response induced by LPS [168]. Frenatin 2.1S activates the cytotoxic capacity of NK cells [169], thereby acting as an antitumor immunotherapeutic agent. Frenatin 2D stimulates cytokine-mediated stimulation of adaptive immune response [170]. AMPs that predominantly possess immunomodulatory activity seem to have little antimicrobial and hemolytic activity [171].

### 3.8.5 Spermicidal peptides

Besides killing pathogens causing sexually transmitted disease, magainins [172] and dermaseptins [173] were observed to possess spermicidal activity. These peptides restricted sperm motility and cell viability in a dose-dependent manner. The exact mechanism is not known. However, the presence of these peptides altered the permeability of the plasma membrane leading to cell death. Maximins display spermicidal action by forming pores via detergent-like disruption mechanisms [174]. The spermicidal activity of these anuran peptides could be promising for the development of safer and natural contraceptives.

## 3.9 Limitations and challenges

The microbicidal action of cationic peptides and their role in innate immunity was recognized way back in the late 19th century, which drove the development of AMPs as possible new therapeutics. Antibiotic resistance is one of the critical health challenges today. Microbes develop resistance to antibiotics due to their abuse and improper medication practices. Certain species show cross-resistance to antibiotics, wherein they are more resistant to similar kinds of antibiotics. Strikingly, antibiotic-resistant bacteria show a high rate of collateral sensitivity to AMPs, whereas cross-resistance is relatively rare [175]. The main charatcteristics of innate immunity peptides are that they target the nonprotein component *i.e.* the bacterial membrane, and also possess auxiliary activities such as immunomodulatory and antiinflammatory activities [176]. Their effectiveness, biodegradability, and versatile mode of action are notable. The ability to kill the dividing and nondividing cells, biofilms, and being less prone to induce resistance are some of the striking attributes of AMPs [123,177,178]. However, they have failed the researchers as far as their clinical applications are concerned. This is due to several limitations and hindrances that need to be overcome before AMPs could find clinical applications.

### 3.9.1 Stability

The AMPs characterized by in vitro activity studies need to be tested for their stability in vivo as multiple proteolytic enzymes exist in the serum. Though effective in killing microbes, most AMPs are susceptible to proteolytic degradation, which reduces their half-life [179]. In vivo stability happens to be one of the key requirements of an effective AMP as a drug. Peptide stability is correlated with amino acid composition and lipophilicity of AMPs. Cyclic peptides or disulfide-containing peptides, specific amino acid-rich peptides, tend to be more stable due to their complex structure. Peptide cyclization appears to be an interesting strategy for serum stability without compromising the antimicrobial activity [180]. The incorporation of nonnatural amino acids also provides protection against proteolytic degradation. As AMPs display their antibacterial activity largely via membrane perturbation, the all D-peptides often show comparable activity but very high serum stability. The additional nonantibiotic features, however, are severely affected.

### 3.9.2 Toxicity

One of the major hindrances in realizing peptides as drugs is their poor selectivity. AMPs display cell selectivity that corresponds to the selective inhibitory action on microbial cells and leaving the mammalian cells intact. The mammalian cells, however, are not really spared. Many peptides possess strong hemolytic activity. Besides, the selectivity of an AMP depends on its concentration. An otherwise noncytotoxic AMP can show significant cytotoxicity at higher concentrations. This attribute can lead to deleterious effects [181].

### 3.9.3 Salt sensitivity

Salt sensitivity poses an obstacle to the development of AMPs as therapeutic agents. Retention of activity under physiological salt concentrations is vital for efficient antimicrobial activity in vivo. Lee et al. reported that histidine-rich AMPs from tunicate, clavinins show better tolerance to high salt concentration compared to magainin-1 [182]. Kim and coworkers investigated the self-assembly of an α-helical peptide $[RLLR]_5$ [183]. They found that the activity of the peptide is severely compromised in the presence of salt. This is attributed to the disruption of the peptide's helical conformation. Capping of the peptide with helix-capping motifs was shown to stabilize the helical conformation, thereby causing retention of activity. The structural stability, therefore, is important for both antimicrobial action and peptide's salt tolerance [184].

### 3.9.4 Aggregation propensity

As most AMPs are amphipathic in nature, they have a natural tendency to self-assemble in water to bury their hydrophobic patches. Aggregation is particularly pronounced in the β-sheet peptides as they can readily self-assemble to form long β-sheet-rich fibrillar assemblies [185]. Higher aggregation propensity turns out to be a challenge for the peptide synthesis as well. The peptides tend to aggregate during synthesis giving poor yields, thereby making synthesis a costlier affair. However, strategies have been devised to reduce aggregation during synthesis [186].

## 3.10 Conclusion

Peptides with improved stability, low cytotoxicity to mammalian cells, and high microbicidal activity happen to be the ideal antimicrobial therapeutics. Several strategies are being actively considered to overcome the limitations that restrain the success of AMPs. Active research is being carried out to optimize the AMPs characteristics for clinical applications, that is, to improve the antimicrobial activity and reduce the toxicity against healthy human cells [187–189]. Proteolytic cleavage of peptides can be decreased by protecting their C- and N- termini through acetylation or amidation. Substitution of L-amino acids with their D enantiomers can make the peptides resistant to proteolysis, without adversely affecting their antimicrobial activity [190,191]. $\alpha/\beta$-Substituted $\alpha$-amino acids or even $\beta$-amino acids are other similar approaches that result in overcoming peptide hydrolysis. Reducing the sequence and systematically substituting each residue with other coded amino acids could yield peptide candidates with improved antibacterial activity. Such strategies, however, might compromise the other auxiliary activities. PEGylation of proteins and peptides increases their molecular mass, thus shielding these small molecules from proteolytic enzymes [192,193]. These approaches could enhance the bioavailability of peptides and improve their biodistribution and rate of clearance. Despite the great deal of efforts that have been put in to realize the clinical applications of AMPs, only a small number of them are in the different stages of clinical trials, and only a handful in clinical applications, mostly for topical applications. There is still a long way to go before the AMPs gain the respect that conventional antibiotics keep enjoying.

## References

[1] M. Pasupuleti, A. Schmidtchen, M. Malmsten, Antimicrobial peptides: key components of the innate immune system, Critical Reviews in Biotechnology 32 (2012) 143–171.
[2] S.C. Mansour, O.M. Pena, R.E.W. Hancock, Host defense peptides: front-line immunomodulators, Trends in Immunology 35 (2014) 443–450.
[3] Y. Lai, R.L. Gallo, AMPed up immunity: how antimicrobial peptides have multiple roles in immune defense, Trends in Immunology 30 (2009) 131–141.
[4] M.L. Mangoni, A.M. McDermott, M. Zasloff, Antimicrobial peptides and wound healing: biological and therapeutic considerations, Experimental Dermatology 25 (2016) 167–173.
[5] A.L. Hilchie, K. Wuerth, R.E. Hancock, Immune modulation by multifaceted cationic host defense (antimicrobial) peptides, Nature Chemical Biology 9 (2013) 761–768.
[6] World Health Organization, Global Antimicrobial Resistance Surveillance System (GLASS) Report, Early Implementation (2020).
[7] M. Mahlapuu, J. Håkansson, L. Ringstad, C. Björn, Antimicrobial peptides: an emerging category of therapeutic agents, Frontiers in Cellular and Infection Microbiology 6 (2016) 194.
[8] P. Vlieghe, V. Lisowski, J. Martinez, M. Khrestchatisky, Synthetic therapeutic peptides: science and market, Drug Discovery Today 15 (2010) 40–56.

[9] R.J. Dubos, Studies on a bactericidal agent extracted from a soil bacillus: I. Preparation of the agent. Its activity in vitro, The Journal of Experimental Medicine 70 (1939) 1–10.
[10] R.J. Dubos, Studies on a bactericidal agent extracted from a soil bacillus: II. Protective effect of the bactericidal agent against experimental pneumococcus infections in mice, Journal of Experimental Medicine, 70, 1939, pp. 11–17.
[11] R.D. Hotchkiss, R.J. Dubos, Fractionation of the bactericidal agent from cultures of a soil bacillus, Journal of Biological Chemistry 132 (1940) 791–792.
[12] R.D. Hotchkiss, R.J. Dubos, Bactericidal fractions from an aerobic sporulating bacillus, Journal of Biological Chemistry 136 (1940) 803–804.
[13] R.D. Hotchkiss, R.J. Dubos, The isolation of bactericidal substances from cultures of *Bacillus brevis*, Journal of Biological Chemistry 141 (1941) 155–162.
[14] G.F. Gause, M.G. Brazhnikova, Gramicidin S and its use in the treatment of infected wounds, Nature 154 (1944) 703.
[15] C.R. Jones, M. Kuo, Hang, W.A. Gibbons, Multiple solution conformations and internal rotations of the decapeptide gramicidin S, Journal of Biological Chemistry 254 (1979) 10307–10312.
[16] P.J. Loll, E.C. Upton, V. Nahoum, N.J. Economou, S. Cocklin, The high resolution structure of tyrocidine A reveals an amphipathic dimer, Biochimica et Biophysica Acta 1838 (2014) 1199–1207.
[17] S. Pahr, C. Constantin, N.G. Papadopoulos, S. Giavi, M. Mäkelä, A. Pelkonen, et al., α-Purothionin, a new wheat allergen associated with severe allergy, Journal of Allergy and Clinical Immunology 132 (2013) 1000–1003. e1004.
[18] P. Hughes, E. Dennis, M. Whitecross, D. Llewellyn, P. Gage, The cytotoxic plant protein, β-purothionin, forms ion channels in lipid membranes, Journal of Biological Chemistry 275 (2000) 823–827.
[19] F.J. Colilla, A. Rocher, E. Mendez, γ-Purothionins: amino acid sequence of two polypeptides of a new family of thionins from wheat endosperm, FEBS Letters 270 (1990) 191–194.
[20] S. Oard, F. Enright, Expression of the antimicrobial peptides in plants to control phytopathogenic bacteria and fungi, Plant Cell Reports 25 (2006) 561–572.
[21] G. Kiss, H. Michl, On the venomous skin secretion of the orange speckled frog Bombina variegata, Toxicon 1 (1962) 33–39.
[22] M. Simmaco, G. Kreil, D. Barra, Bombinins, antimicrobial peptides from Bombina species, Biochimica et Biophysica Acta (BBA) - Biomembranes 1788 (2009) 1551–1555.
[23] H. Boman, Insect responses to microbial infections, Microbial Control of Pests and Plant Diseases, 1970–1980, Academic Press, New York and London, 1981, pp. 769–784.
[24] H. Steiner, D. Hultmark, Å. Engström, H. Bennich, H.G. Boman, Sequence and specificity of two antibacterial proteins involved in insect immunity, Nature 292 (1981) 246–248.
[25] T. Ganz, M.E. Selsted, D. Szklarek, S.S. Harwig, K. Daher, D.F. Bainton, et al., Defensins. Natural peptide antibiotics of human neutrophils, The Journal of Clinical Investigation 76 (1985) 1427–1435.
[26] M. Zasloff, Magainins, a class of antimicrobial peptides from Xenopus skin: isolation, characterization of two active forms, and partial cDNA sequence of a precursor, Proceedings of the National Academy of Sciences of the United States of America 84 (1987) 5449–5453.
[27] G. Wang, X. Li, Z. Wang, APD3: the antimicrobial peptide database as a tool for research and education, Nucleic Acids Research 44 (2015) D1087–D1093.

[28] F.H. Waghu, R.S. Barai, P. Gurung, S. Idicula-Thomas, CAMPR3: a database on sequences, structures and signatures of antimicrobial peptides, Nucleic Acids Research 44 (2015) D1094–D1097.
[29] L. Fan, J. Sun, M. Zhou, J. Zhou, X. Lao, H. Zheng, et al., DRAMP: a comprehensive data repository of antimicrobial peptides, Scientific Reports 6 (2016) 24482.
[30] F.H. Waghu, S. Idicula-Thomas, Collection of antimicrobial peptides database and its derivatives: applications and beyond, Protein Science 29 (2020) 36–42.
[31] P.M. Hwang, H.J. Vogel, Structure-function relationships of antimicrobial peptides, Biochemistry and Cell Biology 76 (1998) 235–246.
[32] H. Duclohier, Peptaibiotics and peptaibols: an alternative to classical antibiotics? Chemistry & Biodiversity 4 (2007) 1023–1026.
[33] D.M.E. Bowdish, D.J. Davidson, M.G. Scott, R.E.W. Hancock, Immunomodulatory activities of small host defense peptides, Antimicrobial Agents and Chemotherapy 49 (2005) 1727–1732.
[34] P. Kumar, J.N. Kizhakkedathu, S.K. Straus, Antimicrobial peptides: diversity, mechanism of action and strategies to improve the activity and biocompatibility in vivo, Biomolecules 8 (2018) 4.
[35] N. Sitaram, R. Nagaraj, The therapeutic potential of host-defense antimicrobial poeptides, Current Drug Targets 3 (2002) 259–267.
[36] M.A. Riley, J.E. Wertz, Bacteriocins: evolution, ecology, and application, Annual Review of Microbiology 56 (2002) 117–137.
[37] A. Simons, K. Alhanout, R.E. Duval, Bacteriocins, antimicrobial peptides from bacterial origin: overview of their biology and their impact against multidrug-resistant bacteria, Microorganisms 8 (2020) 639.
[38] K.L. Brown, R.E. Hancock, Cationic host defense (antimicrobial) peptides, Current Opinion in Immunology 18 (2006) 24–30.
[39] A. Tossi, L. Sandri, A. Giangaspero, Amphipathic, α-helical antimicrobial peptides, Peptide Science 55 (2000) 4–30.
[40] K. Saikia, N. Chaudhary, Antimicrobial peptides from C-terminal amphipathic region of E. coli FtsA, Biochimica et Biophysica Acta (BBA)-Biomembranes 1860 (2018) 2506–2514.
[41] K. Saikia, N. Chaudhary, Interaction of MreB-derived antimicrobial peptides with membranes, Biochemical and Biophysical Research Communications 498 (2018) 58–63.
[42] H. Sato, J.B. Feix, Peptide–membrane interactions and mechanisms of membrane destruction by amphipathic α-helical antimicrobial peptides, Biochimica et Biophysica Acta (BBA) - Biomembranes 2006 (1758) 1245–1256.
[43] R. Lai, H. Liu, W.H. Lee, Y. Zhang, An anionic antimicrobial peptide from toad Bombina maxima, Biochemical and Biophysical Research Communications 295 (2002) 796–799.
[44] H. Frederick, R.D. Sarah, A.P. David, Anionic antimicrobial peptides from eukaryotic organisms, Current Protein & Peptide Science 10 (2009) 585–606.
[45] K.P. Sai, M.V. Jagannadham, M. Vairamani, N.P. Raju, A.S. Devi, R. Nagaraj, et al., Tigerinins: novel antimicrobial peptides from the Indian FrogRana tigerina, Journal of Biological Chemistry 276 (2001) 2701–2707.
[46] D. Romeo, B. Skerlavaj, M. Bolognesi, R. Gennaro, Structure and bactericidal activity of an antibiotic dodecapeptide purified from bovine neutrophils, The Journal of Biological Chemistry 263 (1988) 9573–9575.
[47] M. Trabi, D.J. Craik, Circular proteins—no end in sight, Trends in Biochemical Sciences 27 (2002) 132–138.

[48] G. Wang, Structural biology of antimicrobial peptides by NMR spectroscopy, Current Organic Chemistry 10 (2006) 569–581.
[49] B. Bechinger, The structure, dynamics and orientation of antimicrobial peptides in membranes by multidimensional solid-state NMR spectroscopy, Biochimica et Biophysica Acta (BBA) - Biomembranes 1462 (1999) 157–183.
[50] S. Narasimhaiah, N. Ramakrishnan, Host-defense antimicrobial peptides: importance of structure for activity, Current Pharmaceutical Design 8 (2002) 727–742.
[51] R.E. Hancock, Peptide antibiotics, Lancet 349 (1997) 418–422.
[52] P.B. Timmons, D. O'Flynn, J.M. Conlon, C.M. Hewage, Structural and positional studies of the antimicrobial peptide brevinin-1BYa in membrane-mimetic environments, Journal of Peptide Science 25 (2019) e3208.
[53] D. Marion, M. Zasloff, A. Bax, A two-dimensional NMR study of the antimicrobial peptide magainin 2, FEBS Letters 227 (1988) 21–26.
[54] O. Lequin, F. Bruston, O. Convert, G. Chassaing, P. Nicolas, Helical structure of dermaseptin B2 in a membrane-mimetic environment, Biochemistry 42 (2003) 10311–10323.
[55] D. Hultmark, Å. EngstrÖM, H. Bennich, R. Kapur, H.G. Boman, Insect immunity: isolation and structure of cecropin D and four minor antibacterial components from *Cecropia Pupae*, European Journal of Biochemistry 127 (1982) 207–217.
[56] J. Johansson, G.H. Gudmundsson, M.E. Rottenberg, K.D. Berndt, B. Agerberth, Conformation-dependent antibacterial activity of the naturally occurring human peptide LL-37, Journal of Biological Chemistry 273 (1998) 3718–3724.
[57] J.P.S. Powers, R.E.W. Hancock, The relationship between peptide structure and antibacterial activity, Peptides 24 (2003) 1681–1691.
[58] I.A. Edwards, A.G. Elliott, A.M. Kavanagh, M.A.T. Blaskovich, M.A. Cooper, Structure–activity and toxicity relationships of the antimicrobial peptide tachyplesin-1, ACS Infectious Diseases 3 (2017) 917–926.
[59] D.M. Hoover, O. Chertov, J. Lubkowski, The structure of human β-defensin-1: new insights into structural properties of β-defensins, Journal of Biological Chemistry 276 (2001) 39021–39026.
[60] C. Landon, P. Sodano, C. Hetru, J. Hoffmann, M. Ptak, Solution structure of drosomycin, the first inducible antifungal protein from insects, Protein Science 6 (1997) 1878–1884.
[61] C.V. Jennings, K.J. Rosengren, N.L. Daly, M. Plan, J. Stevens, M.J. Scanlon, et al., Isolation, solution structure, and insecticidal activity of kalata B2, a circular protein with a twist: do Möbius strips exist in nature? Biochemistry 44 (2005) 851–860.
[62] S.S. Nair, J. Romanuka, M. Billeter, L. Skjeldal, M.R. Emmett, C.L. Nilsson, et al., Structural characterization of an unusually stable cyclic peptide, kalata B2 from *Oldenlandia affinis*, Biochimica et Biophysica Acta (BBA) - Proteins and Proteomics 1764 (2006) 1568–1576.
[63] N. Sitaram, C. Subbalakshmi, R. Nagaraj, Indolicidin, a 13-residue basic antimicrobial peptide rich in tryptophan and proline, interacts with Ca(2+)-calmodulin, Biochemical and Biophysical Research Communications 309 (2003) 879–884.
[64] S. Cociancich, A. Dupont, G. Hegy, R. Lanot, F. Holder, C. Hetru, et al., Novel inducible antibacterial peptides from a hemipteran insect, the sap-sucking bug Pyrrhocoris apterus, Biochemical Journal 300 (1994) 567–575.
[65] P. Casteels, C. Ampe, F. Jacobs, M. Vaeck, P. Tempst, Apidaecins: antibacterial peptides from honeybees, The EMBO Journal 8 (1989) 2387–2391.
[66] F. Oppenheim, T. Xu, F. McMillian, S. Levitz, R. Diamond, G. Offner, et al., Histatins, a novel family of histidine-rich proteins in human parotid secretion. Isolation, characterization,

primary structure, and fungistatic effects on *Candida albicans*, Journal of Biological Chemistry 263 (1988) 7472−7477.
[67] M. Graf, M. Mardirossian, F. Nguyen, A.C. Seefeldt, G. Guichard, M. Scocchi, et al., Proline-rich antimicrobial peptides targeting protein synthesis, Natural Product Reports 34 (2017) 702−711.
[68] G. Renato, Z. Margherita, B. Monica, P. Elena, M. Monica, Pro-rich antimicrobial peptides from animals: structure, biological functions and mechanism of action, Current Pharmaceutical Design 8 (2002) 763−778.
[69] T.J. Falla, D.N. Karunaratne, R.E.W. Hancock, Mode of action of the antimicrobial peptide indolicidin, Journal of Biological Chemistry 271 (1996) 19298−19303.
[70] N. Ilić, M. Novković, F. Guida, D. Xhindoli, M. Benincasa, A. Tossi, et al., Selective antimicrobial activity and mode of action of adepantins, glycine-rich peptide antibiotics based on anuran antimicrobial peptide sequences, Biochimica et Biophysica Acta (BBA) - Biomembranes 2013 (1828) 1004−1012.
[71] S.W. Radermacher, V.M. Schoop, H.J. Schluesener, Bactenecin, a leukocytic antimicrobial peptide, is cytotoxic to neuronal and glial cells, Journal of Neuroscience Research 36 (1993) 657−662.
[72] W. Bellamy, M. Takase, H. Wakabayashi, K. Kawase, M. Tomita, Antibacterial spectrum of lactoferricin B, a potent bactericidal peptide derived from the N-terminal region of bovine lactoferrin, Journal of Applied Bacteriology 73 (1992) 472−479.
[73] R. Bals, Epithelial antimicrobial peptides in host defense against infection, Respiratory Research 1 (2000) 141−150.
[74] R.E. Hancock, G. Diamond, The role of cationic antimicrobial peptides in innate host defences, Trends in Microbiology 8 (2000) 402−410.
[75] K. De Smet, R. Contreras, Human antimicrobial peptides: defensins, cathelicidins and histatins, Biotechnology Letters 27 (2005) 1337−1347.
[76] M. Tajbakhsh, A. Karimi, F. Fallah, M. Akhavan, Overview of ribosomal and non-ribosomal antimicrobial peptides produced by gram positive bacteria, Cellular and Molecular Biology 63 (2017) 20−32.
[77] L. Whitmore, J.K. Chugh, C.F. Snook, B.A. Wallace, The peptaibol database: a sequence and structure resource, Journal of Peptide Science 9 (2003) 663−665.
[78] L. Whitmore, B.A. Wallace, The peptaibol database: a database for sequences and structures of naturally occurring peptaibols, Nucleic Acids Research 32 (2004) D593−D594.
[79] J.K. Chugh, B.A. Wallace, Peptaibols: models for ion channels, Biochemical Society Transactions 29 (2001) 565−570.
[80] D.R. Storm, a K.S. Rosenthal, P.E. Swanson, Polymyxin and related peptide antibiotics, Annual Review of Biochemistry 46 (1977) 723−763.
[81] Y.-X. Li, Z. Zhong, W.-P. Zhang, P.-Y. Qian, Discovery of cationic nonribosomal peptides as Gram-negative antibiotics through global genome mining, Nature Communications 9 (2018) 3273.
[82] D.P. Mankelow, B.A. Neilan, Non-ribosomal peptide antibiotics, Expert Opinion on Therapeutic Patents 10 (2000) 1583−1591.
[83] A.E. Shinnar, K.L. Butler, H.J. Park, Cathelicidin family of antimicrobial peptides: proteolytic processing and protease resistance, Bioorganic Chemistry 31 (2003) 425−436.
[84] E.M. Kościuczuk, P. Lisowski, J. Jarczak, N. Strzałkowska, A. Jóźwik, J. Horbańczuk, et al., Cathelicidins: family of antimicrobial peptides. A review, Molecular Biology Reports 39 (2012) 10957−10970.
[85] B. Skerlavaj, R. Gennaro, L. Bagella, L. Merluzzi, A. Risso, M. Zanetti, Biological characterization of two novel cathelicidin-derived peptides and identification of

structural requirements for their antimicrobial and cell lytic activities, Journal of Biological Chemistry 271 (1996) 28375−28381.
[86] G. Wang, Structures of human host defense cathelicidin LL-37 and its smallest antimicrobial peptide KR-12 in lipid micelles, Journal of Biological Chemistry 283 (2008) 32637−32643.
[87] T. Takahashi, N.N. Kulkarni, E.Y. Lee, L.-J. Zhang, G.C.L. Wong, R.L. Gallo, Cathelicidin promotes inflammation by enabling binding of self-RNA to cell surface scavenger receptors, Scientific Reports 8 (2018) 4032.
[88] K. Kuroda, K. Okumura, H. Isogai, E. Isogai, The human cathelicidin antimicrobial peptide LL-37 and mimics are potential anticancer drugs, Frontiers in Oncology 5 (2015).
[89] E. Piktel, K. Niemirowicz, U. Wnorowska, M. Wątek, T. Wollny, K. Głuszek, et al., The Role of cathelicidin LL-37 in cancer development, Archivum Immunologiae et Therapiae Experimentalis 64 (2016) 33−46.
[90] J. Wang, M. Cheng, I.K.M. Law, C. Ortiz, M. Sun, H.W. Koon, Cathelicidin suppresses colon cancer metastasis via a P2RX7-dependent mechanism, Molecular Therapy - Oncolytics 12 (2019) 195−203.
[91] Y.Q. Tang, M.E. Selsted, Characterization of the disulfide motif in BNBD-12, an antimicrobial beta-defensin peptide from bovine neutrophils, Journal of Biological Chemistry 268 (1993) 6649−6653.
[92] M.E. Selsted, S.S. Harwig, Determination of the disulfide array in the human defensin HNP-2. A covalently cyclized peptide, The Journal of Biological Chemistry 264 (1989) 4003−4007.
[93] A.J. Ouellette, Paneth cell alpha-defensin synthesis and function, Current Topics in Microbiology and Immunology 306 (2006) 1−25.
[94] R. Lehrer, T. Ganz, Defensins of vertebrate animals, Current Opinion in Immunology 14 (2002) 96−102.
[95] A. Patil, A.L. Hughes, G. Zhang, Rapid evolution and diversification of mammalian α-defensins as revealed by comparative analysis of rodent and primate genes, Physiological Genomics 20 (2004) 1−11.
[96] J. Jarczak, E.M. Kościuczuk, P. Lisowski, N. Strzałkowska, A. Jóźwik, J. Horbańczuk, et al., Defensins: natural component of human innate immunity, Human Immunology 74 (2013) 1069−1079.
[97] D. Li, L. Zhang, H. Yin, H. Xu, J.S. Trask, D.G. Smith, et al., Evolution of primate α and θ defensins revealed by analysis of genomes, Molecular Biology Reports 41 (2014) 3859−3866.
[98] M.E. Selsted, Theta-defensins: cyclic antimicrobial peptides produced by binary ligation of truncated alpha-defensins, Current Protein and Peptide Science 5 (2004) 365−371.
[99] W. Wang, A.M. Cole, T. Hong, A.J. Waring, R.I. Lehrer, Retrocyclin, an antiretroviral θ-defensin, is a lectin, The Journal of Immunology 170 (2003) 4708−4716.
[100] B. Stec, Plant thionins—the structural perspective, Cellular and Molecular Life Sciences 63 (2006) 1370−1385.
[101] H. Bohlmann, K. Apel, Thionins, Annual Review of Plant Physiology and Plant Molecular Biology 42 (1991) 227−240.
[102] A. Balls, W. Hale, T. Harris, A crystalline protein obtained from a lipoprotein of wheat flour, Cereal Chemistry 19 (1942) 279−288.
[103] L. Stuart, T. Harris, Bactericidal and fungicidal properties of a crystalline protein isolated from unbleached wheat flour, Cereal Chemistry 19 (1942) 288−300.

[104] W.A. Hendrickson, M.M. Teeter, Structure of the hydrophobic protein crambin determined directly from the anomalous scattering of sulphur, Nature 290 (1981) 107–113.
[105] H.U. Stotz, J.G. Thomson, Y. Wang, Plant defensins: defense, development and application, Plant Signaling & Behavior 4 (2009) 1010–1012.
[106] M.E. Selsted, M.J. Novotny, W.L. Morris, Y.Q. Tang, W. Smith, J.S. Cullor, Indolicidin, a novel bactericidal tridecapeptide amide from neutrophils, The Journal of Biological Chemistry 267 (1992) 4292–4295.
[107] D.J. Schibli, L.T. Nguyen, S.D. Kernaghan, Ø. Rekdal, H.J. Vogel, Structure-function analysis of tritrpticin analogs: potential relationships between antimicrobial activities, model membrane interactions, and their micelle-bound NMR structures, Biophysical Journal 91 (2006) 4413–4426.
[108] C. Lawyer, S. Pai, M. Watabe, P. Borgia, T. Mashimo, L. Eagleton, et al., Antimicrobial activity of a 13 amino acid tryptophan-rich peptide derived from a putative porcine precursor protein of a novel family of antibacterial peptides, FEBS Letters 390 (1996) 95–98.
[109] L. Day, D.G. Bhandari, P. Greenwell, S.A. Leonard, J.D. Schofield, Characterization of wheat puroindoline proteins, The FEBS Journal 273 (2006) 5358–5373.
[110] B. Agerberth, J.Y. Lee, T. Bergman, M. Carlquist, H.G. Boman, V. Mutt, et al., Amino acid sequence of PR-39: isolation from pig intestine of a new member of the family of proline-arginine-rich antibacterial peptides, European Journal of Biochemistry 202 (1991) 849–854.
[111] H.-J. Huang, C.R. Ross, F. Blecha, Chemoattractant properties of PR-39, a neutrophil antibacterial peptide, Journal of Leukocyte Biology 61 (1997) 624–629.
[112] Y.R. Chan, R.L. Gallo, PR-39, a Syndecan-inducing antimicrobial peptide, binds and affects p130Cas*, Journal of Biological Chemistry 273 (1998) 28978–28985.
[113] J. Li, M. Post, R. Volk, Y. Gao, M. Li, C. Metais, et al., PR39, a peptide regulator of angiogenesis, Nature Medicine 6 (2000) 49–55.
[114] H.G. Boman, B. Agerberth, A. Boman, Mechanisms of action on *Escherichia coli* of cecropin P1 and PR-39, two antibacterial peptides from pig intestine, Infection and Immunity 61 (1993) 2978–2984.
[115] F.G. Oppenheim, Y.C. Yang, R.D. Diamond, D. Hyslop, G.D. Offner, R.F. Troxler, The primary structure and functional characterization of the neutral histidine-rich polypeptide from human parotid secretion, Journal of Biological Chemistry 261 (1986) 1177–1182.
[116] R.F. Troxler, G.D. Offner, T. Xu, J.C. Vanderspek, F.G. Oppenheim, Structural relationship between human salivary histatins, Journal of Dental Research 69 (1990) 2–6.
[117] E.J. Helmerhorst, W. van't Hof, P. Breeuwer, E.I. Veerman, T. Abee, R.F. Troxler, et al., Characterization of histatin 5 with respect to amphipathicity, hydrophobicity, and effects on cell and mitochondrial membrane integrity excludes a candidacidal mechanism of pore formation*, Journal of Biological Chemistry 276 (2001) 5643–5649.
[118] J.M. Willey, W.Avd Donk, Lantibiotics: peptides of diverse structure and function, Annual Review of Microbiology 61 (2007) 477–501.
[119] O. McAuliffe, R.P. Ross, C. Hill, Lantibiotics: structure, biosynthesis and mode of action, FEMS Microbiology Reviews 25 (2001) 285–308.
[120] J.N. Hansen, Nisin as a model food preservative, Critical Reviews in Food Science and Nutrition 34 (1994) 69–93.
[121] A. Hollmann, M. Martínez, M.E. Noguera, M.T. Augusto, A. Disalvo, N.C. Santos, et al., Role of amphipathicity and hydrophobicity in the balance between hemolysis

and peptide—membrane interactions of three related antimicrobial peptides, Colloids and Surfaces B: Biointerfaces 141 (2016) 528–536.
[122] J. Lei, L. Sun, S. Huang, C. Zhu, P. Li, J. He, et al., The antimicrobial peptides and their potential clinical applications, American Journal of Translational Research 11 (2019) 3919–3931.
[123] C.D. Fjell, J.A. Hiss, R.E. Hancock, G. Schneider, Designing antimicrobial peptides: form follows function, Nature reviews. Drug discovery 11 (2011) 37–51.
[124] W.C. Wimley, K. Hristova, Antimicrobial peptides: successes, challenges and unanswered questions, The Journal of Membrane Biology 239 (2011) 27–34.
[125] K. Wüthrich, NMR with proteins and nucleic acids, Europhysics News 17 (1986) 11–13.
[126] N.J. Baxter, M.P. Williamson, Temperature dependence of 1H chemical shifts in proteins, Journal of Biomolecular NMR 9 (1997) 359–369.
[127] R.M. Epand, R.F. Epand, Lipid domains in bacterial membranes and the action of antimicrobial agents, Biochimica et Biophysica Acta 1788 (2009) 289–294.
[128] S. Riedl, D. Zweytick, K. Lohner, Membrane-active host defense peptides—challenges and perspectives for the development of novel anticancer drugs, Chemistry and Physics of Lipids 164 (2011) 766–781.
[129] V. Teixeira, M.J. Feio, M. Bastos, Role of lipids in the interaction of antimicrobial peptides with membranes, Progress in Lipid Research 51 (2012) 149–177.
[130] R.F. Epand, J.E. Pollard, J.O. Wright, P.B. Savage, R.M. Epand, Depolarization, bacterial membrane composition, and the antimicrobial action of ceragenins, Antimicrobial Agents & Chemotherapy 54 (2010) 3708–3713.
[131] J. Wiesner, A. Vilcinskas, Antimicrobial peptides: the ancient arm of the human immune system, Virulence 1 (2010) 440–464.
[132] N. Sitaram, R. Nagaraj, Interaction of antimicrobial peptides with biological and model membranes: structural and charge requirements for activity, Biochimica et Biophysica Acta (BBA) - Biomembranes 1462 (1999) 29–54.
[133] M. Burian, B. Schittek, The secrets of dermcidin action, International Journal of Medical Microbiology 305 (2015) 283–286.
[134] G. Baumann, P. Mueller, A molecular model of membrane excitability, Journal of Supramolecular Structure 2 (1974) 538–557.
[135] L.M. Gottler, A. Ramamoorthy, Structure, membrane orientation, mechanism, and function of pexiganan—a highly potent antimicrobial peptide designed from magainin, Biochimica et Biophysica Acta (BBA) - Biomembranes 2009 (1788) 1680–1686.
[136] H.W. Huang, Action of antimicrobial peptides: two-state model, Biochemistry 39 (2000) 8347–8352.
[137] K.J. Hallock, D.-K. Lee, A. Ramamoorthy, MSI-78, an analogue of the magainin antimicrobial peptides, disrupts lipid bilayer structure via positive curvature strain, Biophysical Journal 84 (2003) 3052–3060.
[138] D.I. Fernandez, A.P. Le Brun, T.C. Whitwell, M.-A. Sani, M. James, F. Separovic, The antimicrobial peptide aurein 1.2 disrupts model membranes via the carpet mechanism, Physical Chemistry Chemical Physics 14 (2012) 15739–15751.
[139] W.C. Wimley, Describing the mechanism of antimicrobial peptide action with the interfacial activity model, ACS Chemical Biology 5 (2010) 905–917.
[140] M. Cudic, L. Otvos Jr., Intracellular targets of antibacterial peptides, Current Drug Targets 3 (2002) 101–106.
[141] A. Bera, S. Singh, R. Nagaraj, T. Vaidya, Induction of autophagic cell death in Leishmania donovani by antimicrobial peptides, Molecular and Biochemical Parasitology 127 (2003) 23–35.

[142] K.A. Brogden, Antimicrobial peptides: pore formers or metabolic inhibitors in bacteria? Nature Reviews Microbiology, 3, 2005, pp. 238−250.
[143] C.B. Park, H.S. Kim, S.C. Kim, Mechanism of action of the antimicrobial peptide buforin II: buforin II kills microorganisms by penetrating the cell membrane and inhibiting cellular functions, Biochemical and Biophysical Research Communications 244 (1998) 253−257.
[144] C. Muñoz-Camargo, V.A. Salazar, L. Barrero-Guevara, S. Camargo, A. Mosquera, H. Groot, et al., Unveiling the multifaceted mechanisms of antibacterial activity of buforin II and frenatin 2.3S peptides from skin micro-organs of the Orinoco Lime Treefrog (*Sphaenorhynchus lacteus*), International Journal of Molecular Sciences 19 (2018) 2170.
[145] T. Munshi, A. Sparrow, B.W. Wren, R. Reljic, S.J. Willcocks, The antimicrobial peptide, bactenecin 5, supports cell-mediated but not humoral immunity in the context of a mycobacterial antigen vaccine model, Antibiotics 9 (2020) 926.
[146] C.-H. Hsu, C. Chen, M.-L. Jou, A.Y.-L. Lee, Y.-C. Lin, Y.-P. Yu, et al., Structural and DNA-binding studies on the bovine antimicrobial peptide, indolicidin: evidence for multiple conformations involved in binding to membranes and DNA, Nucleic Acids Research 33 (2005) 4053−4064.
[147] H. Brötz, G. Bierbaum, A. Markus, E. Molitor, H.G. Sahl, Mode of action of the lantibiotic mersacidin: inhibition of peptidoglycan biosynthesis via a novel mechanism? Antimicrobial Agents and Chemotherapy 39 (1995) 714−719.
[148] A. Yonezawa, J. Kuwahara, N. Fujii, Y. Sugiura, Binding of tachyplesin I to DNA revealed by footprinting analysis: significant contribution of secondary structure to DNA binding and implication for biological action, Biochemistry 31 (1992) 2998−3004.
[149] H. Gusman, J. Travis, E.J. Helmerhorst, J. Potempa, R.F. Troxler, F.G. Oppenheim, Salivary histatin 5 is an inhibitor of both host and bacterial enzymes implicated in periodontal disease, Infection and Immunity 69 (2001) 1402−1408.
[150] B. Deslouches, Y.P. Di, Antimicrobial peptides with selective antitumor mechanisms: prospect for anticancer applications, Oncotarget 8 (2017) 46635−46651.
[151] D. Gaspar, A.S. Veiga, M.A.R.B. Castanho, From antimicrobial to anticancer peptides. A review, Frontiers in Microbiology 4 (2013).
[152] L. Wang, C. Dong, X. Li, W. Han, X. Su, Anticancer potential of bioactive peptides from animal sources (review), Oncology Reports 38 (2) (2017) 637−651.
[153] Q. Long, L. Li, H. Wang, M. Li, L. Wang, M. Zhou, et al., Novel peptide dermaseptin-PS1 exhibits anticancer activity via induction of intrinsic apoptosis signalling, Journal of Cellular and Molecular Medicine 23 (2019) 1300−1312.
[154] H. van Zoggel, Y. Hamma-Kourbali, C. Galanth, A. Ladram, P. Nicolas, J. Courty, et al., Antitumor and angiostatic peptides from frog skin secretions, Amino Acids 42 (2012) 385−395.
[155] M.S. Libério, G.A. Joanitti, R.B. Azevedo, E.M. Cilli, L.C. Zanotta, A.C. Nascimento, et al., Anti-proliferative and cytotoxic activity of pentadactylin isolated from Leptodactylus labyrinthicus on melanoma cells, Amino Acids 40 (2011) 51−59.
[156] C. Wang, L.-L. Tian, S. Li, H.-B. Li, Y. Zhou, H. Wang, et al., Rapid cytotoxicity of antimicrobial peptide tempoprin-1CEa in breast cancer cells through membrane destruction and intracellular calcium mechanism, PLoS ONE 8 (2013) e60462.
[157] P. Martin, Wound healing−aiming for perfect skin regeneration, Science 276 (1997) 75.
[158] L.H. Elgarhy, M.M. Shareef, S.M. Moustafa, Granulysin expression increases with increasing clinical severity of psoriasis, Clinical and Experimental Dermatology 40 (2015) 361−366.

[159] R.A. Dorschner, V.K. Pestonjamasp, S. Tamakuwala, T. Ohtake, J. Rudisill, V. Nizet, et al., Cutaneous injury induces the release of cathelicidin anti-microbial peptides active against group A streptococcus, Journal of Investigative Dermatology 117 (2001) 91–97.

[160] A. Duplantier, M. van Hoek, The human cathelicidin antimicrobial peptide LL-37 as a potential treatment for polymicrobial infected wounds, Frontiers in Immunology 4 (2013).

[161] E.J. Veldhuizen, V.A. Schneider, H. Agustiandari, A. van Dijk, J.L. Tjeerdsma-van Bokhoven, F.J. Bikker, et al., Antimicrobial and immunomodulatory activities of PR-39 derived peptides, PLoS ONE 9 (2014) e95939.

[162] J.M. Conlon, V. Musale, S. Attoub, M.L. Mangoni, J. Leprince, L. Coquet, et al., Cytotoxic peptides with insulin-releasing activities from skin secretions of the Italian stream frog Rana italica (Ranidae), Journal of Peptide Science 23 (2017) 769–776.

[163] G.-X. Mo, X.-W. Bai, Z.-J. Li, X.-W. Yan, X.-Q. He, M.-Q. Rong, A novel insulinotropic peptide from the skin secretions of Amolops loloensis Frog, Natural Products and Bioprospecting 4 (2014) 309–313.

[164] V. Musale, B. Casciaro, M.L. Mangoni, Y.H.A. Abdel-Wahab, P.R. Flatt, J.M. Conlon, Assessment of the potential of temporin peptides from the frog Rana temporaria (Ranidae) as anti-diabetic agents, Journal of Peptide Science 24 (2018) e3065.

[165] J.M. Conlon, M. Mechkarska, Y.H. Abdel-Wahab, P.R. Flatt, Peptides from frog skin with potential for development into agents for Type 2 diabetes therapy, Peptides 100 (2018) 275–281.

[166] C. Chessa, C. Bodet, C. Jousselin, M. Wehbe, N. Lévêque, M. Garcia, Antiviral and immunomodulatory properties of antimicrobial peptides produced by human keratinocytes, Frontiers in Microbiology 11 (2020).

[167] C.S.F.C. Popov, B.S. Magalhães, B.J. Goodfellow, A.L. Bocca, D.M. Pereira, P.B. Andrade, et al., Host-defense peptides AC12, DK16 and RC11 with immunomodulatory activity isolated from *Hypsiboas raniceps* skin secretion, Peptides 113 (2019) 11–21.

[168] B. Zeng, J. Chai, Z. Deng, T. Ye, W. Chen, D. Li, et al., Functional characterization of a novel lipopolysaccharide-binding antimicrobial and anti-inflammatory peptide in vitro and in vivo, Journal of Medicinal Chemistry 61 (2018) 10709–10723.

[169] J.M. Pantic, I.P. Jovanovic, G.D. Radosavljevic, N.M. Gajovic, N.N. Arsenijevic, J.M. Conlon, et al., The frog skin host-defense peptide frenatin 2.1S enhances recruitment, activation and tumoricidal capacity of NK cells, Peptides 93 (2017) 44–50.

[170] J.M. Conlon, M. Mechkarska, J.M. Pantic, M.L. Lukic, L. Coquet, J. Leprince, et al., An immunomodulatory peptide related to frenatin 2 from skin secretions of the Tyrrhenian painted frog Discoglossus sardus (Alytidae), Peptides 40 (2013) 65–71.

[171] J.M. Pantic, I.P. Jovanovic, G.D. Radosavljevic, N.N. Arsenijevic, J.M. Conlon, M.L. Lukic, The potential of frog skin-derived peptides for development into therapeutically-valuable immunomodulatory agents, Molecules 22 (2017) 2071.

[172] K.V.R. Reddy, S.K. Shahani, P.K. Meherji, Spermicidal activity of Magainins: in vitro and in vivo studies, Contraception 53 (1996) 205–210.

[173] A. Zairi, C. Serres, F. Tangy, P. Jouannet, K. Hani, In vitro spermicidal activity of peptides from amphibian skin: dermaseptin S4 and derivatives, Bioorganic & Medicinal Chemistry 16 (2008) 266–275.

[174] R. Lai, Y.T. Zheng, J.H. Shen, G.J. Liu, H. Liu, W.H. Lee, et al., Antimicrobial peptides from skin secretions of Chinese red belly toad Bombina maxima, Peptides 23 (2002) 427–435.

[175] V. Lázár, A. Martins, R. Spohn, L. Daruka, G. Grézal, G. Fekete, et al., Antibiotic-resistant bacteria show widespread collateral sensitivity to antimicrobial peptides, Nature Microbiology 3 (2018) 718–731.

[176] M. Zasloff, Antimicrobial peptides of multicellular organisms, Nature 415 (2002) 389–395.
[177] A. Peschel, H.G. Sahl, The co-evolution of host cationic antimicrobial peptides and microbial resistance, Nature Reviews. Microbiology 4 (2006) 529–536.
[178] Y. Huang, L. He, G. Li, N. Zhai, H. Jiang, Y. Chen, Role of helicity of α-helical antimicrobial peptides to improve specificity, Protein & Cell 5 (2014) 631–642.
[179] J. Wang, V. Yadav, A.L. Smart, S. Tajiri, A.W. Basit, Stability of peptide drugs in the colon, European Journal of Pharmaceutical Sciences 78 (2015) 31–36.
[180] K. Anne, T. Marianna, Recent advances in the development of antimicrobial peptides (AMPs): attempts for sustainable medicine? Current Medicinal Chemistry 25 (2018) 2503–2519.
[181] K. Matsuzaki, Control of cell selectivity of antimicrobial peptides, Biochimica et Biophysica Acta (BBA) - Biomembranes 2009 (1788) 1687–1692.
[182] I.H. Lee, Y. Cho, R.I. Lehrer, Effects of pH and salinity on the antimicrobial properties of clavanins, Infection and Immunity 65 (1997) 2898–2903.
[183] I.Y. Park, J.H. Cho, K.S. Kim, Y.-B. Kim, M.S. Kim, S.C. Kim, Helix stability confers salt resistance upon helical antimicrobial peptides*, Journal of Biological Chemistry 279 (2004) 13896–13901.
[184] J.P. Tam, Y.-A. Lu, J.-L. Yang, Correlations of cationic charges with salt sensitivity and microbial specificity of cystine-stabilized β-strand antimicrobial peptides*, Journal of Biological Chemistry 277 (2002) 50450–50456.
[185] G. Bellesia, J.-E. Shea, Effect of β-sheet propensity on peptide aggregation, The Journal of Chemical Physics 130 (2009) 145103.
[186] I. Coin, M. Beyermann, M. Bienert, Solid-phase peptide synthesis: from standard procedures to the synthesis of difficult sequences, Nature Protocols 2 (2007) 3247–3256.
[187] Y. Huang, J. Huang, Y. Chen, Alpha-helical cationic antimicrobial peptides: relationships of structure and function, Protein & Cell 1 (2010) 143–152.
[188] H.M. Han, R. Gopal, Y. Park, Design and membrane-disruption mechanism of charge-enriched AMPs exhibiting cell selectivity, high-salt resistance, and anti-biofilm properties, Amino Acids 48 (2016) 505–522.
[189] N. Sitaram, K.P. Sai, S. Singh, K. Sankaran, R. Nagaraj, Structure-function relationship studies on the frog skin antimicrobial peptide tigerinin 1: design of analogs with improved activity and their action on clinical bacterial isolates, Antimicrobial Agents and Chemotherapy 46 (2002) 2279–2283.
[190] T.G. Castro, N.M. Micaêlo, M. Melle-Franco, Modeling the secondary structures of the peptaibols antiamoebin I and zervamicin II modified with D-amino acids and proline analogues, Journal of Molecular Modeling 23 (2017) 313.
[191] K. Hamamoto, Y. Kida, Y. Zhang, T. Shimizu, K. Kuwano, Antimicrobial activity and stability to proteolysis of small linear cationic peptides with D-amino acid substitutions, Microbiology and Immunology 46 (2002) 741–749.
[192] J.M. Harris, R.B. Chess, Effect of pegylation on pharmaceuticals, Nature Reviews. Drug Discovery 2 (2003) 214–221.
[193] R. Manteghi, E. Pallagi, G. Olajos, I. Csóka, Pegylation and formulation strategy of anti-microbial peptide (AMP) according to the quality by design approach, European Journal of Pharmaceutical Sciences 144 (2020) 105197.

# Purification and characterization of antimicrobial peptides

A.R. Sarika and Arunan Chandravarkar
Kerala State Council for Science, Technology and Environment, Thiruvananthapuram, Kerala, India

## 4.1 Purification techniques

The antimicrobial peptides (AMPs) are evolutionary ancient tools which are widely distributed throughout the animal and plant kingdom. These small cationic peptides are multifunctional as effectors of innate immunity and they have a fundamental role in the evolution of complex multicellular organisms [1,2]. They are present on the skin and mucosas of animals and in certain plants and have direct activity against bacteria, viruses, fungi, and even parasites [3–5]. The antimicrobial peptides database (APD http://aps.unmc.edu/AP/main.html) contains a total of 2619 AMPs from various sources and the number is increasing with the addition of new molecules from other sources, and chemically modified versions. While bacteria, fungi, and viruses develop resistance to drugs, AMPs are promising weapons to these agents [4].

These peptides from different sources are extracted using several techniques. The technique varies and it depends on the nature of source. In the case of invertebrates, the main source of AMP is hemolymph, so before purification starts, the crude peptide mixture is centrifuged to separate the plasma, but in the case of very small animals the protein/peptide extraction is performed from the total body of the animals. In the case of cultured cells, the medium can be submitted directly to purification without any concentration. As the separation techniques of AMPs from their natural source are beyond the scope of this chapter, readers are encouraged to refer to these reviews [6–12].

The level of purification is decided by the final intended use of the peptide. While therapeutic and in vivo studies demand 100% purity, about 98% to 99% purity is enough for characterization studies. For antibody production or N-terminal sequencing or chemical modification, the purity at the level of 95% is reasonable, provided the remaining impurities are inert. Though high-performance liquid chromatography (HPLC) alone can give pure peptide when the peptide is synthesized by chemical process, more than one purification step is required to get desired purity when the peptide/protein is isolated from their natural sources. The following methods are generally used for the purification of AMPs.

## 4.1.1 Solid-phase extraction on C18 column

This is the first step in the purification stage of many AMPs and one of the major methods of first step of purification besides the ion-exchange chromatography (IEC) [13–18]. The sample is acidified with trifluoroacetic acid (TFA) or acetic acid before it is loaded onto the C18 column. The column is eluted sequentially with low, medium, and high percentage of acetonitrile in water. Various salts, sugars, and water-soluble impurities including hydrophilic peptides and proteins are removed in the initial washing stage itself. The hydrophobic AMPs are retained on the column which are then eluted at high acetonitrile washing. Hetru and Bulet described detailed protocol on using solid-phase extraction (SPE) in the separation of AMP from invertebrate [7].

## 4.1.2 Ion-exchange chromatography

IEC is a process wherein the peptide solution after isolated from the natural source is brought in contact with an ion exchange resin, and the active ions on the resin are replaced by ions of similar charge from the peptide solution. This is also one of the first step of purification of crude peptide. The selection of SPE or IEC or both will depend on the knowledge of the peptide characteristics. Both cation and anion exchange columns are available, and since most of the AMPs are cationic in nature, cation exchange resins are appropriate supports in the first level of purification. The AMP extract is passed through the cation-exchange column (Fig. 4.1), washed with a phosphate buffer containing NaCl at concentration of 50–100 mM (this step will remove the anionic contaminants), and the peptide is eluted with a gradient of NaCl from 100 to 1000 mM [11]. Ion-exchange columns have proven particularly useful in a multistep protocol for peptide separations, particularly prior to a final reversed-phase HPLC (RP-HPLC) purification and desalting step [19]. Since ion exchange is an adsorption technique, it can also be used in either positive or negative capture modes. Depending on the pH or conductivity of the sample, the target may adsorb while the contaminant is unretained. When cationic peptides are loaded on an anion exchanger, the impurities having anionic nature are bound on the anion resin eluting

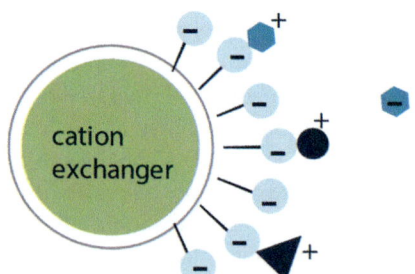

**Figure 4.1** Principle of ion-exchange chromatography. Cationic peptides are bound to the negatively charged matrix, while anionic and neutral peptides are eluted out of the column.

the cationic peptides. DEAE Sephadex is a weak anion exchanger, and its working pH is 2—9; QAE Sephadex is a strong anion exchanger, and its working pH is 2—12; and CM-Sephadex and SP-Sephadex are weak cation and strong cation exchangers, respectively, and their working pH ranges are 6—10 and 4—13, respectively [19—21].

Proteins are zwitterionic and thus can carry either a net positive or a net negative charge. In theory, each protein/peptide could be purified using either a cation or an anion exchanger by varying the pH of the buffer solution used as the mobile phase. However, in practice, proteins are not stable at every pH. Hence, protein stability and buffer choice therefore dictate ion exchange media a choice for protein purification.

### 4.1.3 Gel permeation chromatography

This is an extremely useful method to peptide separations, possibly as the first step in the resolution of a complex peptide mixture. In this method the separation is achieved based on the relative size of the molecules (Fig. 4.2). The stationary phase is a porous polymer matrix. The pores are filled with the solvent which is used as the mobile phase. Molecules above the pore size of the matrix is totally excluded from the pores and smaller molecules are trapped inside the pores. Molecules which are completely trapped and that are completely excluded are not separated. Molecules which are similar chemical nature are eluted from the pores in the order of their relative molecular mass. Most of the stationary phase used in high-performance gel permeation chromatography (GPC) are weakly anionic in nature [22]. So, when AMP which are mostly cationic in nature is passed through GPC it strongly adheres to the stationary phase. Using salt of increasing ionic strength can help in eluting the peptide from the column. Usually for the purification of AMPs, Sephadex G-100 and 0.02 M sodium acetate buffer containing 0.1—0.4 M salts are used and column elution is also conducted by using the same buffer [10]. GPC is

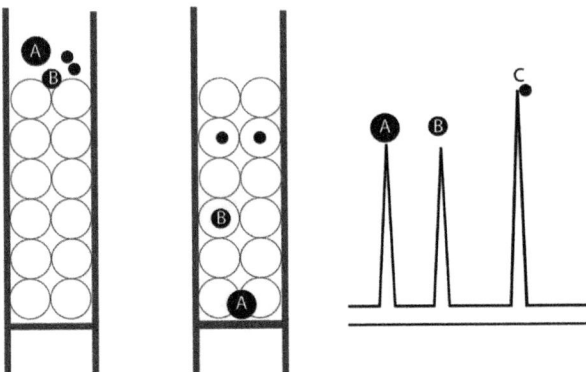

**Figure 4.2** Principle of gel permeation chromatography. Peptides having different sizes are eluted at different times while passing through the pores of stationary phase.

generally used to desalting of ions from the crude peptide/protein, so it is used just after salt precipitation of peptide from its source just like solid phase extraction method.

## 4.1.4 Affinity chromatography

Unlike other methods of separation technique, affinity chromatography (AC) is very specific to peptide/proteins [23]. The principle of AC is that a ligand which shows specific binding affinity to a protein/peptide is bound to a gel matrix and the crude peptide is applied on the top of the column with suitable buffered eluent (Fig. 4.3). The protein/peptide when in contact with the ligand becomes irreversibly bound and the impurities are washed through the column. The pH or composition of the eluent is changed so that the protein/peptide−ligand interaction is weakened and thereby the peptide is separated. AC has considerable potential and it is extensively used in biochemical studies. It provides an elegant method for high yield purification in a single step under mild condition of pH and ionic strength.

This method has been utilized efficiently for isolation of AMP-18 from human seminal plasma [20]. In membrane AC, the conventional AC is modified for efficient screening of peptides. The highly efficient process of membrane-chromatographic separation is based on the use of thin layers of finely organized and well-controlled microporous polymeric stationary phases in the form of rigid disks of 2 to 3 mm thickness [24]. The advantage of this method over conventional AC is time saving and recovery activity. Membrane AC has been used for screening AMP from *Jatropha curcas* [25].

## 4.1.5 Membrane filtration

Advances in material sciences and membrane manufacturing technology have helped this method to develop into an important separation technique for peptides.

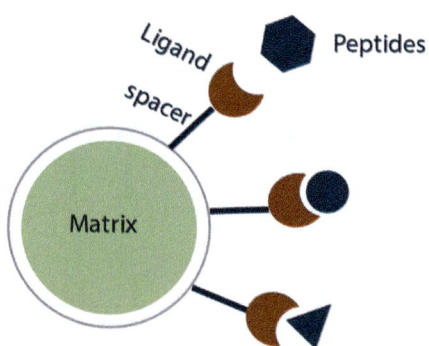

**Figure 4.3** Principle of affinity chromatographic binding. Peptides which can specifically bind to the ligand are attached and other peptides are eluted. Later by changing the pH or composition of the eluent, the bound peptide is released and collected.

Membrane-based separation is performed at room temperature, hence very suitable for heat-sensitive peptides [26].

In normal conditions, the peptides have 3–20 amino acid residues, and membranes with a molecular weight cut-off (MWCO) at 1–10 kDa are suitable for the fractionation of bioactive peptides with desired molecular weights. Membranes with a low MWCO at approximately <1 kDa are used to concentrate the peptides [6]. The membrane filtration in combination with electrophoresis-electro-membrane is used for separation of bioactive peptides [27]. However, the short-life of membrane makes the method membrane filtration less attractive.

### 4.1.6 High-performance liquid chromatography

The development of HPLC packings and instrumentation has revolutionized the efficiency and speed of separation peptides and proteins. HPLC methods are derived from the open column methods that are described above. The performance of open columns is low and they are time consuming. Other disadvantages of classical liquid chromatography are as follows:

1. The column packing procedures are tedious and the columns are used only once and thus the technique is very expensive.
2. The particle used in traditional liquid chromatography is large and hence the efficiency in separation is low and the analysis time is very lengthy.
3. The detection of peptides that elutes from the column is achieved by manual process, and this makes it very labor intensive and time consuming.

All the methods used in open column (gel permeation, reversed phase, and ion-exchange) are adapted in HPLC. The introduction of HPLC has greatly influenced in the peptide chemistry and it helped in speedy discovery and characterization of newer peptides and proteins. High performance in HPLC is achieved by using smaller particle-sized matrix, and this increases surface area of the stationary phase and thereby enhances separation. High pressure allows high flow rate which helps in fast separation of peptides. Using binary or ternary solvent system in a controlled and programmed way, high-resolution separation can be achieved. HPLC has the advantage that the columns are reusable, the sample introduction to the stationary phase is automated and integrated detection techniques helps in quantification of the eluting molecules. These positive features combined with improved efficiency and accuracy makes HPLC one of the top tools in separation techniques. The importance of HPLC is evidenced by a vast growth in published scientific papers, which cite the technique as the chosen method of analysis, and in peptide chemistry this has become the main analytical tool today.

#### 4.1.6.1 High-performance gel permeation chromatography

The separation in GPC is achieved by selective diffusion of molecules based on their size, within the pores of the 3-D lattice of the stationary phase. Usual Gel Permeation matrix such as dextran or styrene-divinylbenzene (DVB) matrix that are used in open column are seldom used in high-performance GPC, because of their poor mechanical strength and their high swelling behavior in organic and aqueous media. Organic gels formed from copolymerization of 2-hydroxyethyl methacrylate and ethyl dimethacrylate and styrene-DVB-based supports on rigid packings-porous silica are used in high-performance gel permeation chromatography (HPGPC) as stationary phase. TSK gel-based columns are widely used in the HPLC as column (TOSOH BIOSCIENCE, Separation Report No. 28). Besides its use in separation of peptides, HPGPC is used in molecular weight determination of peptides and proteins [28–31].

### 4.1.6.2 Cation-exchange high-performance liquid chromatography

Cation-exchange chromatography, more specifically, uses a negatively charged ion-exchange resin with an affinity for molecules having net positive surface charges. Cation-exchange chromatography is used both for preparative and analytical purposes and can separate a large range of molecules from amino acids and nucleotides to large proteins. Many known AMPs are cationic in nature, and this property makes cation exchange an attractive mode of separation. Compared to GPC, cation-exchange method offers higher resolution, but less compared to reversed phase chromatography. Another disadvantage is that peptides that are separated using cation exchange need to be desalted before they are used.

### 4.1.6.3 Reversed-phase high-performance liquid chromatography

RP-HPLC has become a main analytical technique in the separation and analysis of peptides and proteins. RP-HPLC is able to separate peptides of nearly identical sequences. Insulin-like growth factor with an oxidized Met and its nonoxidized analogs [32] and interleukin-2 muteins [33] have been separated using RP-HPLC. RP-HPLC is used to purify micro quantities of peptides for sequencing [34] and to purify milligram to kilogram scale for various purposes therapeutic assays [35,36]. When large quantities of peptides are to be purified, preparative HPLC is used. The flow rate of mobile phase and the columns are so optimized that the chromatographic profile obtained in analytical HPLC matches with that obtains from preparative HPLC. Preparative HPLC is frequently used to purify peptides in milligram and gram scale quantities [37–40].

The peptides adsorb on to the hydrophobic surface of the column and remain absorbed till the organic components of the mobile phase reach a critical concentration to elute the peptide/protein molecule. Different types of reverse phase columns used are C4, C8, and C18 depending on the nature of the hydrophobicity of the peptide or protein targeted. RP-HPLC separation is resulted from subtle interaction between peptides and hydrophobic environment of the reversed phase. Hence it is best to try different reversed phase columns to get the best result. In general, C18 columns are generally preferred for peptides and small protein less than 5 kDa and protein/peptides over 5 kDa are best separated using C4 column.

The desorption of peptide from the reversed phase column is affected with aqueous solvents containing organic solvent (modifiers) and ion-pair reagent/buffer. The ion-pair agent sets the eluent pH and interact with peptide molecules thereby enhance the separation. Some of the organic modifiers are as follows. (1) *Acetonitrile*: it is most commonly used. The merits of using acetonitrile are: its low viscosity, hence low back-pressure develops on the column; the solvent is low boiling; hence it can easily be removed from the peptide solution; and its absorption at low UV wavelength where the eluting peptides are monitored is very low. (2) *Isopropanol*: it is used for very hydrophobic peptides. Its high viscosity makes it less preferable mobile phase in HPLC. Using a combination of acetonitrile and isopropyl alcohol in 1:1 ratio can surpass this disadvantage to certain extent. (3) *Ethanol*: it is a good RP-HPLC solvent in separating peptides/proteins especially membrane-spanning proteins.

Ion-pairing reagents/buffers enhance the separation of peptides. TFA is the most commonly used ion-pairing reagent used in RP-HPLC. Normally 0.1%–0.5% (w/v) TFA is used. Formic acid (10%–60%) is another reagent which is used for very hydrophobic peptides. Heptaflurobutyric acid (HFTBA) is effective in separating basic proteins [40] and triethylamine phosphate (TEAP) is used for preparative separations [39].

## 4.2 Characterization techniques

After isolation and purification of intended peptides, it is essential to get an insight into the structure of the AMPs. Various fractions isolated from HPLC can be used for antimicrobial assay studies to get in information about the required activity of each fraction in the crude mixture. While this gives only the component in the crude that is responsible for its biological activity, it does not offer the amino acid composition or sequence of the peptides (primary structure). Some of the important techniques used for characterization of AMPs are (1) amino acid analysis; (2) sequencing; (3) GPC; (4) two-dimensional poly acrylamide gel electrophoresis (2D-PAGE); and (5) mass spectrometry. Besides these techniques, infra-red spectroscopy, circular dichroism, and nuclear magnetic resonance spectroscopy are extensively used to reveal the secondary and tertiary structures of the peptides.

### 4.2.1 Amino acid analysis

In this method, the peptide is acid digested to separate it to form individual amino acids. The amino acids thus formed are derivatized using suitable reagent and the mixture is subjected to HPLC separation [41–46]. The retention time and peak area of different amino acids eluted are compared with a standard run and the amino acid composition of the peptide is determined.

Acid sensitive amino acids like tryptophan, cysteine, tyrosine, etc., cannot be detected by this method. Asparagine and glutamine are amide derivatives of aspartic acid and glutamic acid, respectively. During acid hydrolysis, asparagine and glutamine are converted to respective acids. Thus the amount determined for aspartic acid represents the total of aspartic acid and asparagine and similarly for glutamic acid and glutamine.

### 4.2.2 Sequencing—Edman procedure

In the Edman procedure, the reagent phenylisothiocynate (PITC) couples with the terminal alpha amino group of a peptide or protein which is bound to a stationary phase by C-terminus to form a phenylthiocarbamyl (PTC) adduct. Under anhydrous acidic conditions, the N-terminal amino acid residue adduct is selectively cleaved from the peptide chain as a heterocyclic derivative. The cleaved amino acid derivative is separated from the residual peptide by extraction with an organic solvent and then converted to a more stable isomer and identified. The procedure is repeated for the next amino acid and the whole process is completely automated [47,48]. Edman sequencing works satisfactorily for peptides of 60−150 residues, and for longer peptides the peptides are shortened enzymatically and subjected to Edman sequencing.

### 4.2.3 Two dimensional—poly acrylamide gel electrophoresis

Two-dimensional gel electrophoresis (2DGel) is a most widely used method for the detection and analysis of proteins. This technology separates the samples by two consecutive techniques: isoelectric focusing, which discriminates proteins based on their isoelectric point, followed by sodium dodecyl sulfate polyacrylamide gel electrophoresis (SDS-PAGE), which discriminates proteins based on their molecular weight [49,50]. Due to the highly cationic behavior, conventional SDS-PAGE analyses often do not work with AMPs [51]. Instead, acidic acryl amide electrophoresis was initially used for checking purity of the AMPs [52]. The Tricine−SDS-PAGE system [53] has been now successfully used for determination of the AMP's size [54]. It is important to choose electrophoresis conditions which allow sufficient loading of SDS to the highly cationic AMPs and minimize formation of dimers and oligomers. Tricine buffers containing urea were found to be optimal. Due to its high sensitivity, silver staining is preferred for detection of AMPs in the gel, because only nanogram amounts of AMPs are required [51].

### 4.2.4 Mass spectrometry

Mass spectrometry (MS) is a central analytical technique in protein and peptide research due to its sensitivity, versatility, and speed. In recent years a wide range of mass spectrometry-based analytical tools have emerged which helps the characterization of peptide and protein with utmost precision and speed. Earlier mass spectrometry was restricted to small, thermostable, and easily volatile organic compounds. The development of electron spray ionization (ESI) and matrix-assisted laser desorption ionization (MALDI) enabled soft ionization of protein and peptides without completely degrading it. MS can provide accurate molecular weights for peptides and proteins with masses up to 500 kDa using only a few picomoles. Detailed reviews on mass spectrometry for peptides are well documented [55−57].

Mass spectrometer is used to measure the molecular mass of peptide/protein and sequence them. In the case of measuring the mass of the peptide/protein, the spectrometer acts as a balance and in sequencing the specific molecular ions after its molecular mass measurement is subjected to fragmentation through collision. This kind of mass spectrometry is called tandem mass spectrometry (MS/MS). By analyzing the masses of the fragments, the structural details of the mother peptide can be drawn and unlike Edman sequencing, this technique provides sequence information of peptide/protein even if they are present in a complex mixture. Today MS/MS is routinely used in sequencing and has replaced Edman sequencing almost completely.

Two ionization techniques used for characterization of proteins and peptides are ESI and MALDI. While a quadrupole mass analyzer is used in ESI, a time-of-flight (TOF) mass analyzer is typically used in MALDI. Both are very sensitive methods, and require picomole to subpicomole ranges of samples and allow determination of wide range of molecular mass. Both methods work best with clean samples, though MALDI can work even with dirty samples. For peptides having molecular mass 5−50 kDa, ESI MS gives accurate results and therefore preferred and for higher molecular mass MALDI is preferred. MALDI produces predominantly single peak and this makes the identification of peptide rather easily from a crude peptide sample, whereas ESI produces multiply charged ions, and hence ESI MS shows several peaks. ESI MS can be coupled with HPLC (LC MS) and this has now a powerful analytical tool in peptide research.

### 4.2.4.1 Sequence by tandem mass spectrometry

In MS/MS two consecutive stages of mass detection is used; in the first state the precursor ion—the actual peptide molecule is identified based on its mass/charge ratio and in the second stage analyzing the product ions formed by spontaneous or induced fragmentation of the precursor ion. Peptide fragment ions are produced in the mass spectrometer primarily by cleavage of the amide bonds that join two amino acid residues. The most commonly observed fragment ions and their nomenclature are presented in Fig. 4.4. The fragmentation of peptides in MS has been well described [58−63]. A detailed account of protein and peptide analysis by mass spectrometry has been reported elsewhere [64,65].

From the fragments that are obtained, the sequence is arrived by two methods: database search or de novo sequencing. In the former method, the mass spectra data of the peptide to be sequenced is submitted to a database search and run to find a match with the known peptide sequence. The peptide with highest matching score is selected. In de novo sequencing different algorithms are used to interpret the structure of the peptide [66−68].

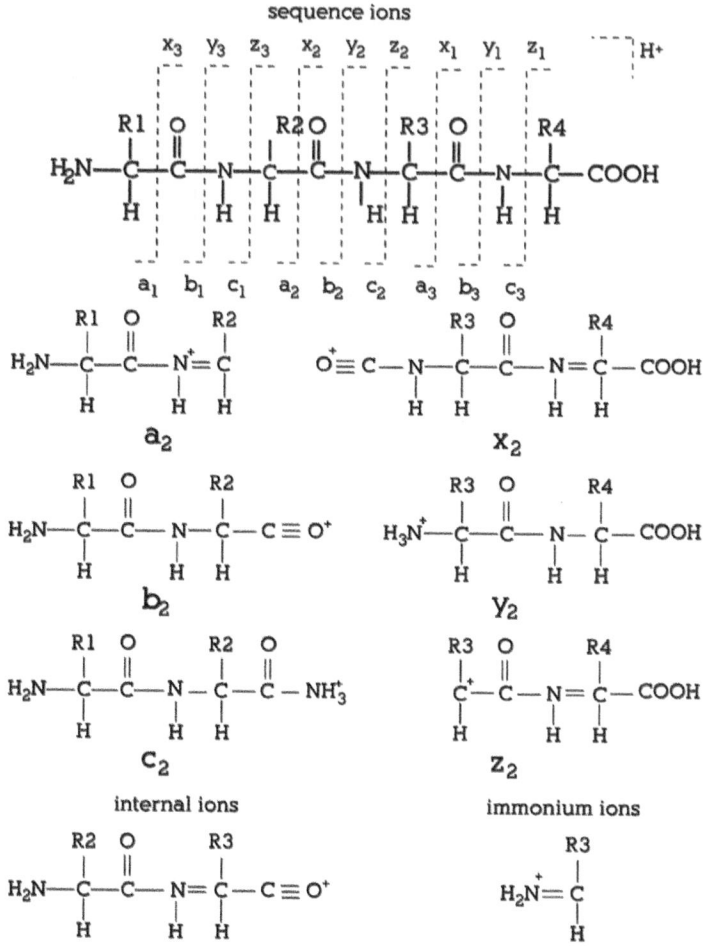

**Figure 4.4** Fragmentation pattern of peptide during mass spectrometry (MS/MS). R—side chain of amino acid. The subscript indicates the number of amino acid residues in the fragment. If the charge is retained on N-terminus, the fragments are called a, b, or c and if the charge is retained on C-terminus the fragments are called $x$, $y$, or $z$. For detection the fragment should carry at least one charge. Internal ions are formed by double backbone fragmentation (typically a combination of b and y cleavages). Sometimes immonium ions can be formed by a combination of a and y cleavages.

# References

[1] S. Gordon, S.H.E. Kaufmann, R. Medzhitov (Eds.), The Innate Immune Response to Infection, ASM Press, Washington, 2004.
[2] T. Ganz, Defensins: antimicrobial peptides of innate immunity, Nature Reviews Immunology 3 (2003) 710—720.

[3] R.E. Hancock, R.I. Lehrer, Cationic peptides: a new source of antibiotics, Trends in Biotechnology 16 (1998) 82–88.
[4] M. Zasloff, Antimicrobial peptides of multicellular organisms, Nature 415 (2002) 389–395.
[5] K.S. Brogden, M. Ackermann, P.B. McCray Jr., B.F. Tack, Antimicrobial peptides in animals and their role in host defensins, International Journal of Antimicrobial Agents 22 (2003) 465–478.
[6] X. Wang, H. Yu, R. Xing, P. Li, Characterization, preparation and purification of marine bioactive peptides, BioMed Research International (2017) 1–15.
[7] C. Hetru, P. Bulet, Strategies for the Isolation and Characterization of Antimicrobial Peptides of Invertebrates, vol 78, Humana Pres Inc, Totowa, NJ, 1997, pp. 35–49.
[8] T. Ganz, A.M. Cole, Human antimicrobial peptides: analysis and application, Biotechniques 29 (2000) 822–831.
[9] J.M. Conlon, A. Sonnevend, Antimicrobial peptides in frog skin secretions, in: A. Giuliani, A.C. Rinaldi (Eds.), Antimicrobial Peptides: Methods and Protocols, vol. 618, Springer, 2010, pp. 3–314.
[10] M. Rai, R. Pandit, S. Gaikwad, G. Kovics, Antimicrobial peptides as natural biopreservative to enhance the shelf-life of food, Journal of Food Science and Technology 53 (2016) 3381–3393.
[11] E.V. Pingitore, E. Salvucci, F. Sesma, M.E. Nader-Macías, Different strategies for purification of antimicrobial peptides from lactic acid bacteria (LAB), in: A. Mendez-Vilas (Ed.), Communicating Current Research and Educational Topics and Trends in Applied Microbiology, 2007, pp. 557–568.
[12] T. Swee-Seong, Z.H. Prodhan, S.K. Biswas, C.-F. Le, S.D. Sekaran, Antimicrobial peptides from different plant sources: isolation, characterisation, and purification, Phytochemistry 154 (2018) 94–100.
[13] C. Lowenberger, P. Bulet, M. Charlet, C. Hetru, B. Hodgeman, B.M. Christensen, et al., Insect immunity: isolation of three novel inducible antibacterial defensins from the vector mosquito *Aedes aegypti*, Insect Biochemistry and Molecular Biology 25 (1995) 867–873.
[14] J. Dimarcq, D. Hoffmann, M. Meister M, P. Bulet, R. Lanot, J.-M. Reichhart, et al., Characterization and transcriptional profiles of a Drosophila gene encoding an insect defensin, European Journal of Biochemistry 221 (1994) 201–209.
[15] S. Cociancich, M. Goyffon, F. Bontems, P. Bulet, F. Bouet, A. Menez, et al., Purification and characterization of a scorpion defensin, a 4kDa antibacterial peptide presenting structural similarities with insect defensins and scorpion toxins, Biochemistry and Biophysical Research Communications 194 (1993) 17–22.
[16] A.L. Battison, R. Summerfield, A. Patrzykat, Isolation and characterisation of two antimicrobial peptides from haemocytes of the American lobster *Homarus americanus*, Fish & Shellfish Immunology 25 (2008) 181–187.
[17] M. Conlon, B. Abraham, A. Sonnevend, T. Jouenne, P. Cosette, J. Leprince, et al., Purification and characterization of antimicrobial peptides from the skin secretions of the carpenter frog *Rana virgatipes* (Ranidae, Aquarana), Regulatory Peptides 131 (2005) 38–45.
[18] J.M. Conlon, M. Milena, E. Ahmed, J. Leprince, H. Vaudry, J.D. King, et al., Purification and properties of antimicrobial peptides from skin secretions of the Eritrea clawed frog *Xenopus clivii* (Pipidae), Comparative Biochemistry and Physiology, Part C 153 (2011) 350–354.
[19] C.T. Mant, Y. Chen, Z. Yan, T.V. Popa, J.M. Kovacs, J.B. Mills, et al., HPLC analysis and purification of peptides, in: G. Fields (Ed.), Methods in Molecular Biology, vol. 386, Humana Press, Totowa, NJ, 2007, pp. 1–53.

[20] E. Andersson, O.E. Sørensen, B. Frohm, N. Borregaard, A. Egesten, J. Malm, Isolation of human cationic antimicrobial protein-18 from seminal plasma and its association with prostasomes, Human Reproduction 17 (2002) 2529–2534.

[21] S.W. Hwang, J.H. Lee, H.B. Park, S.H. Pyo, J.E. So, H.S. Lee, et al., A simple method for the purification of an antimicrobial peptide in recombinant *Escherichia coli*, Molecular Biotechnology 18 (2001) 193–197.

[22] C.T. Mant, J.M. Parker, R.S. Hodges, Size-exclusion high-performance liquid chromatography of peptides. Requirement for peptide standards to monitor column performance and non-ideal behaviour, Journal of Chromatography 397 (1987) 99–112.

[23] J. Turkova, Affinity chromatography, Journal of Chromatography 91 (1974) 267–291.

[24] H. Zou, Q. Luo, D. Zhou, Affinity membrane chromatography for the analysis and purification of proteins, Journal of Biochemical and Biophysical Methods 49 (2001) 199–240.

[25] J. Xiao, H. Zhang, L. Niu, X. Wang, Efficient screening of a novel antimicrobial peptide from *Jatropha curcas* by cell membrane affinity chromatography, Journal of Agricultural and Food Chemistry 59 (2011) 1145–1151.

[26] R. Qilong, X. Huabin, B. Zongbi, S. Baogen, Y. Qiwei, Y. Yiwen, et al., Recent advances in separation of bioactive natural products, Chinese Journal of Chemical Engineering 21 (2013) 952–967.

[27] G. Bargeman, G.-H. Koops, J. Houwing, I. Breebaart, H.C. Horst, M. Wessling, et al., The development of electro-membrane filtration for the isolation of bioactive peptides: the effect of membrane selection and operating parameters on the transport rate, Desalination 149 (2002) 369–374.

[28] G.B. Irvine, C. Shaw, High-performance gel permeation chromatography of proteins and peptides on columns of TSK-G2000-SW and TSK-G3000-SW: a volatile solvent giving separation based on charge and size of polypeptides, Analytical Biochemistry 55 (1986) 141–148.

[29] W.O. Richter, B. Jacob, P. Schwandt, Molecular weight determination of peptides by high-performance gel permeation chromatography, Analytical Biochemistry 133 (1983) 288–291.

[30] G.D. Swergold, C.S. Rubin, High-performance gel-permeation chromatography of polypeptides in a volatile solvent: rapid resolution and molecular weight estimations of proteins and peptides on a column of TSK-G3000-PW, Analytical Biochemistry 131 (1983) 295–300.

[31] G.B. Irvine, Molecular weight estimation for native proteins using high performance size exclusion chromatography, in: J.M. Walker (Ed.), The Protein Protocols Handbook, Springer, 2009, pp. 1029–1037.

[32] M. Hartmanis, A. Engstrom, Occurrence of methionine sulfoxide during production of human insulin-like growth factor (IGF-1), in: Second Symposium of the Protein Society, Abstract Number 502, 1988.

[33] M. Kunitani, D. Johnson, L.R. Snyder, Model of protein conformation in the reversed-phase separation of interleukin-2 muteins, Journal of Chromatography 371 (1986) 313–332.

[34] C. Elicone, M. Lui, S. Geromanos, H. Erdjument-Bromage, P. Tempst, Microbore reversed-phase high-performance liquid chromatographic purification of peptides for combined chemical sequencing-laser-desorption mass spectrometric analysis, Journal of Chromatography 676 (1) (1994) 121–137.

[35] V. Price, D. Mochizuki, C.J. March, D. Cosman, M.C. Deeley, R. Klinke, et al., Expression, purification and characterization of recombinant murine granulocyte-macrophage colony-stimulating factor and bovine interleukin-2 from yeast, Gene 55 (1987) 287–293.

[36] E.P. Kroef, R.A. Owens, E.L. Campbell, R.D. Johnson, H.I. Marks, Production scale purification of biosynthetic human insulin by reversed-phase high-performance liquid chromatography, Journal of Chromatography 461 (1989) 45−61.
[37] J. Rivier, R.A. McClintock, R. Galyean, H. Anderson, Reversed-phase high-performance liquid chromatography: preparative purification of synthetic peptides, Journal of Chromatography. A 288 (1984) 303−328.
[38] C.A. Hoeger, R. Galyean, R.A. McClintock, J.E. Rivier, Practical aspects of preparative reversed-phase chromatography of synthetic peptides, in: T. Colin, R.S.H. Mant (Eds.), High-Performance Liquid Chromatography of Peptides and Proteins, Taylor & Francis, 2017, pp. 753−764.
[39] C. Hoeger, R. Galyean, J. Boublik, R. McClintock, J. Rivier, Preparative reverse phase high performance liquid chromatography: effects of buffer pH on the purification of synthetic peptides, Biochromatography 2 (1987) 134−142.
[40] M.C. McCroskey, V.E. Groppi, J.D. Pearson, Separation and purification of S49 mouse lymphoma histones by reversed-phase high-performance liquid chromatography, Analytical Biochemistry 163 (1987) 427−432.
[41] M.P. Bartolomeo, F. Maisano, Validation of a reversed-phase HPLC method for quantitative amino acid analysis, Journal of Biomolecular Techniques 17 (2006) 131−137.
[42] R.F. Ebert, Amino acid analysis by HPLC: optimized conditions for chromatography of phenylthiocarbamyl derivatives, Analytical Biochemistry 154 (2) (1986) 431−435.
[43] M. Alaiz, J.L. Navarro, J. Girón, E. Vioque, Amino acid analysis by high-performance liquid chromatography after derivatization with diethyl ethoxymethylenemalonate, Journal of Chromatography 591 (1992) 181−186.
[44] R.L. Heinrikson, S.C. Meredith, Amino acid analysis by reverse-phase high-performance liquid chromatography: precolumn derivatization with phenylisothiocyanate, Analytical Biochemistry 136 (1984) 65−74.
[45] C. Cooper, N. Packer, K. Williams (Eds.), Amino Acid Analysis Protocol: Methods in Molecular Biology, vol. 159, Humana Press Inc., Totowa, NJ, 2001.
[46] S.M. Rutherfurd, G.S. Gilani, Amino acid analysis, in: Current Protocols in Protein Science, 2009, Chapter 11: Unit 11.9. doi:10.1002/0471140864.ps1109s58.
[47] H.D. Niall, Automated edman degradation: the protein sequenator, in: C.H.W. Hirs, S.N. Timasheff (Eds.), Methods in Enzymology, Part D: Enzyme Structure, vol. 27, Elsevier, 1973, pp. 942−1010.
[48] R.A. Laursen, Solid-phase Edman degradation: an automatic peptide sequencer, European Journal of Biochemistry 20 (1971) 89−102.
[49] B.A. Petriz, O.L. Franco, Application of cutting-edge proteomics technologies for elucidating host-bacteria interactions, Advances in Protein Chemistry and Structural Biology 95 (2014) 1−24.
[50] A. Görg, W. Weiss, M.J. Dunn, Current two-dimensional electrophoresis technology for proteomics, Proteomics 4 (2004) 3665−3685.
[51] J.M. Schröder, Purification of antimicrobial peptides from human skin. A.C.R. in: A. Giuliani (Ed.), Methods in Molecular Biology: Antimicrobial Peptides, vol. 618, Springer Science, 2010, pp. 15−30.
[52] T. Ganz, J.A. Metcalf, J.I. Gallin, L.A. Boxer, R.I. Lehrer, Microbicidal/cytotoxic proteins of neutrophils are deficient in two disorders: Chediak-Higashi syndrome and "specific" granule deficiency, Journal of Clinical Investigation, 82, 1988, pp. 552−556.
[53] H. Schägger, G.V. Jagow, Tricine-sodium dodecyl sulfate-polyacrylamide gel electrophoresis for the separation of proteins in the range from 1 to 100 kDa, Analytical Biochemistry 166 (1987) 368−379.

[54] J. Harder, J. Bartels, E. Christophers, J.M. Schröder, Isolation and characterization of human beta-defensin-3, a novel human inducible peptide antibiotic, Journal of Biological Chemistry 276 (2001) 5707–5713.
[55] A.P. Jonsson, Mass spectrometry for protein and peptide characterisation, Cellular and Molecular Life Sciences 58 (2001) 868–884.
[56] B. Domon, R. Aebersold, Mass spectrometry and protein analysis, Science 312 (5771) (2006) 212–217.
[57] M. Mann, R.C. Hendrickson, A. Pandey, Analysis of proteins and proteomes by mass spectrometry, Annual Reviews of Biochemistry 70 (2001) 437–473.
[58] D.F. Hunt, J.R. Yates, J. Shabanowitz, S. Winston, C.R. Hauer, Protein sequencing by tandem mass spectrometry, Proceedings of National Academy of Sciences USA 83 (1986) 6233–6237.
[59] K. Biemann, Contributions of mass spectrometry to peptide and protein structure, Biomedical & Environmental Mass Spectrometry 16 (1988) 99–111.
[60] R.S. Johnson, S.A. Martin, K. Biemann, Collision-induced fragmentation of (M + H) + ions of peptides. Side chain specific sequence ions, International Journal of Mass Spectrometry and Ion Processes 86 (1988) 137–154.
[61] K. Biemann, Sequencing of peptides by tandem mass spectrometry and high-energy collision-induced dissociation, in: J.A. McCloskey (Ed.), Methods in Enzymology-Mass Spectrometry, vol. 193, Elsevier Inc, 1990, pp. 455–479.
[62] P.A. Schindler, A. Van Dorsselaer, A.M. Falick, Analysis of hydrophobic proteins and peptides by electrospray ionization mass spectrometry, Analytical Biochemistry 213 (2) (1993) 256–263.
[63] A.M. Falick, W.M. Hines, K.F. Medzihradszky, M.A. Baldwin, B.W. Gibson, Low-mass ions produced from peptides by high-energy collision-induced dissociation in tandem mass spectrometry, Journal of the American Society for Mass Spectrometry 4 (1993) 882–893.
[64] E.J. Zaluzec, D.A. Gage, J.T. Watson, Matrix-assisted laser desorption ionization mass spectrometry: applications in peptide and protein characterization, Protein Expression and Purification 6 (1995) 109–123.
[65] G. Zhang, R.S. Annan, S.A. Carr, T.A. Neubert, Overview of peptide and protein analysis by mass spectrometry, Current Protocols in Molecular Biology 10 (2014) 1–30.
[66] A.L. Yergey, J.R. Coorssen, P.S. Backlund, P.S. Blank, G.A. Humphrey, J. Zimmerberg, et al., De Novo sequencing of peptides using MALDI/TOF-TOF, Journal of the American Society for Mass Spectrometry 13 (2002) 784–791.
[67] A. Shevchenko, I. Chernushevic, A. Shevchenko, M. Wilm, M. Mann, "De novo" sequencing of peptides recovered from in-gel digested proteins by nanoelectrospray tandem mass spectrometry, Molecular Biotechnology 20 (2002) 107–118.
[68] J. Seidler, N. Zinn, M.E. Boehm, W.D. Lehmann, De novo sequencing of peptides by MS/MS, Proteomics 10 (2010) 634–649.

# Antimicrobial lipopeptides of bacterial origin—the molecules of future antimicrobial chemotherapy

P. Prajosh[1], H. Shabeer Ali[1,2], Renu Tripathi[2] and K. Sreejith[1]
[1]Department of Biotechnology and Microbiology, Kannur University, Kannur, Kerala, India, [2]Division of Molecular Parasitology and Immunology, CSIR—Central Drug Research Institute, Lucknow, Uttar Pradesh, India

## 5.1 Introduction

Antibiotics are meant for eradicating infectious diseases of microbial origin. The conventional antibiotics such as β-lactams classes, carbapenems, cephalosporins, aminoglycosides, fluoroquinolones, and macrolides have led the battle against infectious diseases for several decades. Their extended application in the field of food, poultry, cosmetics, and several other industries has contributed to the development of drug resistance in pathogens. Improper applications of antibiotics have further contributed resistance against all the treatment strategies developed so far [1]. The drug resistance among microbes has increased the demand for novel drug candidates that can overcome the drug resistance mechanisms exerted by the pathogens.

In the past few decades, the vast group of antimicrobial agents collectively known as antimicrobial peptides (AMPs) received considerable attention as the alternate antimicrobial agents to counteract drug-resistant microbes [2]. It is surprising that peptides are not limited as antimicrobials, their primary role as anticancer and antiviral agents are the key properties to be highlighted. The class AMP includes structurally diverse and dynamic molecules that contribute to diversity in structure and function. It is not surprising to say that the AMP class includes synthetic as well as engineered peptides along with naturally modified peptides. In most instances peptides of microbial origin are usually found as conjugates with other chemical counterparts such as fatty acids and sugars moieties. In addition to the structural conjugation, presence of unnatural amino acids, cyclization or special structural confirmations of such peptides are the other distinguished features. Such complexities in their structure make them more resistant to the action of hydrolytic chemicals and enzymes [3]. The AMPs class comprises of structurally diverse lipopeptides (LPs), glycolipids, glycoproteins, lipoproteins, or mixtures thereof [4–9].

The focus area of the present chapter includes the AMP subclass "lipopeptides" (LPs) of bacterial origin. LPs are mainly produced by bacterial and fungal genera with their unique nonribosomal peptide synthesis pathway. The structurally diverse LPs are formed by the conjugation of cyclic or linear peptides with a long hydrocarbon tail or other lipophilic molecules [10,11]. Enormous resistance to the action of hydrolytic enzymes and broad biological activities of the LPs makes them suitable for the development of next generation therapeutic agents [12–15].

## 5.2 Lipopeptides

LPs are short peptides conjugated with a lipid tail [16]. The combination of amino acids in the peptide region and the length of the attached lipid tail may significantly affect biological activities of LPs. As mentioned earlier, at least some of the amino acids possessed by the peptide are unique and unusual. Presence of alternating combinations of D and L forms of amino acids is the primary distinguishing feature of such peptides. In addition, the chemical nature of the lipid tail and its length also significantly contribute to functional specificities [17]. Bacteria and fungi are the primary producers of cyclic and linear LPs with important functions such as surfactant, antibacterial, antifungal, antiviral, and even antitumor properties [18,19].

The peptides made of 7 or 10 amino acids linked to β-hydroxy or β-amino fatty acids are the main classes of cyclic LPs produced by *Bacillus* genera. Iturin class of LPs is rare type of LPs with $C_{14}$ to $C_{17}$ carbon long fatty acid tail with an amine group in third position. Similarly, 3-amino-9-methyldecanoic acid has also been reported in LPs produced by *Myxobacteria*. The specific ring structure of fengycins is formed by an ester bond between a third position tyrosine residue and the C-terminal residue.

### 5.2.1 Types of lipopeptides produced by different bacterial genera

The important bacterial genera that are known to produce LPs are *Bacillus*, *Paenibacillus* sp. [20], *Pseudomonas* sp. [11,20,21], *Actinomycetes* [22], *Serratia* [23], and *Propionibacterium* [24]. Among *Bacillus* species, *B. subtilis* strain produces a vast array of well-characterized peptide antibiotics through ribosomal pathway. It includes subtilosin A, subtilin, sublancin, and TasA. On the other hand, peptide antibiotics like bacilysin, mycobacillin, bacillaene, rhizocticins, and LPs from the iturin, surfactin, and fengycin families are synthesized by the nonribosomal pathway and/or polyketide synthases of the bacterium [25]. Some of the well-characterized peptide antibiotics are discussed below.

#### 5.2.1.1 Daptomycin

Daptomycin is a tridecapeptide composed of some nonproteinogenic amino acids derived from *Streptomyces roseosporus*. The N-terminus of the decapeptide region is attached to a decanoyl fatty acid side chain and a lactone core region formed as a result of the cyclization of the Thr 4 -OH onto the C-terminal carboxylate (Fig. 5.1) [26–28]. The molecule was discovered in the early 1980s and approved by US FDA for clinical use during 2003. Probably, daptomycin is the first LP antibiotic that is being effectively used for the treatment of skin infections, endocarditis, and bacteremia.

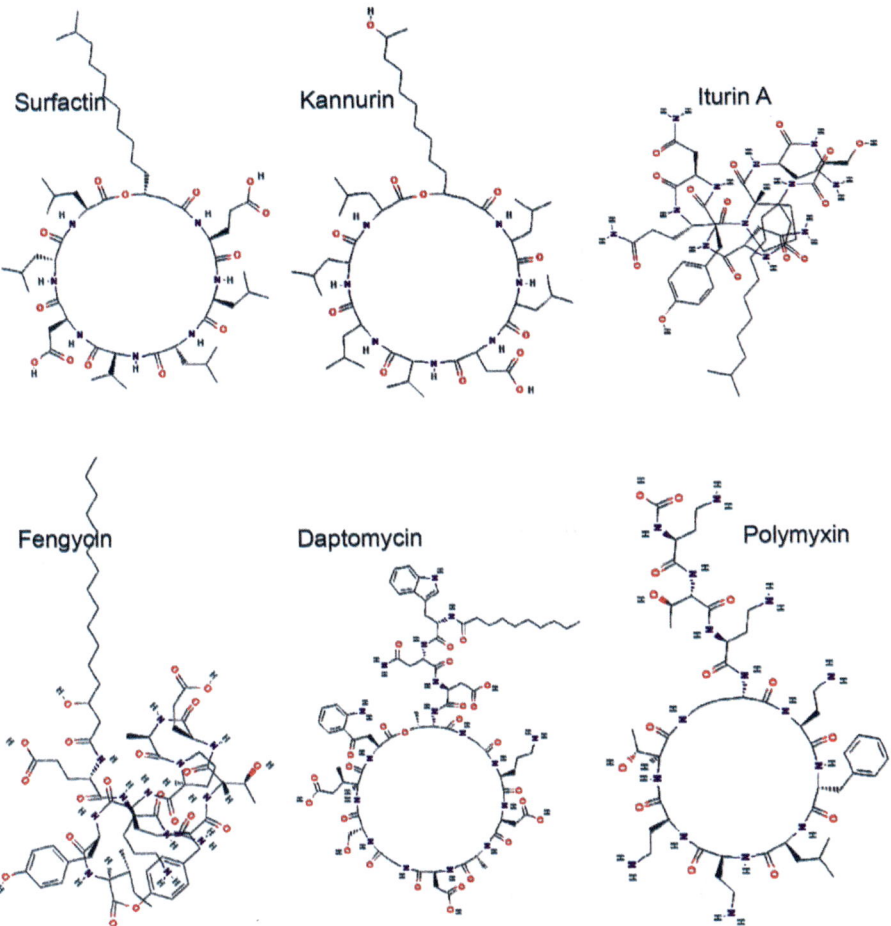

**Figure 5.1** Schematic diagrams (downloaded from pubchem database) of some important lipopeptides.

### 5.2.1.2 Polymyxins

The cationic polypeptides "Polymyxins" comprises of a heptapeptide region and a tripeptide side chain acylated at its N-terminus by a long chain fatty acid tail. Polymyxin B and Polymyxin E (Colistin) are produced by the soil bacterium *Bacillus polymyxa* and these are the well-studied antibiotics of polymyxin family. In addition, Polymyxin M (mattacin) is another member of the family produced by *Paenibacillus kobensis* [29].

### 5.2.1.3 Surfactin

Surfactins are well known for their powerful surfactant property and broad-spectrum antimicrobial activity. The structure of surfactin involves a peptide part

((L)Glu-(L)Leu-(D) Leu-(L)Val-(L)Asp-(D)Leu-(L)Leu) attached with the β-hydroxy fatty acid (12−16 carbon atoms) and cyclized via lactone bond [30]. The amino acids at positions 2, 4, and 7 can be varied [31] for different subfamilies. The combination of amino acids in the heptapeptide region and the variation in lipid tail length may significantly affect the functional specificity of the molecule. Based on the difference in lipid tail length, the surfactin family members are named as Esperin, Lichenysin, Pumilacidin, and Surfactin. Based on the differences in lipid tail length, surfactins can generally be described as C13-surfactin, C14-surfactin, and C15-surfactin. As mentioned above, C15-surfactin has the stronger surfactant property when compared with other members [32,33] and also found to exhibit higher hemolytic activities [34]. Based on the differences in amino acid sequences, the surfactin family is further classified as Surfactin A, Surfactin B, and Surfactin C [35].

### 5.2.1.4 Kannurin

Kannurin is a surfactin-like cyclic LP produced by the soil bacterium *Bacillus cereus* AK1. The structure of the molecule involves a heptapeptide region (amino acid sequence LLLVDLL) attached with the β-hydroxy fatty acid tail via an ester linkage [36]. The β-hydroxy fatty acid tail length may vary from C12 to C15. On the basis of the difference in fatty acid tail length, the different members were named as Kannurin A, Kannurin B, Kannurin C, and Kannurin D. The Kannurin A with short lipid tail was more selective toward bacteria, whereas Kannurin D with long lipid tail was more selective against fungi [17]. In the recent study, it was observed that Kannurin exists as linear forms in solution and was designated as Kannurin $C_L$ and $D_L$.

### 5.2.1.5 Lichenysin

Lichenysin is a surfactin-like LP produced by *Bacillus licheniformis*. The molecule exhibits structural and functional similarities with surfactin class of LPs. Stability over wide pH range, temperature, and salt concentration is the key highlight [16,37−39]. As in the case of other LPs, the variations in amino acid sequences and lipid tail contribute to the formation of various subclasses such as Lichenysin A, Lichenysin B, and Lichenysin C.

Lichenysin A is produced by *B. licheniformis* BAS50. The peptide region of the molecule is formed of "(L)Gln-(L)Leu-(D)Leu-(L)Val-(L)Asp-(D)Leu-(L)Ile" attached with the β-hydroxy fatty acid of C12−C15 carbons long [40]. The sequential arrangement of amino acid present in Lichenycin B ((L)Glu-(L)Leu-(D)Leu-(L)Val-(L)Asp-(D)Leu-(L)Leu) produced by *B. licheniformis* JF-2 [41] is slightly different from Lichenycin A and the β-hydroxy fatty acid attached with the heptapeptide region is of 15 carbons long. Lichenycin C is closely related to Lichenycin B with its heptapeptide region, whereas the β-hydroxy fatty acid tail length ranges from C14 to C15 [42].

### 5.2.1.6 Iturin

Iturin produced by *Bacillus subtilis* is a lactam containing heptapeptide with "Asp or Asn" amino acid at position 1, which is attached with the lipid tail to form a cyclic structure. The presence of a β-amino fatty acid as a replacement for of β-hydroxy fatty acid makes it different from surfactin family. Like the lipid tail of surfactin, iturin also exhibits variation in lipid tail length that ultimately contributes to the formation of different structural variables [43]. Based on the variation in lipid tail length, iturin family consists of multiple structural variants namely Iturin A (Iturin A1, A7, A4, A2, A8, A6, A9), iturin C (C3, C4, C15), Iturin D, Iturin F (Iturin F1 and F2), Bacillomycin D, Bacillomycin F, Bacillomycin L, and Mycosubtilin [39]. In the above-listed structural variants, the length of the β-amino fatty acid tail ranges from C14 to C17. Beyond the variations in lipid tail length, the changes in sequence of the amino acids present in the peptide region are the main reason for the formation of structural variants.

The structure of Iturin A includes the heptapeptide region with the amino acid sequence (L)Asn-(D)Tyr-(D)Asn-(L)Gln-(L)Pro-(D)Asn-(L)Ser which is attached with the $C_{13}$-$C_{16}$ β-amino fatty acid tail [44]. Iturin C differs from iturin A by the presence of an aspartyl residue instead of an asparaginyl residue [45].

### 5.2.1.7 Mycosubtilin

Mycosubtilin is an isoform of iturin produced by the soil bacterium *B. subtilis*. The heptapeptide region contains the amino acid sequence (L)Asn-(D)Tyr-(D)Asn-(L)Gln-(L)Pro-(D)Ser-(L)Asn which is connected with the β-amino fatty acid with C14−C17 carbons long [46]. Mycosubtilin is a strong antifungal agent that can interact with the sterol alcohol group (ergosterol) of yeast or fungal cell membranes [47].

### 5.2.1.8 Bacillomycin L

Bacillomycin L is an iturin-like LP produced by *B. subtilis* NCIB 8872. The peptide region is made of (L)Asp-(D)Tyr-(D)Asn-(L)Ser-(L)Gln-(D)Ser-(L)Thr, and a β-amino fatty acid tail that ranges from C14 to C16 carbons long. The molecule received attention for its strong antifungal activity, which acts by disrupting fungal cell membrane [48,49].

### 5.2.1.9 Fengycin

Unlike other classes of LPs, fengycin consists of a decapeptide attached with the β-hydroxy fatty acid tail [37]. The decapeptide region contains the amino acids (L)Glu-(D)Orn-(L)Tyr-(D)Thr-(L)Glu-(D)Ala/Val-(L)Pro-(L)Gln-(D)Tyr-(L)Ile, in which the tyrosine residue at third position forms a lactone bond with the isoleucine at position 10 to form a cyclic peptide [50]. The unique Ala-Val dimorphy at sixth position in the decapeptide region contribute to the formation of two major variants, fengycin A and fengycin B, respectively (also known as Plipastatins) [51−54].

Variations in β-hydroxy fatty acid tail length (C14−C18) is another criterion for the classification of structural variants.

When compared with surfactins and iturins, fengycins are strong antifungal LPs with the least hemolytic activity [50,55,56].

Plipastatin (Fengicin C) is produced by *B. cereus*, in which the peptide region contains an ornithine moiety [50,57]. The peptide region is formed of the following amino acid sequence: (L)Glu-(D)Orn-(L)Tyr-(D)allo-Thr-(L)Glu-(D)[X]-(L)Pro-(L)Gln-(D)Tyr-(L)Ile which is attached with C16-C17 β-hydroxy fatty acid tail [58].

### 5.2.1.10   WAP-8294A2 (WAP)

The water-soluble lipodepsipeptide, WAP, derived from the Gram-negative *Lysobacter* sp. was found to be highly effective against methicillin-resistant *S. aureus* (MRSA) and multiple strains of *Propionibacterium acnes*. It was demonstrated that WAP exhibits three- to sevenfold therapeutic efficacy than Vancomycin without any indication of antibiotic cross-resistance [59].

### 5.2.1.11   Tridecaptins

The family "tridecaptins" produced by *B. polymyxa* includes tridecaptins A, B, and C. Each member of the family differs in their fatty acid tail and the amino acid sequences [60]. Tridecaptins are believed to be linear acyl tridecapeptides with antibacterial activity against both Gram-negative and Gram-positive bacteria [61].

### 5.2.1.12   Edeines

Edeines are mixture of related basic peptides synthesized by *B. brevis* Vm4 strain. Edeines were proved to be antimicrobial against a number of bacteria in the Gram-positive and Gram-negative class, yeasts and mold [62], and mycoplasmas [63]. Further application in the field of chemotherapy was halted due to their high toxicity in animals. The unique property of Edeines to eliminate antibiotic resistance determining plasmids from bacterial cells is the key hallmark [64]. Edeines are more powerful to selectively inhibit the biosynthesis of prokaryotic DNA [65] without interfering with eukaryotic DNA synthesis [66]. Edeines are well known for their ability to inhibit protein synthesis [67].

### 5.2.1.13   Bogorol cationic peptides

They are selectively potent against MRSA and vancomycin-resistant *Enterococcus* spp., and exhibit temperate activity against *Escherichia coli* as well [68]. The representative member of the family "bogorol A" isolated from *Brevibacillus laterosporus* was found to be effective against multiple pathogens including *S. aureus, Enterococcus faecalis, Stenotrophomonas maltophilia, Burkholderia cepacian, E. coli,* and *Candida albicans* [69].

## 5.2.1.14 Kurstakin

Kurstakins, another family of lipoheptapeptides, were derived from *B. thuringiensis* with an amino acid order in its peptide region as Thr-Gly-Ala-Ser-His-Gln-Gln. However, the first reported kurstakin was devoid of β-hydroxy fatty acid and considered as a linear molecule [70].

## 5.2.1.15 Gramicidins

The linear pentadecapeptides "Gramicidin S and Tyrocidine" are produced by *B. brevis* strain ATCC 9999 and ATCC 8185, respectively [71].

## 5.2.1.16 Circulocins

Circulocin α, β, γ, and δ are novel LPs isolated from *Bacillus circulans* J2154 [72]. These LPs found to exhibit significant inhibitory effect on drug-resistant bacterial pathogens including piperacillin-resistant *Streptococci* and vancomycin-resistant *Enterococci*.

## 5.2.1.17 Amphomycin (Amp)

Amphomycin derived from *Streptomyces canus* is a cyclic peptide with 11 amino acids attached with a lipid tail [14]. Amphomycin acts by inhibiting glycosylation of glycoproteins in eukaryotes by interfering with the synthesis of dolichol-linked saccharides [73].

## 5.2.1.18 Pseudomonas *antimicrobial peptides*

The overall structure of *Pseudomonas* AMPs is more or less similar to that of Gram-positive bacterial LPs. The general structure involves a fatty acid tail attached with a peptide and form a lactone ring between two amino acids. *Pseudomonas* AMPs were classified as viscosin, amphisin, tolaasin, and syringomycin [74]. The subfamily "viscosin" contains a short peptide region with nine amino acids linked to a 3-hydroxy decanoic fatty acid tail. Amphicin contain a peptide region with 11 amino acids which is linked with 3-hydroxy fatty acid. Beyond these, *Pseudomonas* also produces AMPs such as putisolvin I and II [75].

### 5.2.1.18.1 Viscosin

Viscosin group of LPs produced by *P. fluorescens* SBW25 is the largest and most widely characterized *Pseudomonas* cyclic LP group. Viscosins are encoded by the viscA, viscB, and viscC genes and are nonribosomally synthesized [76]. Unlike other LPs of *Bacillus* origin, viscosin is unique with its peptide region containing nine amino acids conjugated with 3-hydroxydecanoic acid [74]. In the structure of viscosin, a 7-residue macrocycle can be observed which is formed by the ester bond between the C-terminal and the hydroxyl group of the D-allo-Thr (D-aThr) side chain at position 3 of the nonapeptide. The N-terminus of the peptide is attached with the R-isomeric 3-hydroxydecanoic acid (3-HDA). Bruijn et al. [76] have identified the role of viscosin on biofilm formation by *P. fluorescens* SBW25.

### 5.2.1.18.2 Amphisin

Amphisin is another group of LP produced by *Pseudomonas* sp. *strain DSS73*. The structure of amphisin involves a peptide region with the amino acid sequence "(D)Leu-(D)Asp-(D)allo-Thr-(D)Leu-(D)Leu-(D)Ser-(L)Leu-(D)Gln-(L)Leu-(L)Ile-(L)Asp." The peptide forms a lactone ring by the interaction between Thr4 and the C-terminus. The entire peptide is cyclized with the α-hydroxydecanoic acid.

### 5.2.1.18.3 Tolaasin

Tolaasin toxin is produced by *P. tolaasii*, causing brown-blotch disease in mushrooms. It is a lipodepsipeptide having a molecular weight 1985. The structural studies revealed that the structure of Tolaasin contains β-hydroxyoctanoyl-ΔBut-(D)Pro-(D)Ser-(D)Leu-(D)Val-(D)Ser-(D)Leu-(D)Val-(L)Val-(D)Gln-(L)Leu-(D)Val-ΔBut-(D)allo-Thr-(L)Ile-(L)Hse-(D)Dab-(L)Lys (cyclized by lactone ring formation between D-allo-Thr and the C-terminus) [77].

### 5.2.1.18.4 Syringomycin

Syringomycins are necrosis-inducing cyclic lipodepsinonapeptides produced by *P. syringae* and are considered as the important virulence factors of this organism against plants (Fig. 5.2).

## 5.2.2 Structure–activity relationship of lipopeptides

As mentioned above, the composition, sequential order, arrangement, and combination of amino acids in the peptide region of a LP significantly affect its surfactant property [31]. Similarly, the modification of some amino acids to more hydrophobic residues in the peptide region of Surfactin has enhanced the surface activity and reduced the critical micelle concentration [31,37,79,80]. The concept of structure-–activity relationship of LPs is not only limited to the peptide region, the lipid tail also possesses important role. Slight changes in the lipid tail length and composition also determine the potency of the LP. The increased percentage of branched chain fatty acids to a LP found to reduce its surface activity [81].

Studies on the biological activity of AMPs showed that incorporation of hydrophobic amino acids increases their antifungal activity and eventually causes undesirable changes in their secondary structure and amphipathic nature [82–84]. Another study conducted by the conjugation of aliphatic acids also significantly

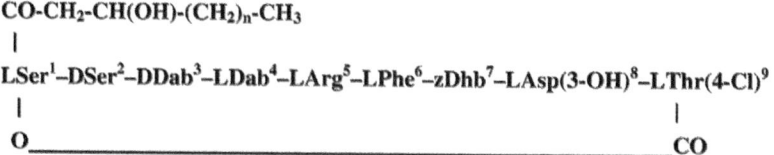

**Figure 5.2** Schematic structure of Syringomycin [78].

increased the antibacterial activity by providing additional antifungal activity [85,86]. Experiments were also proved effective to convert an inactive peptide to a membrane active form by incorporating a palmitoyl group [87]. The experiment conducted by Malina and Shai [88] converged to a set of conclusions that the differences in lipid tail length of a LP determines its selectivity toward bacteria or fungi and also make differences in their hemolytic activity.

The recent research revealed that the structural variants of Kannurin (Kannurin A and B) with shorter lipid tails are selective toward bacteria and are less hemolytic to mammalian RBCs. In contrast, the long lipid tail variants (Kannurin C and D) are highly active against fungi than bacteria [17].

### 5.2.3 Mechanism of action of lipopeptides

Majority of the LPs are believed to act on bacteria or fungi by any of the three proposed mechanisms. The first mechanism describes the inhibition of $(1,3)$-$\beta$-D-Glucan or Chitin synthesis, especially in fungi, thereby preventing cell wall formation [89,90]. The natural LP "Pneumocandins" of the Echinocandin class [91] also act by inhibiting the synthesis of chitin. Since the mammalian system lacks such pathway, this unique mode of action makes the LP a selective antifungal agent.

The second mechanism proposed that the LPs act by lysing the cell membranes [37,92,93]. This mechanism is more applicable in the case of bacteria and several LPs such as Iturins, Bacillomycin, and Surfactin were proven to kill the bacteria by this mechanism. There are some additional mechanisms such as the formation of ion-conducting pores in the lipid bilayer induced by Surfactins were also involved [94]. Carrillo et al. [95] reported the loss of vesicular contents through membrane destabilization or pore formation induced by Surfactin.

### 5.2.3.1 Daptomycin—mode of action

The mode of action of the FDA-approved Daptomycin is the more investigated one. The earlier hypothesis stated that the Daptomycin acts by inhibiting the synthesis of lipoteichoic acid [96,97], further, no significant evidences were obtained to support this hypothesis. In contrary to this, more widely accepted mechanism of Daptomycin action states that the insertion of lipid tail into the bacterial lipid bilayer membrane causes depolarization and $K+$ efflux which ultimately arrest DNA, RNA. and protein synthesis [97–100].

### 5.2.3.2 Polymyxin—mode of action

Polymyxin is another FDA-approved LP antibiotic that adopts a mechanism which is entirely different from the previously described modes of action. It acts by inhibiting the synthesis of Gram-negative outer membrane by making an electrostatic interaction between its lipid tail and the lipopolysaccharide of bacteria [101,102].

### 5.2.3.3 Mode of action for other lipopeptides

The third mechanism proposed by Schneider et al. [103] describes the inhibition of microbial cell wall synthesis by a cyclic LP Friulimicin B derived from the actinomycete *Actinoplanes friuliensis* by complex formation with bactoprenol phosphate. Rest of the other LPs including Surfactin, Fengicin, and Iturin are believed to act by inserting the lipid tail into the bacterial lipid bilayer membrane which in turn destabilizes the membrane and eventually leads to cellular leakage.

## 5.2.4 Antiadhesion and antibiofilm activities of lipopeptides

Hydrophobicity or electrostatic repulsion exerted by the LPs serves the basis for their properties preventing adhesion or biofilm formation [104]. LPs of the Surfactin class are well known for their ability to inhibit microbial attachment to surfaces and also for the removal of biofilm formed on various biotic and abiotic surfaces [105,106]. There are examples for the use of LPs as coating agents on vinyl urethral catheters for the prevention of microbial colonization on indwelling medical devices [107].

## 5.2.5 Natural role of lipopeptides

The major roles of LPs described so far are their role as antagonists against other inhabitants to prevent their motility and attachment to surfaces. For *Bacillus* species, LPs were acting a major role as signal transducers for the growth and cellular differentiation [108].

In natural habitats, the organisms that produce LPs use them as a weapon to acquire a competitive benefit over other microorganisms in their close proximity. Inhibitory effect of LPs on motility of *Bacillus* and *Pseudomonas* has been extensively investigated earlier [109]. The in vitro analysis of LPs produced by *Pseudomonas* and *Bacillus* species exhibited growth-inhibitory and lytic properties against bacteria, fungi, mycoplasmas, oomycetes, and viruses. In 1951 the antiviral effect of viscosin against enveloped viruses was reported [110]. Recent studies revealed the involvement of LPs in *B. subtilis* for forming biofilms. Surfactin also serves as a signaling molecule to elicit matrix formation and cannibalism in biofilms [108,111], whereas other compounds including iturin that are not able to induce potassium leakage are failed to induce multicellularity [111].

## 5.2.6 Lipopeptides in the treatment of multidrug-resistant infections

LPs are the molecules of interest to combat infections caused by multidrug-resistant organisms, because of their efficacy against drug-resistant pathogens [112] and the rare occurrence of resistance against LPs [113]. In this context, WAP-8294A2 exhibits activity against MRSA infections as well as various multidrug-resistant strains of *P. acnes* [59]. In addition to this, several LPs are under investigation for their efficacy against multidrug-resistant pathogens.

Polymyxin E or Colistin was the ancient LP antibiotic used against Gram-negative bacilli, which was discontinued due to severe neurotoxic and nephrotoxic effects. Still there is no active alternative for polymyxin in the treatment of cystic fibrosis. Due to the emergence of carbapenem-resistant pathogens from the past decades, medical experts are started reusing polymyxin E to counterfeit multidrug-resistant Gram-negative bacterial infections [114]. This history signifies the advantage of LP antibiotics over conventional antibiotics in the battle against multidrug-resistant infectious diseases.

Similarly, Battacin or Octapeptin B5, produced by the soil isolate *Paenibacillus tianmuensis* was found to be highly potent against multidrug-resistant and extremely drug-resistant clinical isolates especially in the Gram-negative class. The key highlight of Battacin is its less acute toxicity than polymyxin B [115]. However, Battacin has not been approved for clinical use.

Cadasides A and B subfamily of the acidic LPs is calcium-dependent antibiotic discovered by Wu et al. [116] through metagenomic analysis of soil samples. The Cadasides act by inhibiting the cell wall synthesis of multidrug-resistant pathogens.

Several other researchers have also claimed that the members in the subfamily of Fengycin and Surfactin are efficient in the inhibition of various bacterial pathogens. Most of them have extended application in various fields such as cancer biology. Majority of the LPs reported for activity against multidrug pathogens are in the preliminary stage of drug development pipeline, and equal proportions have withdrawn due to severe toxicities.

## 5.3 Conclusion

In the current scenario, microbial derived LPs with their broad-spectrum activity against pathogens and their diverse structural complexities to resist multiple drug-resistant mechanisms make them an attractive molecule for future antimicrobial therapy. Even though the researches on antimicrobial LPs are progressing worldwide, the number of approved LP antibiotics coming to market is very less. Apart from the few FDA-approved LP antibiotics including Daptomycin and Polymyxin B, there are no other promising candidates available in the market. This is the alarming situation to enhance the research on identification of novel antimicrobial LPs and efforts should also be taken to bring the already reported LP antibiotics in to the market. In this context, deep knowledge on current status of LP antibiotics, their benefits and disadvantages are necessary to explore further in this area.

## References

[1] L. Cantas, S.Q. Shah, L.M. Cavaco, C.M. Manaia, F. Walsh, M. Popowska, A brief multi-disciplinary review on antimicrobial resistance in medicine and its linkage to the global environmental microbiota, Frontiers in Microbiology 4 (2013) 96.

[2] D.B. Diep, I.F. Nes, Ribosomally synthesized antibacterial peptides in Gram positive bacteria, Current Drug Targets 3 (2002) 107–122.
[3] G. Schoenafinger, N. Schracke, U. Linne, M.A. Marahiel, Formylation domain: an essential modifying enzyme for the nonribosomal biosynthesis of linear gramicidin, Journal of the American Chemical Society 128 (2006) 7406–7407.
[4] T.R. Neu, Significance of bacterial surface-active compounds in interaction of bacteria with interfaces, Microbiology Reviews 60 (1996) 151–166.
[5] I.M. Banat, R.S. Makkar, S.S. Cameotra, Potential commercial applications of microbial surfactants, Applied Microbiology and Biotechnology 53 (2000) 495–508.
[6] R.Z. Ron, E. Rosenberg, Natural roles of biosurfactants, Environmental Microbiology 3 (2001) 229–236.
[7] R.M. Maier, Biosurfactants: evolution and diversity in bacteria, Advances in Applied Microbiology 52 (2003) 101–121.
[8] C.N. Mulligan, Environmental applications for biosurfactants, Environmental Pollution 133 (2005) 183–198.
[9] K. Muthusamy, S. Gopalakrishnan, T.K. Ravi, P. Sivachidambaram, Biosurfactants: properties, commercial production and application, Current Science 94 (2008) 736–747.
[10] C.J. Arnusch, R.J. Pieters, E. Breukink, Enhanced membrane pore formation through high-affinity targeted antimicrobial peptides, PLoS One 7 (2012) 39768.
[11] J.M. Raaijmakers, I. De Bruijn, O. Nybroe, M. Ongena, Natural functions of lipopeptides from *Bacillus* and *Pseudomonas*: more than surfactants and antibiotics, FEMS Microbiology Reviews 34 (2010) 1037–1062.
[12] S.S. Cameotra, R.S. Makkar, Recent applications of biosurfactants as biological and immunological molecule, Current Opinion in Microbiology 7 (2004) 262–266.
[13] H. Gross, J.E. Loper, Genomics of secondary metabolite production by *Pseudomonas* spp, Natural Product Reports 26 (2009) 1408–1446.
[14] G. Pirri, A. Giuliani, S.F. Nicoletta, L. Pizzuto, A.C. Rinaldi, Lipopeptides as anti-infectives: a practical perspective, Central European Journal of Biology 4 (2009) 258–273.
[15] A.K. Marr, W.J. Gooderham, R.E. Hancock, Antibacterial peptides for therapeutic use: obstacles and realistic outlook, Current Opinion in Pharmacology 6 (2006) 468–472.
[16] J.M. Bonmatin, O. Laprévote, F. Peypoux, Diversity among microbial cyclic lipopeptides: iturins and surfactins. Activity-structure relationships to design new bioactive agents, Combinatorial Chemistry & High Throughput Screening 6 (2003) 541–556.
[17] H. Shabeer Ali, K. Ajesh, K.V. Dileep, P. Prajosh, K. Sreejith, Structural characterization of Kannurin isoforms and evaluation of the role of β-hydroxy fatty acid tail length in functional specificity, Scientific Reports 10 (2020) 2839.
[18] F.C. Hsieh, T.C. Lin, M. Meng, S.S. Kao, Comparing methods for identifying *Bacillus* strains capable of producing the antifungal lipopeptide iturin A, Current Microbiology 56 (2008) 1–5.
[19] K. Meena, S. Kanwar, Lipopeptides as the antifungal and antibacterial agents: applications in food safety and therapeutics, BioMed Research International (2015) 1–9.
[20] S.A. Cochrane, J.C. Vederas, Lipopeptides from *Bacillus* and *Paenibacillus* spp. A gold mine of antibiotic candidates, Medicinal Research Reviews 36 (2016) 4–31.
[21] T. Janek, M. Łukaszewicz, A. Krasowska, Antiadhesive activity of the biosurfactant pseudofactin II secreted by the Arctic bacterium *Pseudomonas fluorescens* BD5, BMC Microbiology 12 (2012) 24.

[22] L.S. Singh, H. Sharma, N.C. Talukdar, Production of potent antimicrobial agent by actinomycete, Streptomyces sannanensis strain SU118 isolated from phoomdi in Loktak Lake of Manipur, India, BMC Microbiology 14 (2014) 278.
[23] S. Thies, B. Santiago-Schübel, F. Kovačic, F. Rosenau, R. Hausmann, K.E. Jaeger, Heterologous production of the lipopeptide biosurfactant serrawettin W1 in *Escherichia coli*, Journal of Biotechnology 181 (2014) 27–30.
[24] H. Hajfarajollah, B. Mokhtarani, K.A. Noghabi, Newly antibacterial and antiadhesive lipopeptide biosurfactant secreted by a probiotic strain, Propionibacterium *freudenreichii*, Applied Biochemistry and Biotechnology 174 (2014) 2725–2740.
[25] T. Stein, Bacillus *subtilis* antibiotics: structures, syntheses and specific functions, Molecular Microbiology 56 (2005) 845–857.
[26] M. Debono, M. Barnhart, C.B. Carrell, J.A. Hoffmann, J.L. Occolowitz, B.J. Abbott, A21978C, a complex of new acidic peptide antibiotics: isolation, chemistry, and mass spectral structure elucidation, The Journal of Antibiotics 40 (1987) 761–777.
[27] L.H. Lakey, E.J. Lea, B.A. Rudd, H.M. Wright, D.A. Hopwood, A new channel-forming antibiotic from Streptomyces coelicolor A3(2) which requires calcium for its activity, Journal of General Microbiology 129 (1983) 3565–3573.
[28] F.M. Huber, R.L. Pieper, A.J. Tietz, The formation of daptomycin by supplying decanoic acid to *Streptomyces roseosporus* cultures producing the antibiotic complex A21978C, Journal of Biotechnology 7 (1988) 283–292.
[29] N.I. Martin, H. Hu, M.M. Moake, J.J. Churey, R. Whittal, R.W. Worobo, Isolation, structural characterization, and properties of mattacin (polymyxin M), a cyclic peptide antibiotic produced by *Paenibacillus kobensis* M, The Journal of Biological Chemistry 278 (2003) 13124–13132.
[30] K. Arima, A. Kakinuma, G. Tamura, Surfactin, a crystalline peptide lipid surfactant produced by *Bacillus subtilis*: isolation, characterization and its inhibition of fibrin clot formation, Biochemical and Biophysical Research Communications 31 (1968) 488–494.
[31] J.M. Bonmatin, H. Labbé, I. Grangemard, F. Peypoux, R. Maget-Dana, M. Ptak, Production, isolation and characterization of [Leu4]- and [Ile4] surfactins from *Bacillus subtilis*, Letters in Peptide Science 2 (1995) 41–47.
[32] T. Yoneda, T. Tsuzuki, E. Ogata, Y. Fusyo, Surfactin sodium salt: an excellent biosurfactant for cosmetics, Journal of Cosmetic Science 52 (2001) 153–154.
[33] H. Razafindralambo, P. Thonart, M. Paquox, Dynamic and equilibrium surface tensions of surfactin aqueous solutions, Journal of Surfactants and Detergents 7 (2004) 41–46.
[34] M. Kracht, H. Rokos, M. Ozel, M. Kowal, G. Pauli, J. Vater J, Antiviral and hemolytic activities of surfactin isoforms and their methyl ester derivatives, The Journal of Antibiotics 52 (1999) 613–619.
[35] L. Rodrigues, I.M. Banat, J. Teixeira, R. Oliveira, Biosurfactants: potential applications in medicine, Journal of Antimicrobial Chemotherapy 57 (2006) 609–618.
[36] K. Ajesh, S. Sudarslal, C. Arunan, K. Sreejith, Kannurin, a novel lipopeptide from *Bacillus cereus* strain AK1: isolation, structural evaluation and antifungal activities, Journal of Applied Microbiology 115 (2013) 1287–1296.
[37] F. Peypoux, J.M. Bonmatin, J. Wallach, Recent trends in the biochemistry of surfactin, Applied Microbiology and Biotechnology 51 (1999) 553–563.
[38] S. Dufour, M. Deleu, K. Nott, B. Wathelet, P. Thonart, M. Paquot, Hemolytic activity of new linear surfactin analogs in relation to their physico-chemical properties, Biochimica et Biophysica Acta 1726 (2005) 87–95.

[39] M. Ongena, P. Jacques, *Bacillus* lipopeptides: versatile weapons for plant disease biocontrol, Trends in Microbiology 16 (2008) 115–125.

[40] M.M. Yakimov, W.R. Abraham, H. Meyer, G. Laura, P.N. Golyshin, Structural characterization of lichenysin A components by fast atom bombardment tandem mass spectrometry, Biochimica et Biophysica Acta (BBA)—Molecular and Cell Biology of Lipids 1438 (1999) 273–280.

[41] S.C. Lin, M.A. Minton, M.M. Sharma, G. Georgiou, Structural and immunological characterization of a biosurfactant produced by *Bacillus licheniformis* JF-2, Applied and Environmental Microbiology 60 (1994) 31–38.

[42] K. Jenny, O. Käppeli, A. Fiechter, Biosurfactants from *Bacillus licheniformis*: structural analysis and characterization, Applied Microbiology and Biotechnology 36 (1991) 5–13.

[43] S. Hiradate, S. Yoshida, H. Sugie, H. Yada, Y. Fuji, Mulberry anthracnose antagonists (iturins) produced by *Bacillus amyloliquefaciens* RC-2, Phytochemistry 61 (2002) 693–698.

[44] A. Isogai, S. Takayama, S. Murakoshi, A. Suzuki, Structure of β-amino acids in antibiotics iturin, Tetrahedron Letters 23 (1982) 3065–3068.

[45] F. Besson, G. Michel, Isolation and characterization of new iturins: iturin D and iturin E, The Journal of Antibiotics 40 (1987) 437–442.

[46] E.H. Duitman, L.W. Hamoen, M. Rembold, G. Venema, H. Seitz, W. Saenger, The mycosubtilin synthetase of *Bacillus subtilis* ATCC6633: a multifunctional hybrid between a peptide synthetase, an amino transferase, and a fatty acid synthase, Proceedings of the National Academy of Sciences 96 (1999) 13294–13299.

[47] M.N. Nasir, F. Besson, Interactions of the antifungal mycosubtilin with ergosterol-containing interfacial monolayers, Biochimica et Biophysica Acta (BBA) - Biomembranes 1818 (2012) 1302–1308.

[48] B. Zhang, C. Dong, Q. Shang, Y. Cong, W. Kong, P. Li, Purification and partial characterization of Bacillomycin L produced by *Bacillus amyloliquefaciens* K103 from lemon, Applied Biochemistry and Biotechnology 171 (2013) 2262–2272.

[49] F. Peypoux, M.T. Pommier, B.C. Das, F. Besson, L. Delcambe, G. Michel, Structures of bacillomycin D and bacillomycin L peptidolipid antibiotics from *Bacillus subtilis*, The Journal of Antibiotics 37 (1984) 1600–1604.

[50] N. Vanittanakom, W. Loeffler, U. Koch, G. Jung, Fengycin-a novel antifungal lipopeptide antibiotic produced by Bacillus subtilis F-29-3, The Journal of Antibiotics 39 (1986) 888–901.

[51] J. Schneider, K. Taraz, H. Budzikiewicz, M. Deleu, P. Thonart, P. Jacques, The structure of two fengycins from *Bacillus subtilis* S499, Journal of Biosciences 54 (1999) 859–865.

[52] Q. Tang, X. Bie, Z. Lu, F. Lv, Y. Tao, X. Qu, Effects of fengycin from *Bacillus subtilis* fmbJ on apoptosis and necrosis in *Rhizopus stolonifera*, Journal of Microbiology 52 (2014) 675–680.

[53] J. Vater, B. Kablitz, C. Wilde, P. Franke, N. Mehta, S.S. Cameotra, Matrix-assisted laser desorption ionization–time of flight mass spectrometry of lipopeptide biosurfactants in whole cells and culture filtrates of *Bacillus subtilis* C-1 isolated from petroleum sludge, Applied and Environmental Microbiology 68 (2002) 6210–6219.

[54] J. Wang, J. Liu, X. Wang, J. Yao, Z. Yu, Application of electrospray ionization mass spectrometry in rapid typing of fengycin homologues produced by *Bacillus subtilis*, Letters in Applied Microbiology 39 (2004) 98–102.

[55] A. Koumoutsi, X.-H. Chen, A. Henne, H. Liesegang, G. Hitzeroth, P. Franke, Structural and functional characterization of gene clusters directing nonribosomal synthesis of bioactive cyclic lipopeptides in *Bacillus amyloliquefaciens* strain FZB42, Journal of Bacteriology 186 (2004) 1084–1096.

[56] J. Hofemeister, B. Conrad, B. Adler, B. Hofemeister, J. Feesche, N. Kucheryava, Genetic analysis of the biosynthesis of non-ribosomal peptide- and polyketide-like antibiotics, iron uptake and biofilm formation by *Bacillus subtilis* A1/3, Molecular Genetics and Genomics 272 (2004) 363–378.

[57] S. Steller, D. Vollenbroich, F. Leenders, T. Stein, B. Conrad, J. Hofemeister, Structural and functional organization of the fengycin synthetase multienzyme system from *Bacillus subtilis* b213 and A1/3, Chemistry & Biology 6 (1999) 31–41.

[58] T. Nishikiori, H. Naganawa, Y. Muraoka, T. Aoyagi, H. Umezawa, Plipastatins: new inhibitors of phospholipase A2, produced by *Bacillus cereus* BMG302-fF67. III. Structural elucidation of plipastatins, The Journal of Antibiotics 39 (1986) 755–761.

[59] A. Kato, S. Nakaya, N. Kokubo, Y. Aiba, Y. Ohashi, H. Hirata, A new anti-MRSA antibiotic complex, WAP-8294A. I. Taxonomy, isolation and biological activities, The Journal of Antibiotics 51 (1998) 929–935.

[60] T. Kato, H. Hinoo, J. Shoji, The structure of tridecaptin A (studies on antibiotics from the genus Bacillus. XXIV), The Journal of Antibiotics 31 (1978) 652–661.

[61] T. Kato, R. Sakazaki, H. Hinoo, J. Shoji, The structures of tridecaptins B and C (studies on antibiotics from the genus Bacillus. XXV), The Journal of Antibiotics 32 (1979) 305–312.

[62] H. Chmara, E. Borowski, Antibiotic Edeine: VII. Biological activity of edeine A and B, Acta Microbiologica Polonica Pol 17 (1968) 59–66.

[63] J. Briseis, Effect of various inhibitors of protein and deoxyribonucleic acid synthesis on the growth of mycoplasmas, Applied Microbiology 14 (1966) 1049–1050.

[64] J. Borowski, E. Borowski, A. Ciepat, D. Dzierzanowska, P. Jakubicz, A. Smorczewski, Elimination of plasmids determining bacterial antibiotic resistance by edeine, Drugs Under Experimental and Clinical Research 3 (1977) 189–191.

[65] Z. Kurylo-Borowska, On the mode of action of edeine effect of edeine on the bacterial DNA, Biochimica et Biophysica Acta 87 (1964) 305–313.

[66] B. Woynarowska, H. Chmara, E. Borowski, Differential mechanism of action of the antibiotic edeine on prokaryotic and eukaryotic organism points to new basis for selective toxicity, Drugs Under Experimental and Clinical Research 5 (1979) 181–186.

[67] T. Obrig, J. Irvin, W. Culp, B. Hardesty, Inhibition of peptide initiation on reticulocyte ribosomes by edeine, European Journal of Biochemistry 21 (1971) 31–41.

[68] T. Barsby, K. Warabi, D. Sørensen, W.T. Zimmerman, M.T. Kelly, R.J. Andersen, The Bogorol family of antibiotics: template-based structure elucidation and a new approach to positioning enantiomeric pairs of amino acids, The Journal of Organic Chemistry 71 (2006) 6031–6037.

[69] T. Barsby, M.T. Kelly, S.M. Gagné, R.J. Andersen, Bogorol A produced in culture by a marine *Bacillus* sp. reveals a novel template for cationic peptide antibiotics, Organic Letters 3 (2001) 437–440.

[70] Y. Hathout, Y.-P. Ho, V. Ryzhov, P. Demirev, C. Fenselau, Kur takins: a new class of lipopeptides isolated from *Bacillus thuringiensis*, Journal of Natural Products 63 (2000) 1492–1496.

[71] M.M. Nakano, P. Zuber, Molecular biology of antibiotic production in *Bacillus*, Critical Reviews in Biotechnology 10 (1990) 223–240.

[72] H. He, B. Shen, J. Korshalla, G.T. Carter, Circulocins, new antibacterial lipopeptides from *Bacillus circulans*, J2154, Tetrahedron 57 (2001) 1189–1195.
[73] M.S. Kang, J.P. Spencer, A.D. Elbein, Amphomycin inhibition of mannose and GlcNAc incorporation into lipid-linked saccharides, The Journal of Biological Chemistry 253 (1978) 8860–8866.
[74] J.M. Raaijmakers, I. de Bruijn, M.J. de Kock, Cyclic lipopeptide production by plant-associated *Pseudomonas* spp. diversity, activity, biosynthesis, and regulation, Molecular Plant-Microbe Interactions 19 (2006) 699–710.
[75] I. Kuiper, E.L. Lagendijk, R. Pickford, J.P. Derrick, G.E. Lamers, J.E. Thomas-Oates, et al., Characterization of two *Pseudomonas putida* lipopeptide biosurfactants, putisolvin I and II, which inhibit biofilm formation and break down existing biofilms, Molecular Microbiology 51 (2004) 97–113.
[76] I. Bruijn, M. Kock, M. Yang, P. Waard, T. van Beek, J. Raaijmakers, Genome-based discovery, structure prediction and functional analysis of cyclic lipopeptide antibiotics in *Pseudomonas* species, Molecular Microbiology 63 (2007) 417–428.
[77] J.C. Nutkins, R.J. Mortishire-Smith, L.C. Packman, C.L. Brodey, P.B. Rainey, K. Johnstone, Structure determination of tolaasin, an extracellular lipodepsipeptide produced by the mushroom pathogen, *Pseudomonas tolaasii* Paine, Journal of the American Chemical Society 113 (1991) 2621–2627.
[78] M. Anselmi, T. Eliseo, L. Zanetti-Polzi, M.R. Fullone, V. Fogliano, A. Di Nola, Structure of the lipodepsipeptide syringomycin E in phospholipids and sodium dodecylsulphate micelle studied by circular dichroism, NMR spectroscopy and molecular dynamics, Biochimica et Biophysica Acta (BBA) – Biomembranes 1808 (2011) 2102–2110.
[79] F. Peypoux, J.-M. Bonmatin, H. Labbe, I. Grangemard, B.C. Das, M. Ptak, [Ala4] Surfactin, a novel isoform from *Bacillus subtilis* studied by mass and NMR spectroscopies, European Journal of Biochemistry 224 (1994) 89–96.
[80] A. Schneider, T. Stachelhaus, M.A. Marahiel, Targeted alteration of the substrate specificity of peptide synthetases by rational module swapping, Molecular and General Genetics 257 (1983) 308–318.
[81] M.M. Yakimov, K.N. Timmis KN, V. Wray, H.L. Fredrickson, Characterization of a new lipopeptide surfactant produced by thermotolerant and halotolerant subsurface *Bacillus licheniformis* BAS50, Applied and Environmental Microbiology 61 (1995) 1706–1713.
[82] J. Strahilevitz, A. Mor, P. Nicolas, Y. Shai, Spectrum of antimicrobial activity and assembly of dermaseptin-b and its precursor form in phospholipid membranes, Biochemistry 33 (1994) 10951–10960.
[83] M.K. Lee, L. Cha, S.H. Lee, K.S. Hahm, Role of amino acid residues within the disulfide loop of thanatin, a potent antibiotic peptide, Journal of Biochemistry and Molecular Biology 35 (2002) 291–296.
[84] I. Kustanovich, D.E. Shalev, M. Mikhlin, L. Gaidukov, A. Mor, Structural requirements for potent vs selective cytotoxicity for antimicrobial dermaseptin S4 derivatives, The Journal of Biological Chemistry 277 (2002) 16941–16951.
[85] A. Majerle, J. Kidric, R. Jerala, Enhancement of antibacterial and lipopolysaccharide binding activities of a human lactoferrin peptide fragment by the addition of acyl chain, The Journal of Antimicrobial Chemotherapy 51 (2003) 1159–1165.
[86] D. Avrahami, Y. Shai, Conjugation of a magainin analogue with lipophilic acids controls hydrophobicity, solution assembly, and cell selectivity, Biochemistry 41 (2002) 2254–2263.

[87] D. Avrahami, Y. Shai, A new group of antifungal and antibacterial lipopeptides derived from non-membrane active peptides conjugated to palmitic acid, The Journal of Biological Chemistry 279 (2004) 12277−12285.

[88] A. Malina, Y. Shai, Conjugation of fatty acids with different lengths modulates the antibacterial and antifungal activity of a cationic biologically inactive peptide, The Biochemical Journal 390 (2005) 695−702.

[89] J.M. Balkovec, Section review: anti-infectives: lipopeptide antifungal agents, Expert Opinion on Investigational Drugs 3 (1994) 65−82.

[90] M. Debono, R.S. Gordee, Antibiotics that Inhibit fungal cell wall development, Annual Review of Microbiology 48 (1994) 471−497.

[91] M.B. Kurtz, C. Douglas, J. Marrinan, K. Nollstadt, J. Onishi, S. Dreikorn, et al., Increased antifungal activity of L-733, 560, a water-soluble, semisynthetic pneumocandin, is due to enhanced inhibition of cell wall synthesis, Antimicrobial Agents and Chemotherapy 38 (1994) 2750−2757.

[92] R. Maget-Dana, F. Peypoux, Iturins, a special class of pore-forming lipopeptides: biological and physicochemical properties, Toxicology 87 (1994) 151−174.

[93] R. Maget-Dana, M. Ptak, Interactions of surfactin with membrane models, Biophysical Journal 68 (1995) 1937−1943.

[94] J.D. Sheppard, C. Jumarie, D.G. Cooper, R. Laprade, Ionic channels induced by surfactin in planar lipid bilayer membranes, Biochimica et Biophysica Acta (BBA) − Biomembranes 1064 (1991) 13−23.

[95] C. Carrillo, J.A. Teruel, F.J. Aranda, A. Ortiz, Molecular mechanism of membrane permeabilization by the peptide antibiotic surfactin, Biochimica et Biophysica Acta 1611 (2003) 91−97.

[96] M. Boaretti, P. Canepari, M.M. Lleò, G. Satta, The activity of daptomycin on *Enterococcus faecium* protoplasts: indirect evidence supporting a novel mode of action on lipoteichoic acid synthesis, The Journal of Antimicrobial Chemotherapy 31 (1993) 227−235.

[97] P. Canepari, M. Boaretti, M.M. Lleó, G. Satta, Lipoteichoic acid as a new target for activity of antibiotics: mode of action of daptomycin (LY146032), Antimicrobial Agents and Chemotherapy 34 (1990) 1220−1226.

[98] J.A. Silverman, N.G. Perlmutter, H.M. Shapiro, Correlation of daptomycin bactericidal activity and membrane depolarization in *Staphylococcus aureus*, Antimicrobial Agents and Chemotherapy 47 (2003) 2538−2544.

[99] W.E. Alborn, N.E. Allen Jr., D.A. Preston, Daptomycin disrupts membrane potential in growing *Staphylococcus aureus*, Antimicrobial Agents and Chemotherapy 35 (1991) 2282−2287.

[100] N.E. Allen, W.E. Alborn, J.N. Hobbs Jr., Inhibition of membrane potential-dependent amino acid transport by daptomycin, Antimicrobial Agents and Chemotherapy 35 (1991) 2639−2642.

[101] Z.Z. Deris, J. Akter, S. Sivanesan, K.D. Roberts, P.E. Thompson, R.L. Nation, A secondary mode of action of polymyxins against Gram-negative bacteria involves the inhibition of NADH-quinone oxidoreductase activity, The Journal of Antibiotics 67 (2014) 147−151.

[102] Z.Z. Deris, J.D. Swarbrick, K.D. Roberts, M.A.K. Azad, J. Akter, A.S. Horne, Probing the penetration of antimicrobial polymyxin lipopeptides into gram-negative bacteria, Bioconjugate Chemistry 25 (2014) 750−760.

[103] T. Schneider, K. Gries, M. Josten, I. Wiedem
mann, S. Pelzer, H. Labischinski, The lipopeptide antibiotic Friulimicin B inhibits cell wall biosynthesis through complex

formation with bactoprenol phosphate, Antimicrobial Agents and Chemotherapy 53 (2009) 1610–1618.
[104] A.E. Zeraik, M. Nitschke, Biosurfactants as agents to reduce adhesion of pathogenic bacteria to polystyrene surfaces: effect of temperature and hydrophobicity, Current Microbiology 61 (2010) 554–559.
[105] P. Das, S. Mukherjee, R. Sen, Antiadhesive action of a marine microbial surfactant, Colloids and Surfaces, B 71 (2009) 183–186.
[106] F. Rivardo, R.J. Turner, G. Allegrone, H. Ceri, M.G. Martinotti, Anti-adhesion activity of two biosurfactants produced by *Bacillus* spp. prevents biofilm formation of human bacterial pathogens, Applied Microbiology and Biotechnology 83 (2009) 541–553.
[107] J.R. Mireles, A. Toguchi, R.M. Harshey, Salmonella *enterica serovar typhimurium* swarming mutants with altered biofilm-forming abilities: surfactin inhibits biofilm formation, Journal of Bacteriology 183 (2001) 5848–5854.
[108] D. López, M.A. Fischbach, F. Chu, R. Losick, R. Kolter, Structurally diverse natural products that cause potassium leakage trigger multicellularity in *Bacillus subtilis*, Proceedings of the National Academy of Science USA 106 (2009) 280–285.
[109] J. Henrichsen, Bacterial surface translocation: a survey and a classification, Bacteriological Reviews 36 (1972) 478–503.
[110] O. Nybroe, J. Sørensen, Production of cyclic lipopeptides by fluorescent pseudomonads, in: J.-L. Ramos (Ed.), Pseudomonas: vol. 3 Biosynthesis of Macromolecules and Molecular Metabolism, Springer USA, Boston, MA, 2004, pp. 147–172.
[111] D. López, H. Vlamakis, R. Losick, R. Kolter, Cannibalism enhances biofilm development in *Bacillus subtilis*, Molecular Microbiology 74 (2009) 609–618.
[112] M.L. Mangoni, Y. Shai, Short native antimicrobial peptides and engineered ultrashort lipopeptides: similarities and differences in cell specificities and modes of action, Cellular and Molecular Life Sciences 68 (2011) 2267–2280.
[113] H.S. Sader, D.J. Farrell, R.N. Jones, Antimicrobial activity of daptomycin tested against gram-positive strains collected in European hospitals: results from 7 years of resistance surveillance (2003–2009), Journal of Chemotherapy 23 (2011) 200–206.
[114] A.A. Fayad, H. Jennifer, M. Rolf, Octapeptins: lipopeptide antibiotics against multidrug-resistant superbugs, Cell Chemical Biology 25 (2018) 351–353.
[115] C.-D. Qian, X.-C. Wu, Y. Teng, W.-P. Zhao, O. Li, S.-G. Fang, Battacin (Octapeptin B5), a new cyclic lipopeptide antibiotic from *Paenibacillus tianmuensis* active against multidrug-resistant Gram-negative bacteria, Antimicrobial Agents and Chemotherapy 56 (2012) 1458–1465.
[116] C. Wu, Z. Shang, C. Lemetre, M.A. Terne, S.F. Brady, Cadasides, calcium-dependent acidic lipopeptides from the soil metagenome that are active against multidrug-resistant bacteria, Journal of the American Chemical Society 141 (2019) 3910–3919.

# Antimicrobial peptides of fungal origin

S. Shishupala

Department of Microbiology, Davangere University, Davangere, Karnataka, India

## 6.1 Introduction

Fungi are unique group of eukaryotic microorganisms. Their metabolic flexibility is unparalleled among living organisms. Apart from regular primary metabolites they have varieties of special secondary metabolites. Ever since, the first antibiotic penicillin was identified in the fungus *Penicillium notatum* innumerable number of secondary metabolites have been characterized from fungi with significant biological activities. Fungi exist in nature as saprophytes, endophytes, and pathogens and do produce metabolites of significance to humans [1−5].

Antibiotics and mycotoxins are special metabolites produced by fungi with health significance. Though it appears that these metabolites are of secondary importance to the growth of producing species they have plenty to serve the mankind. Antimicrobial compounds are of chief significance considering their utility. Among such compounds antimicrobial peptides have attained importance in terms of various applications. Antimicrobial peptides are either linear or cyclic compounds made up of five to hundred amino acid residues. Different microorganisms, plants, and animals are capable of producing antimicrobial peptides [6−10]. These antimicrobial peptides perform crucial functions in the producing organisms mainly responsible for defense. Several hundreds of antimicrobial peptides have been characterized with important biological activities [10−15]. Ever increasing drug resistance in pathogenic microorganisms induced search for alternate compounds. This resulted in finding antimicrobial peptides with therapeutic value as best possible compounds [16−22]. These peptides are primary defense chemicals in invertebrates and part of innate immunity in other animals. Range of biological activities attributed to antimicrobial peptides makes them unique group of compounds [14,23−25]. The fungi also produce antimicrobial peptides to different levels. The unique metabolic pathways make fungi to produce extensively diverse antimicrobial peptides. Structural and biological diversity of antimicrobial compounds from fungi are exemplary [26−31]. These peptides provide competitive edge to the producing organisms to survive in nature. Various types of antimicrobial peptides from fungi have been characterized with range of biological activities. This chapter discusses antimicrobial peptides of fungal origin with emphasis on their producers, types, and biological activities. Additional importance was given for detection methods. Significance of these peptides in biotechnology is also highlighted.

## 6.2 Fungi-producing antimicrobial peptides

Various fungal species are known to produce antimicrobial peptides. The major genera of fungi involved are *Acremonium, Apiocrea, Boletus, Emericellopsis, Cephalosporium, Mycogone, Sepedonium, Silbella, Trichoderma, Tolypocladium,* and *Verticimonosporium*. Among these fungal genera *Trichoderma* (anamorphic state) is one of the soil fungi producing whitish or greenish yellow growth under cultural conditions (Fig. 6.1). The telemorphic (sexual) state is *Hypocrea*. The species/strains are used extensively in biological control of plant diseases. Enzymes and antimicrobial substances produced by these fungi serve as primary criteria in antagonism. Several species and strains of *Trichoderma* are prolific producers of peptaibol antibiotics [32]. Both filamentous fungi and mushrooms have developed biosynthetic pathways for the production of peptides with significant biological activities.

Fungal peptides were isolated from 17 genera of marine fungi. The fungi were *Acremonium, Ascotricha, Aspergillus, Asteromyces, Ceratodictyon, Clonostachys, Emericella, Exserohilum, Microsporum, Metarrhizium, Penicillium, Scytalidium, Simplicillium, Stachylidium, Talaromyces, Trichoderma,* and *Zygosporium*. Structural elucidation was carried out for 131 peptides using NMR. Antimicrobial and cytotoxic properties were shown by 53% of the peptides and the remaining peptides did not have biological activities. It is suggested that more number of fungi to be screened for biologically active peptides [33]. Peptide-producing fungi are found in different ecological conditions. Explorations of many fungi from different sources have resulted in chemically diverse peptides with range of biological activities. Fungal peptides were also of therapeutic value against many pathogenic species. The peptide-producing fungal species have competitive ability to survive in nature due to these peptides.

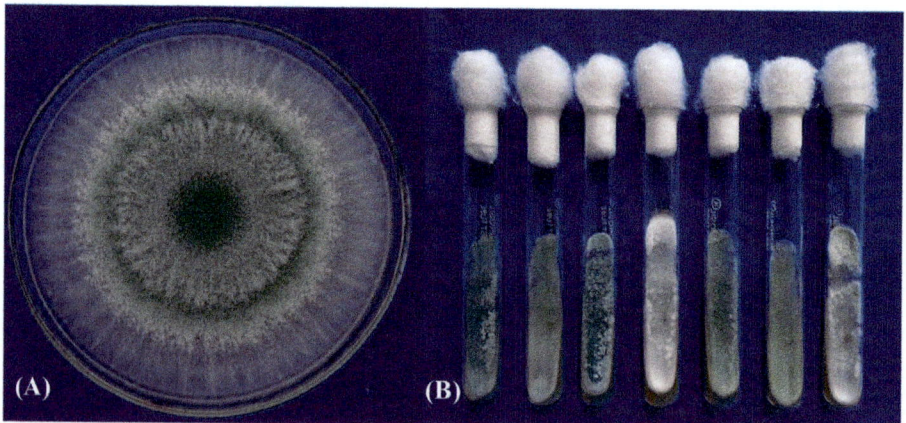

**Figure 6.1** *Trichoderma* sp.—prolific producer of antimicrobial peptides. (A) Colony morphology on Potato dextrose agar. (B) Pure cultures of different isolates.

## 6.3 Fungal peptides

Fungal peptides are categorized as peptaibols, petaibiotics, defensins, efrapeptins, asperripeptides, and others. Their sequence, structures, and functional abilities have been investigated. Peptaibols are linear peptides containing high number of nonstandard amino acids such as α-aminoisobutyric acid (Aib). The amino acid at the C-terminal end is converted to amino alcohol. The number of amino acid residues present in peptaibols range from 5 to 21 and molecular weight ranging from 500 to 2200 Da. They have highly diverse sequence and microheterogeneous groups of polypeptides differing in molecular mass 14 or 16 Da. Such alternation in molecular mass is due to interchange of Aib to Ala or Val, Leu to Val and so on in the peptide sequence. More than 317 peptaibols have been characterized. Wide range of biological activities has been attributed to peptaibols. Extensive chemical diversity of peptaibols is credited to the ability of modular peptaibol synthetases. The species *T. harzianum* itself is known to produce 55 different peptaibols having 11-, 14-, 18-, and 19-residues with lot of subfamilies and groups [5,32,34,35].

*Trichoderma* is abundant producer of peptaibols and petaibiotics. These are linear peptides show incredible diversity and majorly attributed for biocontrol efficiency of the fungus. These peptides exhibit wide range of antimicrobial properties. Mass spectrometry and NMR spectroscopy are the analytical techniques used for the detection of these peptides [5,35]. A novel emercellipsin A was identified from culture of *Emericellopsis alkalina*. This peptaibol was shown to have nine amino acids and antifungal activity against *Candida* and *Aspergillus niger*. The peptide also exhibited antibacterial and cytotoxic activities [36].

*Trichoderma koningiopsis* and *T. gamsii* were detected to have new group of peptaibols. The new group of peptides was named as "Koningiopsin" having 19-residue peptides related to trikoningin. Wide range of biological activities was attributed to these peptides. The sequence and structural diversity were also reported [37]. Species of *Trichoderma* belonging to clade *Longibrachiatum* were investigated to reveal the presence of 20-residue peptabiols related to alamethicins, gliodeliquescins, hypophellins, hyporientalins, longibrachins, metanicins, paracelsins, suzukacillins, saturnisporins, trichobrachins, trichoaureocins, trichocellins, trichokonins, trilongins, and trichosporins. A new subfamily of peptaibols—brevicelsins with 19-residues was also detected. Structural characterization and sequence determination revealed extensive chemical diversity in peptaibols [38].

New peptaibol ampullosporin F and ampullosporin G with five other known peptides were reported from *Sepedonium ampullosporum*. These peptaibols showed antifungal and anticancerous properties. Amino acid sequence variations were also observed between the peptides indicating extensive chemical diversity [39].

Roosebol A—peptaibol with 11-residues was isolated from *Clonostachys rosea*. Combination of NMR and tandem mass analysis revealed the presence of amino acids such as isovaline, α-aminoisobutyric acid, and hydroxyproline. These are characteristic components of peptaibols [40].

*Emercellopsis* spp. from different soil samples were characterized for peptaibol profile with special reference to emercellipsin A. The species tested were *E. alkalina*, *E. maritima*, and *E. terricola* isolated from terrestrial, marine, and saline soils. An anticancerous peptide EmiA with molecular weight of 1032.7 Da was defined using MALDI-TOF-MS [41].

Defensins from fungi are rich in cysteine residue. Antibacterial defensin—Plectasin was isolated from fungus *Pseudoplectania nigrella* known to inhibit bacterial cell wall synthesis. The structural similarities of plectasin with plant and insect defensins have been recorded. Similarly, defensin Copsin was isolated from fungus *Coprinopsis cinerea* showing antibacterial property on gram-positive bacteria. The other fungus *Microsporium canis* was shown to produce Micasin—a defensin-like peptide capable of killing drug-resistant bacteria [42]. Likewise, many antimicrobial peptides have been isolated and characterized from fungi. Distinct mode of action and unique biosynthesis of these fungal peptides make them ideal candidates for biotechnological applications.

## 6.4 Mode of action and biological activities

Antimicrobial peptides have shown variations in their biological activities. Some peptides are having broad spectrum and others have narrow spectrum of activity. Peptides from different sources were compared for their mode of action [43]. Antimicrobial peptides are significant cationic, amphipathic molecules capable of inducing membrane ion channels in the target organisms leading to membrane disruptions [44]. Entomopathogenic and mycoparasitic fungi were found to produce both linear and cyclic peptides used in controlling insects and phytopathogens [45].

Significant biological activities are rendered by fungal peptides. Due to amphipathic nature these peptides induce ion channels in lipid bilayer membrane of target organisms. Wide range of bacteria and fungi get inhibited by fungal peptides. These peptides can also kill cancerous cells. Few fungal peptides were shown to act as inducers of plant growth promotion and activation of plant defense mechanisms [5,46].

Atroviridins—group of peptaibols produced by *Trichoderma atroviride*—were highly effective against Gram-positive bacteria. The growth of methicillin-resistant *Staphylococcus aureus* was inhibited by these fungal peptides. Increase in peptaibol synthetase activity was correlated well with biological activity [47]. Peptides of fungal origin have often been involved in complex action after the entry into the target cell. Broad and specific mode of action by antimicrobial peptides is of consideration [48]. These peptides specifically target the membrane activity however, they have been also showed to have adverse effect on nucleus and protein synthesis of target organisms [49].

Protection of wheat seedlings from *Fusarium culmorum* infection was achieved by metabolites of *Trichoderma harzianum*. The fungal metabolites induced alternation of protein components in seedlings. Oxidative stress induced by pathogen was suppressed [50]. New tripeptides belonging to asterripeptides were isolated from *Aspergillus terreus* found in mangroves. These peptides showed anticancerous and antibacterial activities [51]. The information provided here clearly demonstrated multiple biological activities of fungal peptides. Hence, fungal peptides will have great impact as therapeutic agents.

## 6.5 Mechanisms of synthesis

Most of the fungal peptides are produced without the involvement of ribosomes. They also contain many nonstandard amino acids. The non-ribosomal pathways providing opportunities to synthesize diverse peptides have been studied. Multimodular enzymes such as peptide synthetases are investigated being responsible for the peptide production [52–55]. Non-ribosomal peptide synthetases from *Trichoderma* and *Ganoderma* are characterized and cloned. Nucleotide sequences of the genes and enzyme activity were found to be highly versatile in order to synthesize innumerable varieties of peptides. These enzymes are capable of performing adenylation, thiolation, and condensation of individual monomers in growing peptide chain. Non-ribosomal peptide synthetases are structurally unique enzymes with modules which are units capable of modifying each amino acid being added during peptide synthesis. Peptide synthetases provide protein templates for non-ribosomal assembly of peptides [55]. Analysis of functional abilities in peptide synthetases revealed reasons for variations encountered in the amino acid sequences of antimicrobial peptides. Diverse functional sites of these enzymes are responsible for chemical diversity of fungal peptides.

## 6.6 Detection methods of antimicrobial peptides

Development of novel biochemical techniques paved the way for the detection and analysis of antimicrobial peptides. Rapid progress in biophysical techniques provided opportunity for characterization of the peptides. A separate branch of biology called "Peptiodomics" was introduced which made use of chromatographic techniques and computational analysis [56]. Powerful separation techniques with digital analysis were real strength of detection of charged molecules such as peptides.

Peptaibiotics from *Trichoderma brevicompactum* were analyzed by Liquid Chromatography/Electro-Spray Ionization-Mass Spectrometry (LC/ESI-MS) for the production alamethicins. Many of the novel peptides such as trichocryptins, trichobrevins, trichofesins, and trichocampactins were also detected. These peptides were found to be involved in antagonism. At the same time, diversity in these peptaibiotics may serve as chemotaxonomic markers for species identification [57].

*Trichoderma* conidia/mycelium was subjected to Matrix-Assisted Laser Desorption/Ionization—Time of Flight-Mass Spectrometry (MALDI-TOF-MS). Peptaibols were easily ionizable compounds produced significant peaks of molecular masses with high relative intensities [58–64]. The analysis of mass spectra clearly demonstrated cellular detection of peptaibols from various *Trichoderma* isolates. Signature sequences showing microheterogenity with mass difference of 14 or 16 Da were evident in the spectra (Fig. 6.2). Among 32 isolates of *Trichoderma* presence of 12-, 14-, 17-, and 19-residue

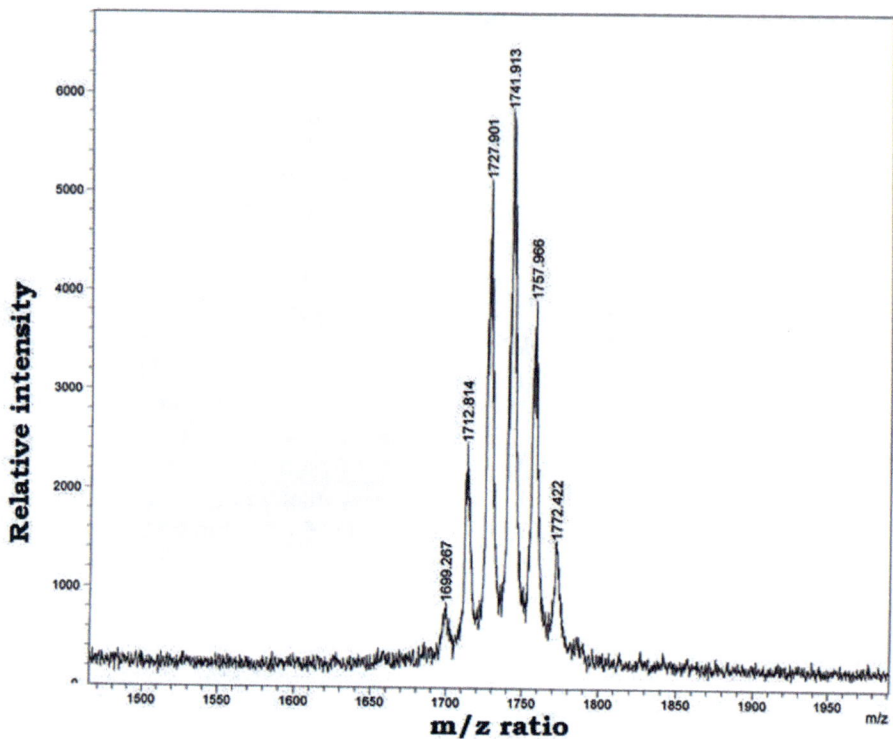

**Figure 6.2** Cellular MALDI-TOF-mass spectrum of *Trichoderma* isolate (KUMB 553) producing 17-residue peptaibol.

peptaibols was revealed. *Trichoderma* isolates showing one, two, or three groups of peptaibols were detected using mass spectra (Figs. 6.3–6.5). Cellular peptidomics having mass analysis was great technique to detect the nature of peptides present in fungi. Some of the peptaibols detected based on molecular mass with the help of peptaibols database are shown in Table 6.1. Similar approach in different species of Trichoderma was able to detect atroviridins, trichovirins, and paracelsins by using intact cells as sample. Amino acid sequences of selected peptaibols from different fungi are given in Table 6.2. The genetic analysis provided information on complexity involved in peptaibol derivatives [30]. Four novel peptaibols from *Nectriopsis* were detected using mass spectrometry. These peptaibols contained threonine in unusual positions. All the peptaibols were characterized and one of them showed anticancerous property [65].

Mass spectrometry has revolutionized fungal peptide research. Different modification of the conditions in mass spectrometric methods can directly provide mass and amino acid sequence of fungal peptides. Extensive use of peptidomics would definitely provide valuable information on fungal peptides.

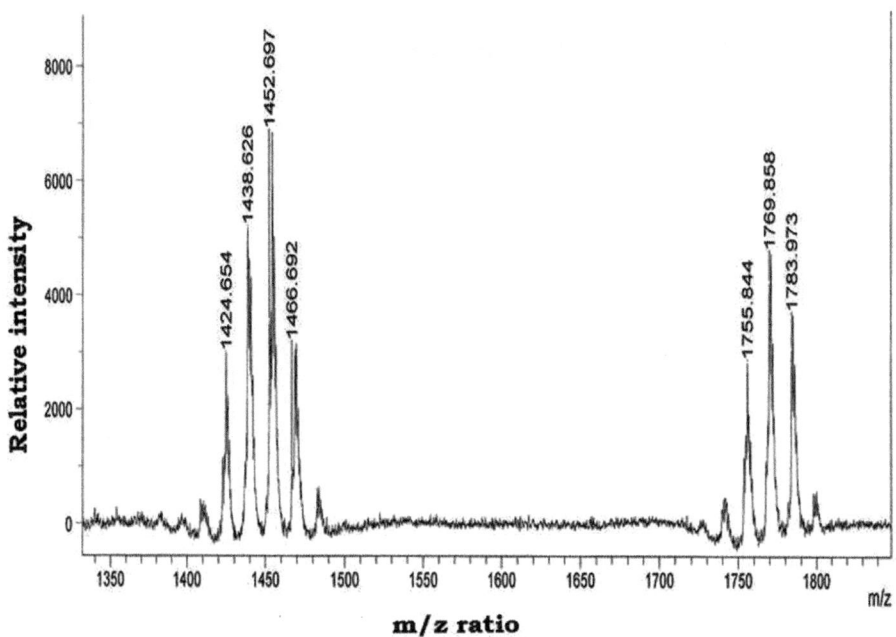

**Figure 6.3** Cellular MALDI-TOF-mass spectrum of *Trichoderma* isolate (KUMB 14) producing 14- and 17-residue peptaibol.

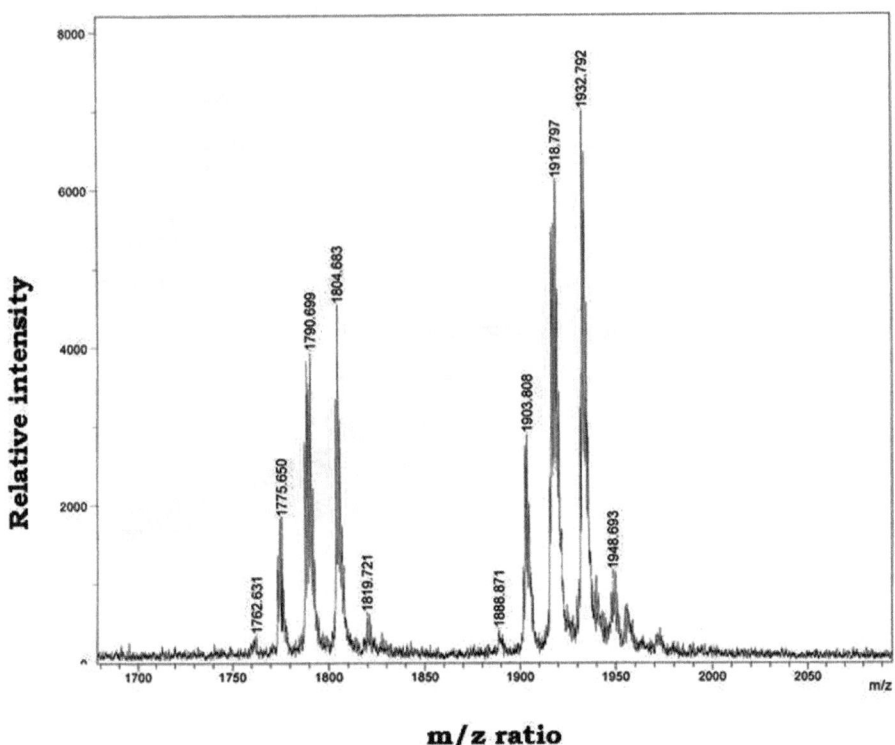

**Figure 6.4** Cellular MALDI-TOF-mass spectrum of *Trichoderma* isolate (KUMB 9) producing 17- and 19-residue peptaibol.

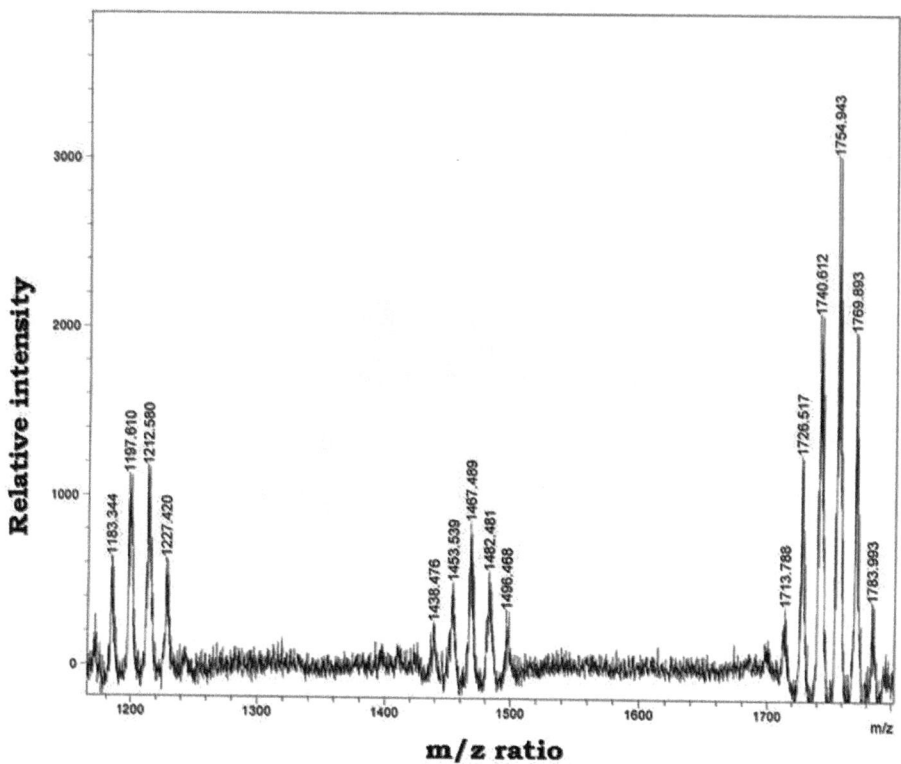

**Figure 6.5** Cellular MALDI-TOF-mass spectrum of *Trichoderma* isolate (KUMB 59) producing 12-, 14-, and 17-residue peptaibol.

**Table 6.1** Peptaibols identified in *Trichoderma* isolates from MALDI-TOF-MS of cells.

| Name[a] | Number of amino acid residues | *Trichoderma* isolate | Calculated molecular weight[a] | Observed molecular weight |
|---|---|---|---|---|
| Trichorozin II | 11 | KUMB 98 | 1174 | 1171 |
| Harzianin HC XV | 14 | KUMB 14 | 1428 | 1425 |
| Trichorozin HA V | 18 | KUMB 61 | 1731 | 1727 |
| Trichokindin VI | 18 | KUMB 9 | 1775 | 1775 |
| Trichorzianine TB VIa | 19 | KUMB 9 | 1937 | 1932 |

[a]Peptaibol database—http://www.cryst.bk.ac.uk/peptaibol.

**Table 6.2** Amino acid sequence of selected peptaibols from different species of fungi.

| Peptaibol | Source fungal species | Amino acid sequence |
|---|---|---|
| Aibellin | *Verticimonosporium eilipticum* | Ac U A U A U A Q U F U G U U P V U U E E NHC (CH2Ph) HCH2NHCH2CH2OH |
| Alamethicin_F-30 | *Trichoderma viride* | Ac U P U A U A Q U V U G L U P V U U E Q F OH |
| Ampullosporin | *Sepedonium ampullosporum* | Ac W A U U L U Q U U U Q L U Q L OH |
| Antiamoebin_I | *Emericellopsis poonensis, E. synnematicola Cephalosporium pimprina* | Ac F U U U J G L U U O Q J O U P F OH |
| Antiamoebin_VI | *Stilbella fimetaria* | Ac F U U U U G L U U O Q U O U P F OH |
| Atroviridin_A | *Trichoderma atroviride* | Ac U P U A U A Q U V U G L U P V U U Q Q F OH |
| Boletusin | *Boletus* spp. | Ac F U A U J L Q G U U A A U P U U U Q W OH |
| Cephaibol_A | *Acremonium tubakii* | Ac F U U U U G L J U O Q J O U P F OH |
| Cervinin_1 | *Mycogone cervina* | Ac L U P U L U P A U P V L OH |
| Chrysospermin_A | *Apiocrea chrysosperma* | Ac F U S U U L Q G U U A A U P U U U Q W OH |
| Harzianin_HB_I | *Trichoderma harzianum* | Ac U N L I U P J L U P L OH |
| Hypelcin_A | *Hypocrea peltata* | Ac U P U A U U Q L U G U U U P V U U Q Q L OH |
| LP237_F8 | *Tolypocladium geodes* | Oc U P F U Q Q U Zor Q A L OH |
| Stilboflavin_A_1 | *Silbella flavipes* | Ac U P U A U A Q U V U G U U P V U U E Q V OH |

*Note*: Non-standard amino acids found in peptaibols are aminoisobutyric acid (Aib; U); ethylnorvaline (EtN or EtNor; Z); isovaline (Iva; J); hydroxyproline (Hyp; O); Ac-acylation of N-terminal, OH- C-terminal amino alcohol.
*Source*: From http://peptaibol.cryst.bbk.ac.uk/home.shtml.

## 6.7 Peptide databases

The genome and proteome sequences of organisms are available in many of the websites. Peptides from fungi and other sources are synthesized by non-ribosomal peptide synthetases. Hence, these peptides are not generally included in the regular protein databases. However, considering the significance and available data on peptides certain databases have been developed (Table 6.3).

**Table 6.3** Databases for antimicrobial peptides from fungi and other sources.

| Sl. no. | Name of the database | Internet access |
|---|---|---|
| 1 | Peptaibol database | http://peptaibol.cryst.bbk.ac.uk/home.shtml |
| 2 | Antimicrobial peptide database (APD3) | https://aps.unmc.edu/ |
| 3 | Collection of antimicrobial peptides database (CAMPR3) | http://www.camp3.bicnirrh.res.in |
| 4 | Database of antimicrobial activity and structure of peptides (DBAASP) | http://csb.cse.yzu.edu.tw/dbAMP/ |
| 5 | Database of antimicrobial peptides (dbAMP) | http://csb.cse.yzu.edu.tw/dbAMP/ |

## 6.7.1 Peptaibol database

The antimicrobial peptides are produced by non-ribosomal synthetic pathways using modular peptide synthetases. Antimicrobial peptides produced by fungi contain non-proteinogenic, unusual amino acids. They are not coded directly by genes and not having genetic code. Hence, they are not generally included in the protein sequence databases available. Peptaibol detection, analysis, sequence and structure determination provided opportunity to have enormous data of this group of antimicrobial peptides. The information collected led to the Peptaibol Database establishment in 1997 at the Department of Crystallography, Birkbeck College, University of London, United Kingdom [66]. Currently, the database has 317 sequences of peptaibols produced by fungi. The database provides information of fungal producers of peptaibols, sequence, crystal structure, NMR data, and other results. It includes only naturally occurring peptaibols. The details are collected from literature worldwide and stored here. The database has easily accessible version of sequence and structures of peptaibols [46,66,67]. Numbers of other databases with extensive information are available. The databases provide sources, classification, structures, and biological activities of antimicrobial peptides [42,68].

Bioinformatics, cheminformatics, and drug-designing approaches are being used in antimicrobial peptide research. This information will be valuable for fungal peptides to be used as therapeutics [69]. Such databases are highly useful to know already characterized peptides and their sequences. Evidence on producing organisms is suitable for screening protocols and sources of organisms in search of novel peptides. These databases can also be valuable in synthesis of new peptides with biological activities.

## 6.8 Biotechnological applications

Immense value of fungal peptides for therapeutic use paved the way for biotechnological applications. A number of techniques were employed to analyze fungal peptides. Efforts were also made to improve biological activities. Site-directed changes of peptide sequences and cloning of peptide synthase genes were also envisaged. Improvement in therapeutic potential of fungal peptides was of major concern. Use of natural peptides as lead molecules for the synthesis of peptides with desired sequence was also reported. Some of the recent developments in this direction are discussed here. Application and delivery of the peptide therapeutics are essential process involved.

Generation of novel peptaibols by incorporation of fluorine was achieved [70]. Biosynthetic pathway of *Trichoderma arundinaceum* was modified by employing precursor-directed biosynthesis where *ortho*- and *meta*-F-phenylalanine incorporation to fermentation medium resulted in biosynthesis of fluorine-containing alamethicin derivatives. Such an approach increased the potential of biological activity playing important role of fluorine compounds in drugs.

Aculeacin A an antibacterial peptide from *Aspergillus aculeatus* was detected. Diversity in structure and functions of antimicrobial peptides from different origin provides an opportunity to design new drugs. Proper understanding of sequence, structure, and biological activities of these peptides offer a prospect to create synthetic antimicrobials [71]. Natural antimicrobial peptides were found to be excellent starting points for designing novel drugs. New generation of antifungal compounds were developed based on structural and functional characteristics of peptide.

More than half of the currently used drugs are non-ribosomal peptides or polyketides of natural products mainly from bacteria and fungi. Non-ribosomal peptides from marine organisms are potential pharmaceutical compounds [72]. Diversity of antimicrobial peptides with respect to biological activities and biocompatibility was assessed. Development of strategies such as chemical modifications and better delivery systems were analyzed and presented [73].

Antimicrobial peptides can be transported using nano-vehicles for increasing fungicidal efficiency in targeted drug delivery [74]. Nanotechnological advancements are definitely helpful in delivering fungal peptides to realize their potential as therapeutic drugs [75]. Genomic level of understanding peptide synthesis in fungi is extremely important for modifications and to improve the efficiency. Whole genome analysis of *Penicillium citreonigrum* led to identification of gene cluster involved in mycotoxin citreoviridin production [76]. Similar methodology is required for understanding genetic basis of fungal peptide production.

Expression of antimicrobial peptide in chloroplast genome of transgenic tobacco plant offered resistance against bacteria and fungi. Disease resistance in plants by achieving peptide expression using chloroplast genome was desirable approach [77]. Entomopathogenic nematodes and bacteria can produce effective antimicrobial peptides which are essential for designing peptidomimetic antimicrobial agents [78]. Such an approach is required in fungal peptide drug designing. Using de novo synthesis strategy antimicrobial peptides can be modified to improve their biological activities. Their uses in drug designing are discussed [49]. Biocontrol efficiency of *Trichoderma* to manage fungal phytopathogens has been attributed to the production of antimicrobial peptides. Some of the compounds produced by biocontrol fungi also improve the plant growth [45,79].

Hence, assessment of *Trichoderma* peptaibols for their antimicrobial and plant growth promoting activity needs to be investigated.

Peptaibol synthetase from *Trichoderma* was cloned and expressed [55,80,81]. The gene coding this enzyme was termed as *tex1* of the size 62.8 kb open reading frame. In this enzyme 18 modules have been identified. Versatility of peptide synthetases is responsible for chemical diversity. Complete understanding of enzyme structure-—function relationship is necessary to make use in peptide synthesis. This needs to be exploited at molecular level [82].

## 6.9 Summary and conclusions

Being unique group organisms, fungi are having flexible biosynthetic pathways. Fungal peptides have gained importance due to their biological activities. Several fungi from different sources are capable of producing antimicrobial peptides. Fungi belonging *Trichoderma* are significant producers of antimicrobial peptides. Many other fungi from soil to marine water have been identified as source for biologically active peptides. Peptaibols and peptaibiotics are extensively produced by certain groups of fungi. These peptides have been sequenced and structures are presented. Extensive chemical diversity has been reported from peptides of fungal origin. In particular, high propensity of unusual amino acids makes them more interesting. Multiple mode of actions is assigned to fungal peptides to make them as useful from therapeutic point of view. Modular non-ribosomal peptide synthetases are involved in the production of such diverse peptides in fungi. Mass spectrometry-based detection methods are found to be highly useful in identification and characterization of fungal peptides. Few databases are available providing information on source, sequence, structure, and biological activities of fungal peptides. Biotechnological approaches in understanding fungal peptides will be helpful for developing fungal peptides as useful drugs. Combinatorial chemistry and cloning peptides synthetases are positive approaches to achieve this goal. Fungal peptides have assumed special status both in human medicine and plant protection methods. Nanotechnological methods have been made use to design the drug delivery systems. Extensive chemical diversity in fungal peptides offers ample opportunity to understand the biology of fungi. It is hoped that these useful peptides can provide desirable source of therapeutic agents. The fermentation technology already being established in fungal culturing should also be useful in establishing industrial production technology

## Acknowledgments

The author is grateful to Davangere University for providing necessary facilities and Indian Institute of Science, Challkere Campus, Chitradurga for providing library facilities for literature survey. The author is grateful to Prof. P. Balaram, Director, Indian

Institute of Science, Bangalore for having introduced to the field of peptaibols, constructive discussion, and providing laboratory facilities. Thanks are also due to Mr. S. Prakash for his technical assistance in procuring mass spectra. The photographic help of Mr. Narendrababu, B.N. is acknowledged.

# References

[1] J. Deacon, Fungal Biology, Blackwell Publishing, Oxford, UK, 2006, p. 371.
[2] S.-E. Kim, B.S. Hwang, J. Song, S.W. Lee, I.-K. Lee, B.-S. Yun, New bioactive compounds from Korean native mushrooms, Mycobiology 41 (2013) 171–176.
[3] M. Staszczak, Fungal secondary metabolites as inhibitors of the ubiquitin–proteasome system, International Journal of Molecular Science 22 (2021) 13309–13339.
[4] T.B. Ng, Peptides and proteins from fungi, Peptides 25 (2004) 1055–1073.
[5] J.F.D.S. Daniel, E.R. Filho, Peptaibols of *Trichoderma*, Natural Product Reports 24 (2007) 1128–1141.
[6] Y. Ku, S. Cheng, A. Gerhardt, M. Cheung, C.A. Contador, L.W. Poon, et al., Secretory peptides as bullets: effector peptides from pathogens against antimicrobial peptides from soybean, International Journal of Molecular Science 21 (2020) 9294–9309.
[7] A. Kombrink, A. Tayyrov, A. Essig, M. Stöckli, S. Micheller, J. Hintze, et al., Induction of antibacterial proteins and peptides in the coprophilous mushroom *Coprinopsis cinerea* in response to bacteria, The ISME Journal 13 (2019) 588–602.
[8] K.G.N. Oshiro, G. Rodrigues, B.E.D. Monges, M.H. Cardoso, O.L. Franco, Bioactive peptides against fungal biofilms, Frontiers Microbiology 10 (2019) 2169–2185.
[9] D.A. Phoenix, S.R. Dennison, F. Harris, Antimicrobial Peptides: Their History, Evolution, and Functional Promiscuity, Antimicrobial Peptides, first ed., Wiley-VCH Verlag GmbH & Co. KGaA, 2013, pp. 1–38.
[10] Y. Huan, Q. Kong, H. Mou, H. Yi, Antimicrobial peptides: classification, design, application and research progress in multiple fields, Frontiers in Microbiology 11 (2020) 582779–582799.
[11] L. Zhang, R.L. Gallo, Antimicrobial peptides, Current Biology 26 (2016) R1–R21.
[12] T. Roncevic, J. Puizina, A. Tossi, Antimicrobial peptides as anti-infective agents in pre-post-antibiotic era? International Journal of Molecular Science 20 (2019) 5713–5744.
[13] G.B.D. Cesare, S.A. Cristy, D.A. Garsin, M.C. Lorenza, Antimicrobial peptides: a new frontier in antifungal therapy, mBio 11 (2020). Available from: https://doi.org/10.1128/mBio.02123-20.
[14] M.F. de Ullivarri, S. Arbulu, E. Garcia-Gutierrez, P.D. Cotter, Antifungal peptides as therapeutic agents, Frontiers in Cellular and Infection Microbiology 10 (2020) 105–126.
[15] T. Mirski, M. Niemcewicz, M. Bartoszcze, R. Gryko, A. Michalski, Utilisation of peptides against microbial infections – a review, The Annals of Agricultural and Environmental Medicine 25 (2018) 205–210.
[16] J. Lei, L. Sun, S. Huang, C. Zhu, P. Li, J. He, et al., The antimicrobial peptides and their potential clinical applications, The American Journal of Translational Research 11 (2019) 3919–3931.
[17] C. Cheng, Z. Hua, Lasso peptides: heterologous production and potential medical application, Frontiers in Bioengineering and Biotechnology 8 (2020) 571165–571180.
[18] J. Mwangi, X. Hao, R. Lai, Z. Zhang, Antimicrobial peptides: new hope in the war against multi-drug resistance, Zoological Research 40 (2019) 488–505.

[19] F. Pierre, D.L. Torre, A. Sidebottom, A. Kambal, X. Zhu, Y. Tao, et al., Peptide YY: a novel Paneth cell antimicrobial peptide that maintains fungal commensalism, bioRxiv (2020). Available from: https://doi.org/10.1101/2020.05.15.096875.

[20] D. Brady, A. Grapputo, O. Romoli, F. Sandrelli, Insect Cecropins, antimicrobial peptides with potential therapeutic applications, International Journal of Molecular Science 20 (2019) 5862−5883.

[21] H.X. Luonga, T.T. Thanha, T.H. Trana, Antimicrobial peptides − advances in development of therapeutic applications, Life Sciences 260 (2020) 118407−118422.

[22] K. Browne, S. Chakraborty, R. Chen, M.D.P. Willcox, D.S. Black, W.R. Walsh, et al., A New era of antibiotics: the clinical potential of antimicrobial peptides, International Journal of Molecular Science 21 (2020) 7047−7069.

[23] T. Ganz, The role of antimicrobial peptides in innate immunity, Integrative and Comparative Biology 43 (2003) 300−304.

[24] A.A. Bahar, R.D. Ren, Antimicrobial peptides, Pharmaceuticals 6 (2013) 1543−1575.

[25] H. Duclohier, G.M. Alder, C.L. Bashford, H. Bruckner, J.K. Chugh, B.A. Wallace, Conductance studies on trichotoxin_A50E and implications for channel structure, Biophysical Journal 87 (2004) 1705−1710.

[26] J. Lee, D.G. Lee, Antimicrobial peptides (AMPs) with dual mechanisms: membrane disruption and apoptosis, Journal of Microbiology and Biotechnology 25 (2015) 759−764.

[27] D. Jakubczyk, F. Dussart, Selected fungal natural products with antimicrobial properties, Molecules 25 (2020) 911−928.

[28] E. Montesinos, Antimicrobial peptides and plant disease control, FEMS Microbiology Letters 270 (2007) 1−11.

[29] C.T. Walsh, Polyketide and non-ribosomal peptide antibiotics: modularity and versatility, Science 303 (2004) 1805−1810.

[30] T. Neuhof, R. Dieckmann, I.S. Druzhinina, C.P. Kubicek, H.V. Dohren, Intact-cell MALDI-TOF mass spectrometry analysis of peptaibol formation by the genus *Trichoderma/Hypocrea*: can molecular phylogeny of species predict peptaibol structures? Microbiology 153 (2007) 3417−3437.

[31] A. Rani, K.C. Saini, F. Bast, S. Vajani, S. Mehariya, S.K. Bhatia, et al., A review on microbial products and their perspective application as antimicrobial agents, Biomolecules 11 (2021) 1860−1887.

[32] S. Shishupala, in: K.R. Sridhar (Ed.), Bioactive peptaibol antibiotics from *Trichoderma*, Frontiers in Fungal Ecology, Diversity and Metabolites, I.K. International Publishing House Pvt. Ltd., New Delhi, 2009, pp. 300−321.

[33] F.S. Youssef, M.L. Ashour, A.N.B. Singab, M. Wink, A comprehensive review of bioactive peptides from marine fungi and their biological significance, Marine Drugs 17 (2019) 559−582.

[34] S. Shishupala, Biomolecular Analysis of Fungi by Matrix-Assisted Laser Desorption-Ionization Mass Spectrometry. Research Project Report, Indian Academy of Sciences, Bangalore, 2001.

[35] T. Degenkolb, H. Bruckner, Peptaibiomics: towards a myriad of bioactive peptides containing Ca-dialkylamino acids? Chemistry & Biodiversity 5 (2008) 1817−1843.

[36] E.A. Rogozhin, V.S. Sadykova, A.A. Baranova, A.S. Vasilchenko, V.A. Lushpa, K.S. Mineev, et al., Lipopeptaibol emericellipsin A with antimicrobial and antitumor activity produced by the extremophilic fungus *Emericellopsis alkalina*, Molecules 23 (2018) 2785−2796.

[37] T. Marik, C. Tyagi, G. Racic, D. Rakk, A. Szekeres, C. Vágvölgyi, et al., New 19-residue peptaibols from *Trichoderma* clade *Viride*, Microorganisms 6 (2018) 85−100.

[38] T. Marik, C. Tyagi, D. Balázs, P. Urbán, Á. Szepesi, L. Bakacsy, et al., Structural diversity and bioactivities of peptaibol compounds from the *Longibrachiatum* clade of the filamentous fungal genus *Trichoderma*, Frontiers in Microbiology 10 (2019) 1434–1471.
[39] Y.T.H. Lam, M.G. Ricardo, R. Rennert, A. Frolov, A. Porzel, W. Brandt, et al., Rare gutamic acid methyl ester peptaibols from *Sepedonium ampullosporum* Damon KSH 534 exhibit promising antifungal and anticancer activity, International Journal of Molecular Science 22 (2021) 12718–12740.
[40] C.-K. Kim, L.R.H. Krumpe, E. Smith, C.J. Henrich, I. Brownell, K.L. Wendt, et al., A, a new peptaibol from the fungus *Clonostachys rosea*, Molecules 26 (2021) 3594–3603.
[41] A.E. Kuvarina, I.A. Gavryushina, M.A. Sykonnikov, T.A. Efimenko, N.N. Markelova, E.N. Bilanenko, et al., Exploring peptaibol's profile, antifungal, and antitumor activity of emericellipsin A of *Emericellopsis* species from soda and saline soils, Molecules 27 (2022) 1736–1751.
[42] A.B. Hafeez, X. Jiang, P.J. Bergen, Y. Zhu, Antimicrobial peptides: an update on classifications and databases, International Journal of Molecular Science 22 (2021) 11691–11742.
[43] N.L.V. Weerden, M.R. Bleackley, M.A. Anderson, Properties and mechanisms of action of naturally occurring antifungal peptides, Cellular and Molecular Life Sciences (2013). Available from: https://doi.org/10.1007/s00018-013-1260-1.
[44] A. Marquette, B. Bechinger, Biophysical investigations elucidating the mechanisms of action of antimicrobial peptides and their synergism, Biomolecules 8 (2018) 18–41.
[45] X. Niu, N. Thaochan, Q. Hu, Diversity of linear non-ribosomal peptide in biocontrol fungi, Journal of Fungi 6 (2020) 61–81.
[46] J.K. Chugh, B.A. Wallace, Peptaibols: models for ion channels, Biochemical Society Transactions 29 (2001) 565–570.
[47] J. Viglas, S. Dobiasová, J. Viktorová, T. Ruml, V. Repiská, P. Olejníková, et al., Peptaibol-containing extracts of *Trichoderma atroviride* and the fight against resistant microorganisms and cancer cells, Molecules 26 (2021) 6025–6043.
[48] T. Li, L. Li, F. Du, L. Sun, J. Shi, M. Long, et al., Activity and mechanism of action of antifungal peptides from microorganisms: a review, Molecules 26 (2021) 3438–3456.
[49] S. Kapil, V. Sharma, D-Amino acids in antimicrobial peptides: a potential approach to treat and combat antimicrobial resistance, Canadian Journal of Microbiology 67 (2021) 119–137.
[50] J. Mironenka, S. Rózalska, P. Bernat, Potential of *Trichoderma harzianum* and its metabolites to protect wheat seedlings against *Fusarium culmorum* and 2,4-D, International Journal of Molecular Science 22 (2021) 13058–13078.
[51] E.V. Girich, A.B. Rasin, R.S. Popov, E.A. Yurchenko, E.A. Chingizova, P.T.H. Trinh, et al., New tripeptide derivatives asperripeptides A−C from Vietnamese mangrove-derived fungus *Aspergillus terreus*, Marine Drugs 20 (2022) 77.
[52] V. Gogineni, M.T. Hamann, Marine natural product peptides with therapeutic potential: chemistry, biosynthesis, and pharmacology, Biochimica et Biophysica Acta 2018 (1862) 81–196.
[53] N. Shokrollahi, C.-L. Ho, N.A.I.M. Zainudin, M.A.B.A. Wahab, M.-Y. Wong, Identification of non-ribosomal peptide synthetase in *Ganoderma boninense* Pat. that was expressed during the interaction with oil palm, Scientific Reports 11 (2021) 16330–16345.
[54] M. Duban, S. Cociancich, V. Leclère, Nonribosomal peptide synthesis definitely working out of the rules, Microorganisms 10 (2022) 577–595.
[55] A. Wiest, D. Grzegorski, B. Xu, C. Goulard, S. Rebuffat, D.J. Ebbole, et al., Identification of peptaibols from *Trichoderma virens* and cloning of a peptaibol synthetase, Journal of Biological Chemistry 277 (2002) 20862–20868.

[56] P. Schulz-Knappe, M. Schrader, H. Zucht, The peptidomics concept, Combinatorial Chemistry & High Throughput Screening 8 (2005) 697–704.
[57] T. Degenkolb, T. Grafenhan, H.I. Nirenberg, W. Gams, H. Bruckner, *Trichoderma brevicompactum* complex: rich source of novel and recurrent plant-protective polypeptide antibiotics (peptaibiotics), Journal of Agriculture and Food Chemistry 54 (2006) 7047–7061.
[58] S. Shishupala, Y.K. Saikumari, P. Balaram, Matrix-assisted laser desorption ionization mass spectral analysis of fungi, in: Asian Congress of Mycology and Plant Pathology, 1–4 October 2002, University of Mysore, Mysore, India.
[59] S. Shishupala, Microbial metabolites with antifungal activity and their applications, in: A.R. Alagawadi, P.U. Krishnaraj, G. Shirnalli, K.S. Jagadeesh (Eds.), In: *Souvenir - Microbes and Human Sustenance*, University of Agricultural Sciences, Dharwad, India, 2003, pp. 78–81.
[60] S. Shishupala, Y.K. Saikumari, P. Balaram, Mass spectral detection of peptaibol antibiotics in *Trichoderma*, in: 44th Annual Conference of Association of Microbiologists of India, 12–14 November 2003, University of Agricultural Sciences, Dharwad, India, pp. 183–184.
[61] S. Shishupala, Biomolecular profiling of fungi using MALDI mass spectrometry, in: National Seminar on Recent Advances in Mycology and 31st Annual Meeting of the Mycological Society of India, 2–3 December 2004, Mangalore University, Mangalore, India.
[62] V. Sabareesh, V.P. Balaram, Tandem electrospray mass spectrometric studies of proton and sodium ion adduct of neutral peptides with modified N- and C-termini: synthetic model peptides and micro heterogeneous peptaibol antibiotics, Rapid Communications in Mass Spectrum. 20 (2006) 618–628.
[63] S. Shishupala, P. Balaram, $N^{15}$ labeled peptide profiling by matrix-assisted laser desorption/ionization mass spectrometry in cells of *Trichoderma*, in: 46th Annual Conference of Association of Microbiologists of India, 8–10 December 2005, Osmania University, Hyderabad, India.
[64] S. Shishupala, Biochemical analysis of fungi using matrix-assisted laser desorption/ionization time-of-flight mass spectrometry (MALDI-TOF-MS), Fungal Diversity Research Series 20 (2008) 327–372.
[65] V.P. Sicaa, E.R. Reesa, H.A. Rajaa, J. Rivera-Cháveza, J.E. Burdetteb, C.J. Pearcec, et al., *In situ* mass spectrometry monitoring of fungal cultures led to the identification of four peptaibols with a rare threonine residue, Phytochemistry 143 (2017) 45–53.
[66] L. Whitmore, B.A. Wallace, The peptaibol database: a database for sequences and structures of naturally occurring peptaibols, Nucleic Acids Research 32 (2004) D593–D594.
[67] L. Whitmore, J.K. Chugh, C.F. Snook, B.A. Wallace, The peptaibol database: a sequence and structure resource, Journal of Peptide Sciences 9 (2003) 663–665.
[68] G. Wang, X. Li, Z. Wang, APD3: the antimicrobial peptide database as a tool for research and education, Nucleic Acids Research 44 (2016) D1087–D1093.
[69] J.D. Romano, N.P. Tatonetti, Informatics and computational methods in natural product drug discovery: a review and perspectives, Frontiers in Genetics 10 (2019) 368–383.
[70] J. Rivera-Chávez, H.A. Raja, T.N. Graf, J.E. Burdette, C.J. Pearce, N.H. Oberlies, Biosynthesis of fluorinated peptaibols using a site-directed building block incorporation approach, Journal of Natural Products 80 (2017) 1883–1892.
[71] M. Bondaryk, M. Staniszewska, P. Zielinska, Z. Urbanczyk-Lipkowska, Natural antimicrobial peptides as inspiration for design of a new generation antifungal compounds, Journal of Fungi 3 (2017) 46–81.

[72] S. Agrawal, D. Acharya, A. Adholeya, C.J. Barrow, S.K. Deshmukh, Non-ribosomal peptides from marine microbes and their antimicrobial and anticancer potential, Frontiers Pharmacology 8 (2017) 828–854.

[73] P. Kumar, J.N. Kizhakkedathu, S.K. Straus, Antimicrobial peptides: diversity, mechanism of action and strategies to improve the activity and biocompatibility *in vivo*, Biomolecules 8 (2018) 4–29.

[74] I. Mela, C.F. Kaminski, Nano-vehicles give new lease of life to existing antimicrobials, Emerging Topics in Life Sciences 4 (2020) 555–566.

[75] M.C. Teixeira, C. Carbone, M.C. Sousa, M. Espina, M.L. Garcia, E. Sanchez-Lopez, et al., Nanomedicines for the delivery of antimicrobial peptides (AMPs), Nanomaterials 10 (2020) 560–582.

[76] T. Okano, N. Kobayashi, K. Izawa, T. Yoshinari, Y. Sugita-Konishi, Whole genome analysis revealed the genes responsible for Citreoviridin biosynthesis in *Penicillium citreonigrum*, Toxins 12 (2020) 125–134.

[77] G. DeGray, K. Rajasekaran, F. Smith, J. Sanford, H. Daniell, Expression of an antimicrobial peptide via the chloroplast genome to control phytopathogenic bacteria and fungi, Plant Physiology 127 (2001) 852–862.

[78] S.D. Mandal, A.K. Panda, C. Murugan, X. Xu, N.S. Kumar, F. Jin, Antimicrobial Peptides: novel source and biological function with a special focus on entomopathogenic nematode/bacterium symbiotic complex, Frontiers in Microbiology 12 (2021) 555022–555037.

[79] R. Tyskiewicz, A. Nowak, E. Ozimek, J. Jaroszuk- Scisel, *Trichoderma*: The current status of its application in agriculture for the biocontrol of fungal phytopathogens and stimulation of plant growth, International Journal of Molecular Sciences 23 (2022) 2329–2356.

[80] J.A. Vizcaino, L. Sanz, R.E. Cardoza, E. Monte, S. Gutierrez, Detection of putative peptide synthetase genes in *Trichoderma* species: application of this method to the cloning of a gene from *T. harzianum* CECT 2413, FEMS Microbiology Letters 244 (2005) 139–148.

[81] C. Chutrakul, J.F. Peberdy, Isolation and characterization of a partial peptide synthetase gene from *Trichoderma asperellum*, FEMS Microbiology Letters 252 (2005) 257–265.

[82] I. Teichert, Fungal RNA editing: who, when, and why? Applied Microbiology and Biotechnology 104 (2020) 5689–5695.

# Insect peptides with antimicrobial effects

Daljeet Singh Dhanjal[1,*], Chirag Chopra[1,*], Sonali Bhardwaj[1,*], Parvarish Sharma[2], Eugenie Nepovimova[3], Reena Singh[1] and Kamil Kuca[3]

[1]School of Bioengineering and Biosciences, Lovely Professional University, Phagwara, Punjab, India, [2]School of Pharmaceutical Sciences, Lovely Professional University, Phagwara, Punjab, India, [3]Department of Chemistry, Faculty of Science, University of Hradec Kralove, Hradec Kralove, Czech Republic

## 7.1 Introduction

In prokaryotes and eukaryotes, antimicrobial peptides (AMPs) are versatile innate defensive components. AMPs are usually 12–50 amino acids (AA) long and are classified into various categories based on AA substitutions, structure, and composition. More than 50% of the total AA consolidate the hydrophobic region of AMPs [1]. AMPs in eukaryotes are produced ribosomally and contribute to the innate immunity of the organism. In contrast, the fungal and bacterial AMPs differ from the former to be synthesized nonribosomally [2]. AMPs are known to possess an array of anticancer activities, antiviral, antifungal, and antibacterial, among others, which make them a suitable candidate as a prophylactic and therapeutic agent. AMP-derived drugs are often employed to treat wound and skin infections in the form of topical preparations [3]. AMPs have an advantage over conventional antibiotics as they are not as susceptible to developing resistance in bacteria as antibiotics. Therefore, they can act as an alternative to antibiotics in the period of antibiotic resistance. AMPs are known to kill a range of pathogens owing to their ability to disintegrate cell membranes [4]. Therefore, the search for AMPs is often directed toward those with the potential to disrupt microbial membranes. Membrane disrupting property of AMPs outstand them as perfect candidates to be employed in combination therapies, especially with traditional antibiotics as AMPs aid in the facilitation of antibiotic molecules into the cell by disrupting the cell membrane, thus allowing the antibiotic to act on its target [1].

It is well versed in the literature that insects are impervious to attack by pathogenic microorganisms. They are known to synthesize a variety of peptides and proteins as a defensive action against bacterial infections. They either directly kill pathogenic microbes and viruses or trigger the immune system to impede them.

---

*Authors Daljeet Singh Dhanjal, Chirag Chopra, and Sonali Bhardwaj have contributed equally to this work.

Antimicrobial Peptides. DOI: https://doi.org/10.1016/B978-0-323-85682-9.00015-5
© 2023 Elsevier Inc. All rights reserved.

These defensive proteins and peptides are produced by innate immune cells such as hemocytes [5]. AMPs are involved in multiple defensive responses such as neutralizing and binding to endotoxins, killing pathogenic microbes, and controlling immune responses during infection. The production of AMPs induces a transient humoral immune response marked by increased levels of AMPs in the hemolymph, which tend to last longer than the initial cellular responses, thus acting as back-up for persistent microbial infections [6]. With further exploration, AMPs are involved in modulating endosymbionts apart from their defensive roles. Despite the ubiquity of AMPs amid eukaryotic organisms, they have only been extensively investigated in insects [7].

## 7.2 The need for antimicrobial peptides

The discovery of antibiotics has led to the prevention and treatment of various diseases and infections. The effectiveness of antibiotics in treatment has brought forth their immense exploitation through overuse, resulting in antibiotic resistance in microorganisms. Due to emerging antibiotic resistance in diverse microbial strains, the efficacy of these antibiotics in the present scenario has declined sufficiently. Some of the most important classes of antibiotics have been rendered ineffective following the development of resistant mechanisms by microorganisms to cope with these drugs. In the current scenario, it has become crucial to search for an alternative to antibiotics [8]. Pharmaceutical industries strive to develop novel antibiotics despite the associated scientific challenges such as exploiting efflux mechanisms and penetration barriers in gram-negative bacteria to reduce the effective concentration of antibiotics. Fluoroquinolones were the last new class of antibiotics introduced to treat gram-negative infections [9]. The development of extensively drug-resistant (XDR), multidrug-resistant (MDR), and pan drug-resistant (PDR) among *Acinetobacter*, *Klebsiella*, and other pathogenic microbes are imposing a significant threat to health and accounting for prevailing death (Fig. 7.1). However, still, pharmaceutical companies are focusing and relying on traditional antibiotic discovery programs for identifying a new class of molecules as potential antimicrobial candidates [10]. Upon observing the challenges, it has become prominent to expand our knowledge and consider the overlooked options such as AMPs for therapeutic purposes to treat and regulate death rates due to microbial infections. Although clinical trials of few AMPs have been conducted, none of them has been approved by the U.S. Food and Drug Administration (FDA), except gramicidin, which is only used for topical application [11]. The hope was raised in this field when the AMP compound pexiganan was proposed by Magainin Pharmaceuticals with an impressive result in early Phase I and II clinical trials to treat diabetic foot ulcers.

Nevertheless, this compound was not approved by FDA as its performance was not up to the mark in contrast to antibiotics available for the treatment of diabetic foot ulcers [12]. The other challenges that set back the application of AMPs are the cost and time-consuming manufacturing procedure of these peptides. In the present scenario,

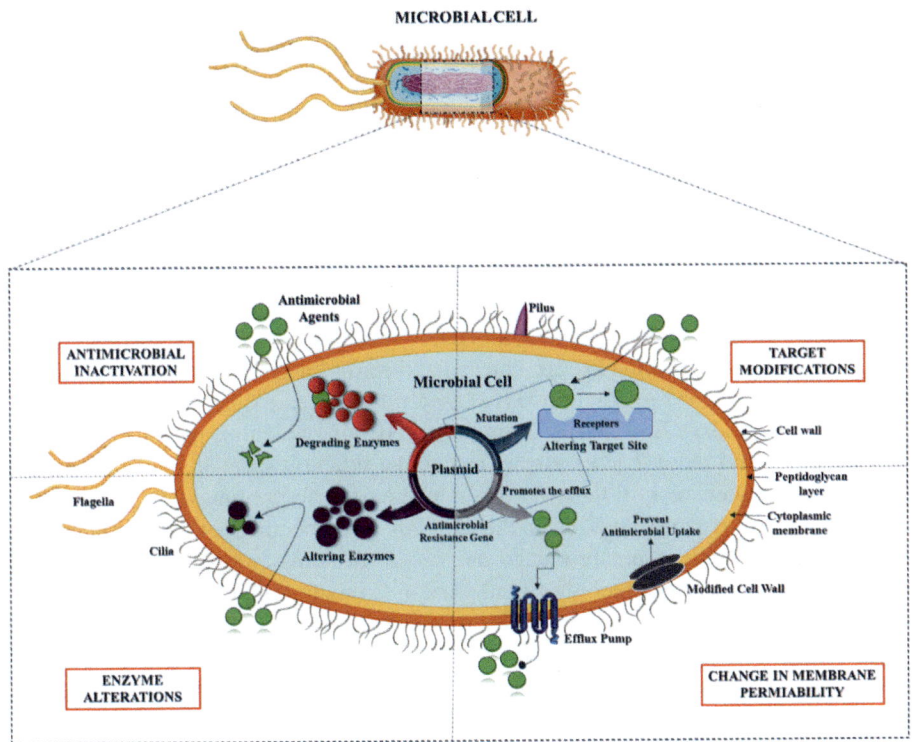

**Figure 7.1** Diagrammatic depiction of antimicrobial resistance mechanism.

antibiotic resistance in pathogenic microbes and the nonavailability of therapeutic drugs to eliminate MDR-pathogenic strains of microbes have prompted the interest of researchers in identifying, manufacturing, and commercializing AMPs [13].

## 7.3 Classification of insect peptides

AMPs are generally categorized as α-helical, β-sheet, or peptides with extended/random-coil structure based on their secondary structure, with most AMPs falling into the initial two categories. AMPs form an α-helical structure through the formation of intramolecular disulfide bridges [14]. In an aqueous solution, α-helical peptides are usually unstructured, but upon interaction with a biological membrane, they acquire an amphipathic helical structure. The β-sheet structure has disulfide bonds that provide stabilization to the structure [15]. The hydrophobic groups in β-sheet peptides transform the peptide chain into a polymer through hydrophobic interaction that increases the peptide chain's affinity toward the cell membrane and penetrates the cell. The structure plays a crucial role in providing the antibacterial potential to AMPs. A balance between certain factors such as length of peptide, hydrophobic character, and charge density is essential to exhibit optimum antibacterial activity. The secondary structure may get affected if any of the factors mentioned above change, such as a shift in the position of positively charged AA or an increase in their number may alter the secondary structure and, thus, the antibacterial activity [16]. Several types of insect-derived AMPs have been discussed in Table 7.1.

### 7.3.1 Attacins

Attacins were first isolated from the hemolymph of bacteria immunized *H. cecropia* pupae. The isolated attacins had a molecular mass of 20−23 kDa and isoelectric points (pI) of 5.7−8.3. Attacins are categorized into two groups: acidic attacins and basic attacins. Attacin F obtained from *H. cecropia* has acidic nature, pI of 9, and is a derivative of attacin E prepared through proteolysis, which on the other hand has neutral nature and pI of 7 [81]. Basic and acidic attacins share high similarities in their peptide sequence, with the only difference being the presence of a more significant number of Asp residues in acidic attacins and encoding of both attacins via separate genes. Attacins are produced as pre-pro-proteins, a signal peptide, a pro-peptide (P domain), and an N-terminal attacin domain, followed by two glycine-rich domains (G1 and G2 domains). N-terminal pro-peptide of attacins comprises of a conserved RXXR motif recognized by furin-like enzymes, which indicates that mature attacins are synthesized via processing of pro-attacins by furin-like enzymes. The pro-peptide of attacin is essential for the synthesis of pro-attacin and is eliminated at or after the trans-Golgi compartment. Pro-attacin does not tend to possess any biological activity. The pro-peptide (P domain) of *Drosophila melanogaster* attacin-C is long and rich in proline residues and shows antibacterial activity against gram-negative [82]. Many lepidopteran species have been reported to synthesize attacins, such as *Hyalophora cecropia, Spodoptera exigua, Heliothis virescens, Hyphantria cunea, Bombyx mori, Trichoplusia ni, Samia cynthia, Helicoverpa armigera*, and *Manduca sexta*. Besides, attacins have also been identified in some dipteran species such as e *Glossina morsitans* and *D. melanogaster* [83]. A large number of attacins show inhibitory activity against *Escherichia coli*. Attacin-A1 obtained from *G. morsitans* has exhibited antibacterial activity against *E. coli* in an in vitro study. In contrast, attacin-B obtained from *H. cunea* has been reported to show antibacterial (*Citrobacter freundii, E. coli*) and antifungal activity (*Candida albicans*) [84,85]. Gram-negative (*Pseudomonas cichorii* and *E. coli*) and gram-positive (*Listeria monocytogenes* and *Bacillus subtilis*) are inhibited by attacins produced by *S. exigua*. At the same time, recombinant *Drosophila* attacin-A has been found to show inhibitory activity against *E. coli* [1,86].

### 7.3.2 Cecropins

Cecropins were first purified from immunized hemolymph of *H. cecropia* pupae and belong to the group of cationic AMPs, usually 31−39 residues long. They have also been isolated from the identified coleopteran, dipteran, and lepidopteran order

**Table 7.1** List of different insect antimicrobial peptides with their antimicrobial potential.

| Insect peptide | Name of peptide | Insect order | Insect species | Antimicrobial activity G+ | Antimicrobial activity G- | Antimicrobial activity F | Amino acid residue (one-letter code) | References |
|---|---|---|---|---|---|---|---|---|
| *Defensins and Defensin-like Peptides* | Defensin 1 | Coleoptera | *Acalolepta luxariosa* | ✓ | ✓ | - | FIxDVLSVEAKGVKLNHAAxGI | [17] |
| | Defensin | Coleoptera | *Allomyrina dichotoma* | ✓ | - | - | VTCDLLSFEAKGFAANHSLCAAHCLAIGRRGGSC ERGVCICRR | [18] |
| | Defensin A, B | Coleoptera | *Anomala cuprea* | ✓ | ✓ | - | VTCDLLSFEAKGFAANHSICAAHCLAIGRKGGSC QNGVCVRN, VTCDLLSFEAKGFAANHSI CAAHCLVIGRKGGACQNGVCVCRN | [19] |
| | Coprisin | Coleoptera | *Copris tripartitus* | ✓ | ✓ | - | VTCDVLSFEAKGIAVNHSACALHCIALRKKGGSCQNGVCVCRN | [20] |
| | Holotricin-I | Coleoptera | *Holotrichia diomphalia* | ✓ | ✓ | - | VTCDLLSLQIKGIAINDSACAAHCLAMRRKGGSCKQGVCVCRN | [21] |
| | Defensin | Coleoptera | *Oryctes rhinoceros* | ✓ | - | - | VTCDLLSFEAKGFAANHSLCAAHCLAIGRRGGSCERGVCICRR | [22] |
| | Tenecin-1 | Coleoptera | *Tenebrio molitor* | ✓ | - | - | VTCDILSVEAKGVKLNDAACAAHCLFRGRSGGYCNGKRVCVCR | [23] |
| | Defensin B, C | Coleoptera | *Zophob atratus* | ✓ | ✓ | - | FTCDVLGFEIAGTKLNSAACGAHCLALGRRGGYCNSKSVCVCR. FTCDVLGFEIAGTKLNSAACGAHCLALGRTGGYCNSKSVCVCR | [24] |
| | Defensin A, B, C | Diptera | *Aedes aegypti* | ✓ | ✓ | - | ATCDLLSGFGVGDSACAAHCIARGNRGGYCNSKKVCVCR, ATCDLL SGFGVGDSACAAHCIARGNRGGYCNSQKVCVCR, ATCDLLSGFG VGDSACAAHCIARRNRGGYCNAKKVCVCRN | [25] |
| | Defensin | Diptera | *Anopheles gambiae* | ✓ | ✓ | ✓ | MKCATIVCTIAVVLAATLLNGSVQAAPQEEAALSGGANLNTLLDELP EETHHAALENYRAKRATCDL ASGFGVGSSLCAAHCIARRYRGGYCNSKAVCVCRN | [26,27] |
| | Defensin A, B | Diptera | *Chironomus plumosus* | ✓ | - | - | LTCDILGSTPACAAHCIAKGYRGGWCDGOSVCNCRR, LTCDVIGST QLCAAHCIAKGYRGGWCDGKSVCNCRR | [28] |
| | Drosomycin, Defensin | Diptera | *Drosophila melanogaster* | ✓ | - | ✓ | MMQIKYLFALFAVLMLVVLGANEADADCLSGRYKGPCAVWDNETCRR VCKEEGRSSGHCSPSLKCWCEGC | [29,30] |
| | Lucifensin | Diptera | *Lucilia sericata* | ✓ | ✓ | - | ATCDLLSGTGVKHSACAAHCLLRGNRGGYCNGRAICVCRN | [31] |
| | Defensin A, B | Diptera | *Phormia terranovae* | ✓ | ✓ | - | ATCDLLSGTGINHSACAAHCLLRGNRGGYCNGKGVCVCRN, ATCD LLSGTGINHSACAAHCLLRGNRGGYCNRKGVCVCRN | [32] |
| | Sapecin, Sapecin B and C | Diptera | *Sarcophaga peregrina* | ✓ | ✓ | - | ATCDLLSGTGINHSACAAHCLLRGNRGGYCNGKAVCVCRN, LTCEIDRSLCLLHCRLK | [33–35] |

(Continued)

Table 7.1 (Continued)

| Insect peptide | Name of peptide | Insect order | Insect species | Antimicrobial activity | | | Amino acid residue (one-letter code) | References |
|---|---|---|---|---|---|---|---|---|
| | | | | G+ | G− | F | | |
| | Defensin | Hemiptera | *Pyrrhocoris apterus* | ✓ | - | - | GYLRAYCSQQKVCRCVQ, ATCDLLSGIGVQHSACALHCVFRG NRGGYCTGKGICVCRN | [36] |
| | Defensin A, B, C | Hemiptera | *Rhodnius prolixus* | ✓ | - | - | ATCDILSFQSQWVTPNHAGCALHCVIKGYKGGQCKITVCHCRR ATCDLFSFRSKWVTPNHAACAAHCLLRGNRGGRCKGTICHCRK, ATCDLLSFSSKWVTPNH AGCAAHCLLRGNRGGHCKGTICHCRK, ATCDLLSLTSKWFTPNHAGCAA HCIFLGNRGGRCVGTVCHCRK | [37] |
| | Royalisin | Hymenoptera | *Apis mellifera* | ✓ | - | - | VTCDLLSFKGQVNDSACAANCLSLGKAGGHCEKGVCICRKTSFKDLWDKYF | [38] |
| | Defensin | Hymenoptera | *Bombus pascuorum* | ✓ | ✓ | ✓ | VTCDLLSIKGVAEHSACAANCLSMGKAGGRCENGICLCRKTTFKELWDKRF | [39] |
| | Defensin | Hymenoptera | *Formica rufa* | ✓ | - | - | FTCDLLSGAGVDHSACAAHCILRGKTGGRCNSDRVCVCRA | [40] |
| | Navidefensin2−2 | Hymenoptera | *Nasonia vitripennis* | ✓ | - | - | MKVLVLAACAVFAGAFGATRIRDGYEDPVFEILGDDIKQDG DNAETVDATDDLSPIKESS DDPTELVQLSYRVRRFSCDVLSFQSKWVSPNHSACAVRCLA QRRKGGKCKNGDCVCR | [41] |
| | Termicin | Isoptera | *Pseudacanthotermes spiniger* | - | - | ✓ | ACNFQSCWATCQAQHSIYFRRAFCDRSQCKCVFVRG | [42] |
| | ARD1 | Lepidoptera | *Archeoprepona demophoon* | - | - | ✓ | DKLIGSCVWGAVNYTSNCNAECKRRGYKGGHCGSFANVNCWCET | [43] |
| | Defensin gallerimycin | Lepidoptera | *Galleria mellonella* | - | - | ✓ | DTLIGSCVVGATNYTSDCNAECKRRGYKGGHCGSFLNVNCWCE | [44] |
| | Heliomicin | Lepidoptera | *Heliothis virescens* | - | - | ✓ | DKLIGSCVWGAVNYTSDCNGECKRRGYKGGHCGSFANVNCWCET | [45] |
| | Defensin | Lepidoptera | *Mamestra brassicae* | ✓ | ✓ | - | MLCLADIRIVASCSAAIKSGYGQQPWLAHVAGPYANSLFDD VPADSYHAAVE YLRLIPASCYLLDGYAAGRDDCRAHCIAPRNRRLYCASYQVCVCRY | [46] |
| | SpliDef | Lepidoptera | *Spodoptera littoralis* | ✓, | ✓ | ✓ | VSCDFEEANEDAVCOEHCLPKGYTYGICVSHTCSCIYIVELIKWYTNTYT | [47] |
| | Defensin | Odonata | *Aeschna cyanea* | ✓ | ✓ | ✓ | GFGCPLDQMQCHRHCQTITGRSGGYCSGPLKLTCTCYR | [48] |

| | | | | | | |
|---|---|---|---|---|---|---|
| Cecropins and Cecropin-like Peptides | Cecropin Sarcotoxin Pd | Coleoptera Coleoptera | *Acalolepta luxuriosa* *Paederus dermatitis* | - ✓ | ✓ ✓ | - ✓ | KNFFKRIEKVGKNIRNAAERSLPTVVGYAGVAKQIGK GWLKKIGKKIERVGQHTRGLGIAQIAANVAATAR | [49] [50] |
| | Cecropin A Cecropin A, B, C | Diptera Diptera | *Aedes aegypti* *Aedes albopictus* | - ✓ | ✓ ✓ | ✓ - | ATXDLLSGFGVGDSAXAAVX GGLKKLGKKLEGVGKRVFKASEKALPVAVGIKALG, KWKVFKKI EKMGRNIRNGIVKA GPAIAVLGEAKAL, GWLKKLGKRIERIGQHTRDATIQGLGIAQQ AANVAATAR | [25] [1] |
| | Cecropin A | Diptera | *Anopheles gambiae* | ✓ | ✓ | ✓ | MNFSKIFIFVVLAVLLLCSQTEAGRLKKLGKKIEGAGKRVFKAAEKA LPVVAGVKALG | [26] |
| | Cecropin A, B, C | Diptera | *Drosophila melanogaster* | ✓ | ✓ | ✓ | MNFYNIFVFVALILAITIGQSEAGWLKIGKKIERVGQHTRDATIQGL GIAQQAANVAATARG, MNFNKIFVFVALILAISLGNSEAGWLRKLGKKIERIGQHTRDASIQ VLGIAQQAANVAATARG, M NFNKIFVFVALILAISLGNSEAGWLRKLGKRIERIGQHTRDASIQV LGIAQQAANVAATARG | [51,52] |
| | Sarcotoxin IA, IB, IC | Diptera | *Sarcophaga peregrina* | ✓ | ✓ | - | GWLKKIGKKIERVGQHTRDATIQGLGIAQQAANVAATAR, GWLK KIGKKIERVGQHTRDATIQVIG VAQQAANVAATAR, GWLRKIGKKIERVGQHTRDATIQVLGI AQQAANVAATAR | [53] |
| | Stomoxyn | Diptera | *Stomoxys calcitrans* | ✓ | ✓ | ✓ | MNFYKYLVVLVLVLVLCLSATQTEARGFRKHFNKLVKVKHTIS ETAHVAKDTAVIAGSGAAVVAATG | [54] |
| | Cecropin D Hinnavin I, II | Lepidoptera Lepidoptera | *Agrius convolvuli* *Artogeia rapae* | ✓ ✓ | ✓ ✓ | - - | WNPFKELERAGQRIRDSIISAAP MNFGELYFLIFACVLALSSVSAAPGWKIGKKLEHMGGQNIRDGLISAG PAVFAVGQAATIYAAAK, MNFGELYFLIFACVLALSSVSAAPKWKIF KKIEIIMGQNIRDGLIKAGPAVQVVGQAATIYKG | [55,56] [57] |
| | Cecropin (A, B), D | Lepidoptera | *Bombyx mori* | ✓ | ✓ | ✓ | MNFVRILSFVFALVLALGAVSAAPEPRWKLFKEKVGRNVRDGLIKAGPAI AVIGQAKSLGK, MNFAKI LSFVFALVLALSMTSAAPEPRWKIFKKIEKMGRNIRDGIVKAGPAIEVLG SAKAIGK, MKFSKIFVFFA IVFATASVSAAPGNFFKDLEKMGQRVRDAVISAAPAVDTLAKAKALGQG | [58–60] |
| | Cecropin D | Lepidoptera | *Helicoverpa armigera* | ✓ | ✓ | - | MNSKIVLFLCVCLVLVSTATAWDFFKELEGAGQRVRDAIISAGPAVDVLTKAKGL YDSSEEKD | [61,62] |
| | Cecropin B Cecropin A, B, C, D, E, F | Lepidoptera Lepidoptera | *Heliothis virescens* *Hyalophora cecropia* | - ✓ | ✓ ✓ | - - | KWKVFKKIEKVGRNIRDGIVKAGPAIAVLGQAN KWKLFKKIEKVGQNIRDGIIKAGPAVAVVGQATQIAK, KWKVFKKIEKMGRNIRDG IVKAGPAIAVLGEAKAIL S, KWKLFKKIEKVGQNIRDGIIKAGPAVAVVGQATQIAK, WNPFKELEKVGQRVRD | [63] [64] |

*(Continued)*

Table 7.1 (Continued)

| Insect peptide | Name of peptide | Insect order | Insect species | G+ | G- | F | Amino acid residue (one-letter code) | References |
|---|---|---|---|---|---|---|---|---|
| | Cecropin A | Lepidoptera | Hyphantria cunea | ✓ | ✓ | - | AVISAGPAVATVAQATAL AK, WNPFKELEKVGQRVRDAVISAGPAVATVAQAT, WNPFKELEKVGQRVRNAV ISAGPAVATVAQAT | [65] |
| | Papiliocin | Lepidoptera | Papilio xuthus | ✓ | ✓ | ✓ | MNFSRILFFMFACFVALASVSAVPEPRWKVFKKIEKVGRHIRDGVIKAGPAITVVG QATAL | [66] |
| | Spodopsin Ia, Ib | Lepidoptera | Spodoptera litura | ✓ | ✓ | - | MNFGKILFFVMACLAALSLTTASPRWKIFKKIEKVGRNVRDGIIKAGPAVAVVGQA ATVVKG | [67,68] |
| | Drosocin, Metchnikowin | Diptera | Drosophila melanogaster | ✓ | ✓ | | RWKVFKKIEKVGRNVRXGIIXAGPAIGVLXQAXAL, RWKVFKKIEKMGRNIRDGIIKAGPAVEVLGSAXAL GKPRPYSPRPTSHPRPIRV, MQLNLGAIFLALLGVMATATSVLAEPHRHQGPIFDTRSP FNPNQPRPGPIY | [69,70] |
| Proline and Proline like Peptide | Pyrrhocoricin Metalnikowin I, IIA, Metalnikowin IIB, Metalnikowin III | Hemiptera Hemiptera | Oncopeltus fasciatus Palomena prasina | ✓ - | ✓ ✓ | - - | VDKPPYLPRP(X/F)PRRIYN(NR) VDKPDYRPRPRPNM, VDKPDYRPRPWPRPN, VDKPDYRPRPWPRPNM VDKPDYRPRPWPRPNM | [71] [72] |
| | Pyrrhocoricin Apidaecin Ia, Ib, II, Abaecin | Hemiptera Hymenoptera | Pyrrhocoris apterus Apis mellifera | ✓ - | ✓ ✓ | - - | VDKGSYLPRPTPPRPIYNRN GNNRPVYIPQPRPPHPRL, GNNRPVYIPQPRPPHPRL, GNNRPIYIPQPRPPHPRL | [36] [73,74] |
| | Apidaecin, Abaecin | Hymenoptera | Bombus pascuorum | ✓ | ✓ | - | GNRPVYIPPPRPPHPRL, FVPYNPPRPGQSKPFSFPGHGPFNPKIQWPYPLPNPGH | [39] |
| | Formaecin Abaecin | Hymenoptera Hymenoptera | Myrmecia gulosa Pteromalus puparum | ✓ - | ✓ ✓ | - - | GRPNPVNNKPTPHPRL YVPLPNVPQPGRRPFPTFP | [75,76] [77] |
| | Lebocins 1/2, 3, 4 | Lepidoptera | Bombyx mori | - | ✓ | - | MYKFLVFSSVLVLFFAQASCQRFIQPTFRPPPTQRPIIRTARQAGQEPLWL YQGDNVPRAPSTAD | [78] |

| Lebocin-A, Lebocin-B, Lebocin-C | Lepidoptera | *Manduca sexta* | ✓ | ✓ | ✓ | HPILPSKIDDVQLDPNRRYVRSVTNPENNEASIEHSHHTVDTGLDQPIESHR NTRDLRFLYP RGKLPVPTPPFNPKPIYIDMGNRYRRHASDDQEELRQYNEHFLIPRDIFQE, MYKFLVFSSVL VLFFAQASCQRFIQPTFRPPTQRPITRTVRQAGQEPLWLYQGDNVPRA PSTADHPILPSKIDDVQLDPNRRYVRSVTNPENNEASIEHSHHTVDIGLDQ PIESHRNTRDLRFLYPRGKLPVPTLPPFNPKPTYIDMGNRYRRHASEDQEE LRQYNEHFLIPRDIFQE, MYKFLVFSSVLVLFFAQASCQRFIQPTYRPPPTR RPIIRTARQAGQEPLWLYQGDNIPRAPSTADHPILPSKIDD VKLDPNRRYVRSVTNPENNEASIESSHHTVDIG LDRPIESHRNTRDLRFWNPREKLPLPTLPPFNPKPTYIDMGNRYRRHASDDQ EELRHHNEHFLIPRDILQD MKLLILLGVALVLLFGESLG QRFSQPTFKLPQGRLTLSRKFRESGNEPLWLYQGDNIPKAP STAEHPFLPSIIDDVKFNPDRRYARSLGTPDHYHGGRHSISRGS QSTGPTHPGYNRRNARSVETLASQEHLSSLPMDSQETLLRGTR, MAKSIFALGVIAVLLITESNCWRSDLPHILPTYKPPRTPSTVIIRTVREAGD KPLWLYQGDDHPRAPSSGDHPVLPPIIDDVKLDPN RRYARSVNEPSSQEHHERFVRSFDSRSSRHHGGSHSTSSGSRDTGATHPG YNRRNS, MMKSVLVLCVVAVLH TAASSGWNKNNGGIILPTFRPPPIWPGITRTVREAGDQPLWLYQGD NHPRAPSSGDHPVLPSIIDDVKLDPNRRYVRSVNEPSSQEHHERFVRSFDS RSSRHHGGSHSTSSGSRDTGATHPGYNRRNS | [79,80] |

G +, gram-positive bacteria; G, gram-negative bacteria; F, fungi; ✓, present; -: not applicable.

of insects. Cecropins are produced as proteins but are activated only upon the elimination of signal peptide [87]. Cecropins are known to exhibit antimicrobial activity against both gram-positive bacteria, gram-negative bacteria, and fungi. Amidation of cecropins occurs at the C-terminus, which is an essential step contributing to enhanced interaction of cecropins with liposomes that ultimately results in antimicrobial activity against a range of microorganisms [88]. For instance, sarcotoxin-IA contains two N-terminal residues, that is, Trp2 and Gly1, which have been reported to be significant in order to exhibit inhibitory activity against *E. coli*. These residues are known to be essential for binding of sarcotoxin-IA to the lipid A of the lipopolysaccharide layer in gram-negative bacteria.

Moreover, *H. cecropi* cecropin-A and papiliocin contains Trp2 and Phe5 as N-terminal residues which are vital for the interaction of cecropins with negatively charged bacterial cell membrane [89]. In addition, cecropins along with their derivatives namely, Shiva and SB-37 have been reported to obstruct the replication of HIV-1 virus. Furthermore, studies evidence that papiliocin can potentially stimulate apoptosis in *C. albicans* [87].

### 7.3.3 Defensins

Defensins have been isolated from almost all living organisms. They are cationic/basic AMPs and comparatively small in size with a molecular mass of approximately 4 kDa. They contain a conserved sequence comprising six cysteine residues, due to which three intramolecular disulfide bridges are formed. Defensins can be categorized into three groups based on structural composition: "classical" defensins, insect defensins, and beta-defensins. Insect defensins are small cationic peptides of 34–51 residues with 6 conserved cysteines. Many peptides rich in cysteine residues belong to the insect defensin family, such as royalisin, sapecins, tenecin-1, spodoptericin, heliomycin, holotricin-1, lucifensin, gallerimycin, coprisin, among others [83]. Insect defensins have been isolated from various orders such as Hymenoptera, Coleoptera, Diptera, Lepidoptera, Hemiptera. Defensins have also been purified from an ancient order of insects, that is, Odonata, which indicates the emergence of defensins from a common ancestor gene. Insect defensins were first isolated from *Sarcophaga peregrina* as sapecins, which comprised 40 amino acids with 6 cysteine sequences.

In contrast, insect defensins as cationic peptides were first reported from the hemolymph of bacteria immunized larvae of *Phormia terranovae* which also consisted of 40 amino acid residues. *P. terranovae* defensin-A and *S. peregrina* sapecin are secreted as pre-pro-proteins comprising of a signal peptide, a pre-peptide, and a mature defensin peptide [90]. The insect defensins contain an N-terminal loop, an α-helix, followed by an antiparallel β-sheet. Two intramolecular disulfide bonds connect the β-sheet and α-helix which forms a "loop-helix-beta-sheet" or "cysteine-stabilized alpha beta (CSαβ)" structure [91]. Insect defensins have been reported to show inhibitory activity against many gram-positive bacteria such as *Staphylococcus aureus, Aerococcus viridians, Bacillus thuringiensis, B. subtilis, Micrococcus luteus* and *Bacillus megaterium* as well as some gram-negative bacteria (*E. coli*) and fungi [92]. Insect defensins form channels cell membrane of bacteria which leads to cell death. Defensins interact with the phospholipid layer of the cell and induce microheterogeneity in the phospholipid membrane which is believed to contribute to the formation of channels in the cell membrane and thus, cell death [93]. A hendecapeptide, a derivative formed from the helical region of sapecin-B has been found to show antimicrobial activity against many bacteria (*S. aureus*, *E. coli*) and some yeasts (*C. albicans*) which indicates the significance of conserved helical structure in providing antimicrobial activity [94].

## 7.3.4 Gloverins

Gloverin is an antibacterial, basic, heat-stable, and glycine-rich protein of size nearly 14 kDa, which was first isolated from the hemolymph of *Hyalophora gloveri* pupae. This peptide is predominantly found in Lepidoptera in species such as *Antheraea mylitta*, *B. mori*, *Diatraea saccharialis*, *Galleria mellonella*, *H. armigera*, *M. sexta*, *Plutella xylostella*, *S. exigua*, and *T. ni* [61]. The amino acid sequence analysis of gloverin peptides has revealed a conserved RXXR motif in the N-terminal part of the pro-gloverins precursor. Gloverins are synthesized from pre-pro-proteins by the action of furin-like enzymes at the N-terminal of pro-regions of protein. *T. ni* pro-gloverin has been reported to show antimicrobial activity against *E. coli*. In contrast, gloverins obtained from *S. exigua* and *M. sexta* have been found to exhibit antimicrobial activity against *Flavobacterium* sp. and *Bacillus cereus*, respectively [83]. Moreover, studies show that *M. sexta* gloverins also possess antimicrobial activity against *Saccharomyces cerevisiae* and *Cryptococcus neoformans* [95].

## 7.3.5 Lebocins

Lebocins, a proline-rich and O-glycosylated 32-residue long peptide, was first isolated from the hemolymph of *E. coli* immunized silkworm viz *B. mori*. The bioinformatic analysis of the amino acid sequence of this peptide showed 41% similarity with proline-rich and non-O-glycosylated peptide abaecin obtained from the honeybee. The investigation of complementary DNA (cDNA) clone of *B. mori* lebocin revealed that lebocin was produced from 179-residue precursor protein, in which an active 32-residue peptide is present near the C-terminal of the precursor [96]. This cDNA encoding for lebocin precursor has been identified in numerous lepidopteran species such as *Antheraea pernyi*, *H. virescens*, *M. sexta*, *Pseudoplusia includens*, *Pieris rapae*, *S. cynthia*, and *T. ni*. All the precursors of lebocin, including *B. mori* precursors, are proline-rich peptides with 22–28 residues and contain 4–6 proline residues N-terminal end of the mature precursor proteins. However, surprisingly, only lebocin precursors of *B. mori* are known to encompass extra 32-residue peptides with 7 proline residues at the C-terminal end [97]. Extensive research in this direction has revealed that lebocins are synthesized via proteolytic removal of the precursor proteins. Molecular analysis of these sequences has elucidated that lebocin precursors contain various conserved regions of RXXR motifs, which can be identified by furin-like enzymes. The antimicrobial activity of lebocins is effective against gram-positive bacteria, gram-negative bacteria, and fungi [98]. *B. mori* lebocins have been reported to show antimicrobial activity against *Acinetobacter* sp. and *E. coli*. In contrast, *M. sexta* lebocin-B and lebocin-C have been reported to show active antimicrobial activity against *B. cereus*, *C. neoformans*, *S. aureus*, *Serratia marcescens*, and *Salmonella typhimurium* [80,96].

### 7.3.6 Moricins

Moricin, a 42-residue long peptide, was first isolated from the hemolymph of *E. coli* immunized larvae of *B. mori* [99]. This AMP is extensively found in lepidopteran insects. cDNAs that encode for this AMP have been identified in *G. mellonella*, *H. armigera*, *Hyblaea puera*, *H. virescens*, *M. sexta*, *S. exigua*, and *Spodoptera litura*. Moricins have been accorded to show antimicrobial activity against both grampositive and gram-negative bacteria. Moricins obtained from *G. mellonella* have shown significant antimicrobial activity against yeast and filamentous fungi [100]. The N-terminal segment of α-helix is found to have amphipathic nature. It is accountable for increased membrane permeability of bacterial cells and cell death.

In contrast, the C-terminal segment of α-helix has hydrophobic nature and is responsible for the antimicrobial activity of moricins. Although moricins share high similarity with cecropins, but it lacks the hinge region. Furthermore, the expression of AMPs gene in insects is controlled by GATA transcription factors and nuclear factor-κB (NF-κB)/Rel. This was confirmed through the assessment of different classes of AMP genes that have GATA and NF-κB binding sites in their promoter region, including attacin as well as moricin genes [101].

## 7.4 Mode of action

There are a variety of mechanisms by which AMPs kill bacteria. These include the targeting of cytoplasmic components, disruption of the cell membrane, and interference with bacterial metabolism. The primary action mechanism recorded for AMPs involves hydrophilic or electrostatic interaction of AMPs with the bacterial cell membrane. AMPs interact with the cell membrane of bacteria to change the permeability of the membrane [102]. Upon interaction of AMP with the cell, it creates a transmembrane potential, altering the osmotic potential. In general, the interaction between the cell membrane and AMPs is directly associated with the antimicrobial potential of AMPs. Currently, a minimum of four action mechanisms explains the membrane activity of AMPs, that is, carpet, barrel-stave, disordered toroidal-pore, and toroidal-pore. The modes mentioned above of action demand a threshold concentration of peptides to show antimicrobial activity [103].

Moreover, AMPs are also known to disrupt DNA and intracellular enzymes during their translocation into pathogens. The action mechanism of AMPs is different and variable, and requires further investigation. To understand the membrane activity of AMPs, certain factors need to be taken into consideration like specific receptors on the membrane and other factors which work in synergism [104]. In general, AMPs of insect origin have a positive charge and encompasses nearly 50% of hydrophobic residues. The positive charge attributes to the interaction of lipophilic and negatively charged cell membrane of bacteria in comparison to the cell membrane of eukaryotes which are dominated by uncharged lipids and zwitterions. This is the reason that AMPs get attracted to the cell membrane of bacteria and upon establishing the contact the hydrophobic residues stimulate the integration and outer leaflet of the cell membrane which then starts to expand and becomes thin, ultimately leading to the formation of pores and causing lysis of bacterial cells [105]. This potential of AMPs to increase the permeability of the membrane was confirmed through an experiment on dye-loaded bacteria exposed to papiliocin and cecropin A which resulted in the leakage of dye from the cell. Various experiments have been conducted to understand the mechanism of AMPs interaction with the cell membrane of bacteria but researchers are still trying to uncover the mechanistic approach which explains how AMPs overcome the barriers such as peptidoglycan-rich cell wall [7]. The binding of AMPs to teichoic acids or anionic lipopolysaccharides is considered to be essential as neutralization of the charge of the cell membrane of bacteria makes them highly resistant to cationic AMPs. Even though a large number of AMPs are known to function by increasing the porosity of membrane, on the contrary, proline-rich AMPs such as bumblebee abaecin interacts with intracellular targets such as protein synthesis apparatus or DnaK of the bacteria. Moreover, attacins have been comprehended to obstruct the protein synthesis involved in the outer membrane of bacterial cells [106].

Furthermore, few AMPs have been reported to inhibit the synthesis of the cell wall by interfering with the lipid phosphatidylethanolamines or corresponding enzymes or by delocalizing the surface proteins of bacterial cells. For instance, cecropin (LSer-Cec6) and defensin (LSer-Def4) obtained from *Lucilia sericata*, when used in combination, have been reported to show effective antibacterial activity. Other AMPs such as insect metalloprotease inhibitors (IMPIs) have also been reported to neutralize precisely virulence-associated microbial metalloproteases [107,108].

## 7.5 Concluding remarks

In recent years, insect AMPs are considered the chief immune effector molecules, and still, a considerable number of AMPs have to be explored yet in this world inhabiting diverse insects. Nowadays, fruit flies and mosquito are used as model organisms, which have enabled us to understand the immunity of insects. Extensive experiments and evidence have now uncovered the complexity of the immune system of insects. In the intervening time, during the evolution process, few molecules and signaling pathways involved in the immune system of insects shared similarities with vertebrates, including humans. Moreover, exploring the immune system of insects has allowed us better to understand the human immune system and its complexity. At present, many bacteria have developed multidrug resistance because of the excessive utilization of antibiotics. Many antibiotic-resistant bacteria have emerged with antibiotic exploitation, which has imposed a significant threat to humanity. Due to this, the exploration of new antimicrobial drugs has become necessary in the medical field. Concerning the above challenges, AMPs have turned out to be an apparent alternative. They kill not only the broad spectrum of bacteria and fungi but also cancer cells, protozoans, and viruses. On associating these AMPs with conventional antibiotics, the action mechanism of these AMPs is found to be unique, and microbes do not develop resistance to them quickly. Besides, these AMPs also do not damage the normal cells of animals. For instance, clavaspirin peptide obtained from *Styela clava* tunicate exhibits antimicrobial activity against *S. aureus* (drug-resistant) without showing any detectable resistance in the bacteria. This ability of insects' AMPs makes them unique peptides that can be used for synthesizing novel antimicrobial drugs.

Moreover, quantitative assessment of these AMPs is rarely executed to assess their synergistic profile, considering which the quantitative assessment of the AMPs should be promoted. Now, demand for peptide drugs is steadily increasing and few AMPs such as Bacitracin, Fuzeon, and polymyxin are commercially being sold in the market. Still, the clinical applications of AMPs are scarce due to some limitations such as instability toward proteases, hemolysis, unknown toxicity and low bioavailability. Besides the extensive studies, we are also not able to completely understand the mechanism of AMP activity and the structure—activity relationship (SAR). Hence, more SAR studies should be conducted to get a deep insight of the AMP activity. In addition, the cellular and molecular mechanism of AMP should also be studied. Lastly, there is a need for the establishment of AMPs insect library which is required for their optimization and improving their antimicrobial activity and toxic properties.

## References

[1] Q. Wu, J. Patočka, K. Kuča, Insect antimicrobial peptides, a mini review, Toxins (Basel) 10 (2018) 461. Available from: https://doi.org/10.3390/toxins10110461.
[2] M. Mahlapuu, J. Håkansson, L. Ringstad, C. Björn, Antimicrobial peptides: an emerging category of therapeutic agents, Frontiers in Cellular and Infection Microbiology 6 (2016) 194. Available from: https://doi.org/10.3389/fcimb.2016.00194.
[3] A. Pfalzgraff, K. Brandenburg, G. Weindl, Antimicrobial peptides and their therapeutic potential for bacterial skin infections and wounds, Frontiers in Pharmacology 9 (2018) 281. Available from: https://doi.org/10.3389/fphar.2018.00281.
[4] J. Lei, L.C. Sun, S. Huang, C. Zhu, P. Li, J. He, et al., The antimicrobial peptides and their potential clinical applications, The American Journal of Translational Research 11 (2019) 3919—3931.
[5] H. Coutinho, K. Lobo, D. Bezerra, I. Lobo, Peptides and proteins with antimicrobial activity, Indian Journal of Pharmacology 40 (2008) 3—9. Available from: https://doi.org/10.4103/0253-7613.40481.

[6] G. Diamond, N. Beckloff, A. Weinberg, K. Kisich, The roles of antimicrobial peptides in innate host defense, Current Pharmaceutical Design 15 (2009) 2377−2392. Available from: https://doi.org/10.2174/138161209788682325.
[7] E. Mylonakis, L. Podsiadlowski, M. Muhammed, A. Vilcinskas, Diversity, evolution and medical applications of insect antimicrobial peptides, Philosophical Transactions of the Royal Society B: Biological Sciences 371 (2016) 20150290. Available from: https://doi.org/10.1098/rstb.2015.0290.
[8] J. Davies, Origins and evolution of antibiotic resistance, Microbiologia (Madrid, Spain) 12 (1996) 9−16. Available from: https://doi.org/10.1128/mmbr.00016-10.
[9] L.D. Högberg, A. Heddini, O. Cars, The global need for effective antibiotics: challenges and recent advances, Trends in Pharmacological Sciences 31 (2010) 509−515. Available from: https://doi.org/10.1016/j.tips.2010.08.002.
[10] S. Basak, P. Singh, M. Rajurkar, Multidrug resistant and extensively drug resistant bacteria: a study, Journal of Pathology 2016 (2016) 1−5. Available from: https://doi.org/10.1155/2016/4065603.
[11] C.H. Chen, T.K. Lu, Development and challenges of antimicrobial peptides for therapeutic applications, Antibiotics 9 (2020) 24. Available from: https://doi.org/10.3390/antibiotics9010024.
[12] J.D. Steckbeck, B. Deslouches, R.C. Montelaro, Antimicrobial peptides: new drugs for bad bugs? Expert Opinion on Biological Therapy 14 (2014) 11−14. Available from: https://doi.org/10.1517/14712598.2013.844227.
[13] A.A. Bahar, D. Ren, Antimicrobial peptides, Pharmaceuticals 6 (2013) 1543−1575. Available from: https://doi.org/10.3390/ph6121543.
[14] M. Pazderková, P. Maloň, V. Zíma, K. Hofbauerová, V. Kopecký, E. Kočišová, et al., Interaction of halictine-related antimicrobial peptides with membrane models, International Journal of Molecular Sciences 20 (2019). Available from: https://doi.org/10.3390/ijms20030631.
[15] A. Giangaspero, L. Sandri, A. Tossi, Amphipathic $\alpha$ helical antimicrobial peptides: a systematic study of the effects of structural and physical properties on biological activity, European Journal of Biochemistry 268 (2001) 5589−5600. Available from: https://doi.org/10.1046/j.1432-1033.2001.02494.x.
[16] M.H. Cardoso, K.G.N. Oshiro, S.B. Rezende, E.S. Cândido, O.L. Franco, The structure/function relationship in antimicrobial peptides: what can we obtain from structural data? Advances in Protein Chemistry and Structural Biology, Academic Press Inc, 2018, pp. 359−384. Available from: https://doi.org/10.1016/bs.apcsb.2018.01.008.
[17] K. Ueda, M. Imamura, A. Saito, R. Sato, Purification and cDNA cloning of an insect defensin from larvae of the longicorn beetle, *Acalolepta luxuriosa*, The Journal Applied Entomology and Zoology 40 (2005) 335−345. Available from: https://doi.org/10.1303/aez.2005.335.
[18] A. Miyanoshita, S. Hara, M. Sugiyama, A. Asaoka, K. Taniai, F. Yukuhiro, et al., Isolation and characterization of a new member of the insect defensin family from a beetle, *Allomyrina dichotoma*, Biochemical and Biophysical Research Communications 220 (1996) 526−531. Available from: https://doi.org/10.1006/bbrc.1996.0438.
[19] H. Yamauchi, Two novel insect defensins from larvae of the cupreous chafer, *Anomala cuprea*: purification, amino acid sequences and antibacterial activity, Insect Biochemistry and Molecular Biology 32 (2001) 75−84. Available from: https://doi.org/10.1016/S0965-1748(01)00082-0.
[20] J.-S. Hwang, J. Lee, Y.-J. Kim, H.-S. Bang, E.-Y. Yun, S.-R. Kim, et al., Isolation and characterization of a defensin-like peptide (Coprisin) from the Dung Beetle, *Copris tripartitus*,

International Journal of Peptide Research (2009). Available from: https://doi.org/10.1155/2009/136284.
[21] S.Y. Lee, H.J. Moon, S. Kawabata, S. Kurata, S. Natori, B.L. Lee, et al., Homologue of *Holotrichia diomphalia*: purification, sequencing and determination of disulfide pairs, Biological and Pharmaceutical Bulletin 18 (1995) 457−459. Available from: https://doi.org/10.1248/bpb.18.457.
[22] H. Saido-Sakanaka, J. Ishibashi, A. Sagisaka, E. Momotani, M. Yamakawa, Synthesis and characterization of bactericidal oligopeptides designed on the basis of an insect anti-bacterial peptide, Biochemistry Journal 338 (1999) 29−33. Available from: https://doi.org/10.1042/bj3380029.
[23] H.J. Moon, S.Y. Lee, S. Kurata, S. Natori, B.L. Lee, Purification and molecular cloning of cDNA for an inducible antibacterial protein from larvae of the coleopteran, *Tenebrio molitor*, Journal of Biochemistry 116 (1994) 53−58. Available from: https://doi.org/10.1093/oxfordjournals.jbchem.a124502.
[24] P. Bulet, S. Cociancich, J.L. Dimarcq, J. Lambert, J.M. Reichhart, D. Hoffmann, et al., Insect immunity. Isolation from a coleopteran insect of a novel inducible antibacterial peptide and of new members of the insect defensin family, Journal of Biological Chemistry 266 (1991) 24520−24525.
[25] C. Lowenberger, P. Bulet, M. Charlet, C. Hetru, B. Hodgeman, B.M. Christensen, et al., Insect immunity: isolation of three novel inducible antibacterial defensins from the vector mosquito, *Aedes aegypti*, Insect Biochemistry and Molecular Biology 25 (1995) 867−873. Available from: https://doi.org/10.1016/0965-1748(95)00043-U.
[26] J. Vizioli, P. Bulet, M. Charlet, C. Lowenberger, C. Blass, H.M. Müller, et al., Cloning and analysis of a cecropin gene from the malaria vector mosquito, *Anopheles gambiae*, Insect Molecular Biology 9 (2000) 75−84. Available from: https://doi.org/10.1046/j.1365-2583.2000.00164.x.
[27] J.M. Meredith, H. Hurd, M.J. Lehane, P. Eggleston, The malaria vector mosquito *Anopheles gambiae* expresses a suite of larval-specific defensin genes, Insect Molecular Biology 17 (2008) 103. Available from: https://doi.org/10.1111/J.1365-2583.2008.00786.X.
[28] X. Lauth, A. Nesin, J.P. Briand, J.P. Roussel, C. Hetru, Isolation, characterization and chemical synthesis of a new insect defensin from *Chironomus plumosus* (Diptera, Insect Biochemistry and Molecular Biology 28 (1998) 1059−1066. Available from: https://doi.org/10.1016/S0965-1748(98)00101-5.
[29] C. Landon, P. Sodano, C. Hetru, J. Hoffmann, M. Ptak, Solution structure of drosomycin, the first inducible antifungal protein from insects, Protein Science 6 (1997) 1878−1884. Available from: https://doi.org/10.1002/pro.5560060908.
[30] J.-L. Imler, P. Bulet, Antimicrobial peptides in Drosophila: structures, activities and gene regulation, Chemical Immunology and Allergy 86 (2005) 1−21. <https://doi.org/10.1159/000086648>.
[31] V. Čeřovský, J. Žďárek, V. Fučík, L. Monincová, Z. Voburka, R. Bém, Lucifensin, the long-sought antimicrobial factor of medicinal maggots of the blowfly *Lucilia sericata*, Cellular and Molecular Life Sciences 67 (2010) 455−466. Available from: https://doi.org/10.1007/s00018-009-0194-0.
[32] J. Lambert, E. Keppi, J.L. Dimarcq, C. Wicker, J.M. Reichhart, B. Dunbar, et al., Insect immunity: isolation from immune blood of the dipteran *Phormia terranovae* of two insect antibacterial peptides with sequence homology to rabbit lung macrophage bactericidal peptides, Proceedings of National Academy of Sciences U. S. A. 86 (1989) 262−266. Available from: https://doi.org/10.1073/pnas.86.1.262.

[33] K. Matsuyama, S. Natori, Molecular cloning of cDNA for sapecin and unique expression of the sapecin gene during the development of *Surcophugu peregrinu*, Journal of Biological Chemistry 263 (1988) 17117−17121.

[34] K. Matsuyama, S. Natori, Purification of three antibacterial proteins from the culture medium of NIH-Sape-4, an embryonic cell line of *Sarcophaga peregrina*, Journal of Biological Chemistry 263 (1988) 17112−17116.

[35] K. Yamada, S. Natori, Purification, sequence and antibacterial activity of two novel sapecin homologues from Sarcophaga embryonic cells: similarity of sapecin B to charybdotoxin, Biochemistry Journal 291 (1993) 275−279. Available from: https://doi.org/10.1042/bj2910275.

[36] S. Cociancich, A. Dupont, G. Hegy, R. Lanot, F. Holder, C. Hetru, et al., Novel inducible antibacterial peptides from a hemipteran insect, the sap-sucking bug *Pyrrhocoris apterus*, Biochemistry Journal 300 (1994) 567−575. Available from: https://doi.org/10.1042/bj3000567.

[37] L. Lopez, G. Morales, R. Ursic, M. Wolff, C. Lowenberger, Isolation and characterization of a novel insect defensin from *Rhodnius prolixus*, a vector of Chagas disease, Insect Biochemistry and Molecular Biology 33 (2003) 439−447. Available from: https://doi.org/10.1016/S0965-1748(03)00008-0.

[38] S. Fujiwara, J. Imai, M. Fujiwara, T. Yaeshima, T. Kawashima, K. Kobayashit, A potent antibacterial protein in royal jelly. Purification and determination of the primary structure of royalisin, Journal of Biological Chemistry 19 (1990) 11333−11337.

[39] J.A. Rees, M. Moniatte, P. Bulet, Novel antibacterial peptides isolated from a European bumblebee, Bombus pascuorum (Hymenoptera, apoidea), Insect Biochemistry and Molecular Biology 27 (1997) 413−422. Available from: https://doi.org/10.1016/S0965-1748(97)00013-1.

[40] S. Taguchi, P. Bulet, J.A. Hoffmann, A novel insect defensin from the ant *Formica rufa*, Biochimie 80 (1998) 343−346. Available from: https://doi.org/10.1016/S0300-9084(98)80078-3.

[41] C. Tian, B. Gao, Q. Fang, G. Ye, S. Zhu, Antimicrobial peptide-like genes in *Nasonia vitripennis*: a genomic perspective, BMC Genomics 11 (2010) 1−19. Available from: https://doi.org/10.1186/1471-2164-11-187.

[42] P. Da Silva, Solution structure of termicin, an antimicrobial peptide from the termite *Pseudacanthotermes spiniger*, Protein Science 12 (2003) 438−446. Available from: https://doi.org/10.1110/ps.0228303.

[43] C. Landon, Lead optimization of antifungal peptides with 3D NMR structures analysis, Protein Science 13 (2004) 703−713. Available from: https://doi.org/10.1110/ps.03404404.

[44] Y.S. Lee, E.K. Yun, W.S. Jang, I. Kim, J.H. Lee, S.Y. Park, et al., Purification, cDNA cloning and expression of an insect defensin from the great wax moth, *Galleria mellonella*, Insect Molecular Biology 13 (2004) 65−72. Available from: https://doi.org/10.1111/j.1365-2583.2004.00462.x.

[45] M. Lamberty, S. Ades, S. Uttenweiler-Joseph, G. Brookhart, D. Bushey, J.A. Hoffmann, et al., Insect immunity. Isolation from the lepidopteran *Heliothis virescens* of a novel insect defensin with potent antifungal activity, Journal of Biological Chemistry 274 (1999) 9320−9326.

[46] M. Mandrioli, S. Bugli, S. Saltini, S. Genedani, E. Ottaviani, Molecular characterization of a defensin in the IZD-MB-0503 cell line derived from immunocytes of the insect *Mamestra brassicae* (Lepidoptera, Biology of the Cell 95 (2003) 53−57. Available from: https://doi.org/10.1016/S0248-4900(02)01219-4.

[47] A.E.M. Seufi, E.E. Hafez, F.H. Galal, Identification, phylogenetic analysis and expression profile of an anionic insect defensin gene, with antibacterial activity, from bacterial-challenged cotton leafworm, *Spodoptera littoralis*, BMC Molecular Biology 12 (2011) 1–14. Available from: https://doi.org/10.1186/1471-2199-12-47.

[48] P. Bulet, S. Cociancich, M. Reuland, F. Sauber, R. Bischoff, G. Hegy, et al., A novel insect defensin mediates the inducible antibacterial activity in larvae of the dragonfly *Aeschna cyanea* (Paleoptera, Odonata), European Journal of Biochemistry 209 (1992) 977–984. Available from: https://doi.org/10.1111/j.1432-1033.1992.tb17371.x.

[49] A. Saito, K. Ueda, M. Imamura, S. Atsumi, H. Tabunoki, N. Miura, et al., Purification and cDNA cloning of a cecropin from the longicorn beetle, *Acalolepta luxuriosa*, Comparative Biochemistry and Physiology Part B: Biochemistry & Molecular Biology 142 (2005) 317–323. Available from: https://doi.org/10.1016/j.cbpb.2005.08.001.

[50] M. Memarpoor-Yazdi, H. Zare-Zardini, A. Asoodeh, A novel antimicrobial peptide derived from the insect *Paederus dermatitis*, The International Journal of Peptide Research and Therapeutics 19 (2013) 99–108. Available from: https://doi.org/10.1007/s10989-012-9320-1.

[51] S. Ekengren, D. Hultmark, Drosophila cecropin as an antifungal agent, Insect Biochemistry and Molecular Biology 29 (1999) 965–972. Available from: https://doi.org/10.1016/S0965-1748(99)00071-5.

[52] B.P. Lazzaro, A.G. Clark, Molecular population genetics of inducible antibacterial peptide genes in *Drosophila melanogaster*, Molecular Biology and Evolution 20 (2003) 914–923. Available from: https://doi.org/10.1093/MOLBEV/MSG109.

[53] M. Okada, S. Natori, Primary structure of sarcotoxin I, an antibacterial protein induced in the hemolymph of *Sarcophaga peregrina* (flesh fly) larvae, Journal of Biological Chemistry 260 (1985) 7174–7177.

[54] N. Boulanger, R.J.L. Munks, J.V. Hamilton, F. Oise Vovelle, R. Brun, M.J. Lehane, et al., Epithelial innate immunity: a novel antimicrobial peptide with antiparasitic activity in the blood-sucking insect *Stomoxys calcitrans*, Journal of Biological Chemistry 277 (2002) 49921–49926. Available from: https://doi.org/10.1074/jbc.M206296200.

[55] W. Kim, H. Koo, A.M. Richman, D. Seeley, J. Vizioli, A.D. Klocko, D.A. O'brochta, , Ectopic expression of a cecropin transgene in the human malaria vector mosquito *Anopheles gambiae* (Diptera: Culicidae): effects on susceptibility to Plasmodium, Journal of Medical Entomology 41 (2004) 447–455. Available from: https://doi.org/10.1603/0022-2585-41.3.447.

[56] I.H. Lee, K.Y. Chang, C.S. Choi, H.R. Kim, Cecropin D-like antibacterial peptides from the sphingid moth, *Agrius convolvuli*, Archives of Insect Biochemistry and Physiology 41 (1999) 178–185. https://doi.org/https://doi.org/10.1002/(SICI)1520-6327 (1999)41:4 < 178::AID-ARCH2 > 3.0.CO;2-W.

[57] S.M. Yoe, C.S. Kang, S.S. Han, I.S. Bang, Characterization and cDNA cloning of hinnavin II, a cecropin family antibacterial peptide from the cabbage butterfly, *Artogeia rapae*, Comparative Biochemistry and Physiology Part B: Biochemistry & Molecular Biology 144 (2006) 199–205. Available from: https://doi.org/10.1016/j.cbpb.2006.02.010.

[58] Y. Yamano, M. Matsumoto, K. Sasahara, E. Sakamoto, I. Morishima, Structure of genes for cecropin A and an inducible nuclear protein that binds to the promoter region of the genes from the silkworm, *Bombyx mori*, Bioscience, Biotechnology, and Biochemistry 62 (1998) 237–241. Available from: https://doi.org/10.1271/bbb.62.237.

[59] J. Yang, S. Furukawa, A. Sagisaka, J. Ishibashi, K. Taniai, T. Shono, et al., cDNA cloning and gene expression of cecropin D, an antibacterial protein in the silkworm, *Bombyx mori*, Comparative Biochemistry and Physiology Part B: Biochemistry & Molecular Biology 122 (1999) 409–414. Available from: https://doi.org/10.1016/S0305-0491(99)00015-2.

[60] K. Taniai, S. Furukawa, T. Shono, M. Yamakawa, Elicitors triggering the simultaneous gene expression of antibacterial proteins of the silkworm, *Bombyx mori*, Biochemical and Biophysical Research Communications 226 (1996) 783−790. Available from: https://doi.org/10.1006/bbrc.1996.1429.

[61] Q. Wang, Y. Liu, H.J. He, X.F. Zhao, J.X. Wang, Immune responses of *Helicoverpa armigera* to different kinds of pathogens, BMC Immunology 11 (2010) 1−12. Available from: https://doi.org/10.1186/1471-2172-11-9.

[62] L. Wang, Z. Li, C. Du, W. Chen, Y. Pang, Characterization and expression of a cecropin-like gene from *Helicoverpa armigera*, Comparative Biochemistry and Physiology Part B: Biochemistry & Molecular Biology 148 (2007) 417−425. Available from: https://doi.org/10.1016/J.CBPB.2007.07.010.

[63] T.D. Lockey, D.D. Ourth, Formation of pores in *Escherichia coli* cell membranes by a cecropin isolated from hemolymph of *Heliothis virescens* larvae, European Journal of Biochemistry 236 (1996) 263−271. Available from: https://doi.org/10.1111/j.1432-1033.1996.00263.x.

[64] D. Hultmark, Å. Engstrom, H. Bennich, R. Kapur, H.G. Boman, Insect immunity: isolation and structure of cecropin D and four minor antibacterial components from *Cecropia Pupae*, European Journal of Biochemistry 127 (1982) 207−217. Available from: https://doi.org/10.1111/j.1432-1033.1982.tb06857.x.

[65] S. Sik Park, S. Woon Shin, D.S. Park, H. Woo Oh, K. Saeng Boo, H.Y. Park, Protein purification and cDNA cloning of a cecropin-like peptide from the larvae of fall webworm (*Hyphantria cunea*), Insect Biochemistry and Molecular Biology 27 (1997) 711−720. Available from: https://doi.org/10.1016/S0965-1748(97)00049-0.

[66] S.R. Kim, M.Y. Hong, S.W. Park, K.H. Choi, E.Y. Yun, T.W. Goo, et al., Characterization and cDNA cloning of a cecropin-like antimicrobial peptide, papiliocin, from the swallowtail butterfly, *Papilio xuthus*, Molecules and Cells 29 (2010) 419−423. Available from: https://doi.org/10.1007/s10059-010-0050-y.

[67] W.L. Cho, Y.C. Fu, C.C. Chen, C.M. Ho, Cloning and characterization of cDNAs encoding the antibacterial peptide, defensin A, from the mosquito, *Aedes aegypti*, Insect Biochemistry and Molecular Biology 26 (1996) 395−402. Available from: https://doi.org/10.1016/0965-1748(95)00108-5.

[68] C.-S. Choi, S.-M. Yoe, E.-S. Kim, K.-S. Chae, H.R. Kim, Purification and characterization of antibacterial peptides, spodopsin Ia and Ib induced in the larval haemolymph of the common cutworm, *Spodoptera iitura*, Animal Cells and Systems (Seoul) 1 (1997) 457−462.

[69] P. Bulet, L. Urge, S. Ohresser, C. Hetru, L. Otvos, Enlarged scale chemical synthesis and range of activity of drosocin, an O-glycosylated antibacterial peptide of Drosophila, European Journal of Biochemistry 238 (1996) 64−69. Available from: https://doi.org/10.1111/j.1432-1033.1996.0064q.x.

[70] E.A. Levashina, S. Ohresser, P. Bulet, J.-M. Reichhart, C. Hetru, J.A. Hoffmann, Metchnikowin, a novel immune-inducible proline-rich peptide from Drosophila with antibacterial and antifungal properties, European Journal of Biochemistry 233 (1995) 694−700. Available from: https://doi.org/10.1111/j.1432-1033.1995.694_2.x.

[71] M. Schneider, A. Dorn, Differential infectivity of two Pseudomonas species and the immune response in the milkweed bug, *Oncopeltus fasciatus* (Insecta: Hemiptera), Journal of Invertebrate Pathology 78 (2001) 135−140. Available from: https://doi.org/10.1006/jipa.2001.5054.

[72] P. Fehlbaum, P. Bulet, S. Chernysh, J.P. Briand, J.P. Roussel, L. Letellier, et al., Structure-activity analysis of thanatin, a 21-residue inducible insect defense peptide

with sequence homology to frog skin antimicrobial peptides, Proceedings of National Academy of Sciences U. S. A. 93 (1996) 1221−1225. Available from: https://doi.org/10.1073/pnas.93.3.1221.

[73] P. Casteels, C. Ampe, F. Jacobs, M. Vaeck, P. Tempst, Apidaecins: antibacterial peptides from honeybees, EMBO Journal 8 (1989) 2387−2391. Available from: https://doi.org/10.1002/j.1460-2075.1989.tb08368.x.

[74] K. Ando, S. Natori, Molecular cloning, sequencing, and characterization of cDNA for sarcotoxin IIA, an inducible antibacterial protein of *Sarcophaga peregrina* (Flesh Fly), Biochemistry 27 (1988) 1715−1721. Available from: https://doi.org/10.1021/bi00405a050.

[75] J.A. Mackintosh, A.A. Gooley, P.H. Karuso, A.J. Beattie, D.R. Jardine, D.A. Veal, A gloverin-like antibacterial protein is synthesized in *Helicoverpa armigera* following bacterial challenge, Developmental and Comparative Immunology 22 (1998) 387−399. Available from: https://doi.org/10.1016/S0145-305X(98)00025-1.

[76] J. Mackintosh, D. Veal, A. Beattie, A. Gooley, Isolation from an ant *Myrmecia gulosa* of two inducible O-glycosylated proline-rich antibacterial peptides, Journal of Biological Chemistry 273 (1998) 6139−6143. Available from: https://doi.org/10.1074/JBC.273.11.6139.

[77] X. Shen, G. Ye, X. Cheng, C. Yu, I. Altosaar, C. Hu, Characterization of an abaecin-like antimicrobial peptide identified from a *Pteromalus puparum* cDNA clone, The Journal of Invertebrate Pathology 105 (2010) 24−29. Available from: https://doi.org/10.1016/j.jip.2010.05.006.

[78] S. Furukawa, K. Taniai, J. Ishibashi, S. Hara, T. Shono, M. Yamakawa, A novel member of lebocin gene family from the silkworm, *Bombyx mori*, Biochemical and Biophysical Research Communications 238 (1997) 769−774. Available from: https://doi.org/10.1006/bbrc.1997.7386.

[79] S. Rayaprolu, Y. Wang, M.R. Kanost, S. Hartson, H. Jiang, Functional analysis of four processing products from multiple precursors encoded by a lebocin-related gene from *Manduca sexta*, Developmental and Comparative Immunology 34 (2010) 638−647. Available from: https://doi.org/10.1016/j.dci.2010.01.008.

[80] X.J. Rao, X.X. Xu, X.Q. Yu, Functional analysis of two lebocin-related proteins from *Manduca sexta*, Insect Biochemistry and Molecular Biology 42 (2012) 231−239. Available from: https://doi.org/10.1016/j.ibmb.2011.12.005.

[81] Å. Engström, P. Engström, Z.-J. Tao, A. Carlsson, H. Bennich, Insect immunity. The primary structure of the antibacterial protein attacin F and its relation to two native attacins from *Hyalophora cecropia*, EMBO Journal 3 (1984) 2065−2070. Available from: https://doi.org/10.1002/j.1460-2075.1984.tb02092.x.

[82] M. Hedengren, K. Borge, D. Hultmark, Expression and evolution of the Drosophila attacin/diptericin gene family, Biochemical and Biophysical Research Communications 279 (2000) 574−581. Available from: https://doi.org/10.1006/bbrc.2000.3988.

[83] H.Y. Yi, M. Chowdhury, Y.D. Huang, X.Q. Yu, Insect antimicrobial peptides and their applications, Applied Microbiology and Biotechnology 98 (2014) 5807−5822. Available from: https://doi.org/10.1007/s00253-014-5792-6.

[84] Y.M. Kwon, H.J. Kim, Y.I. Kim, Y.J. Kang, I.H. Lee, B.R. Jin, et al., Comparative analysis of two attacin genes from *Hyphantria cunea*, Comparative Biochemistry and Physiology Part B: Biochemistry & Molecular Biology 151 (2008) 213−220. Available from: https://doi.org/10.1016/j.cbpb.2008.07.002.

[85] J. Wang, C. Hu, Y. Wu, A. Stuart, C. Amemiya, M. Berriman, et al., Characterization of the antimicrobial peptide attacin loci from *Glossina morsitans*, Insect Molecular Biology 17 (2008) 293−302. Available from: https://doi.org/10.1111/j.1365-2583.2008.00805.x.

[86] K. Bang, S. Park, J.Y. Yoo, S. Cho, Characterization and expression of attacin, an antibacterial protein-encoding gene, from the beet armyworm, *Spodoptera exigua* (Hübner) (Insecta: Lepidoptera: Noctuidae, Molecular Biology Reports 39 (2012) 5151−5159. Available from: https://doi.org/10.1007/s11033-011-1311-3.

[87] D. Brady, A. Grapputo, O. Romoli, F. Sandrelli, Insect cecropins, antimicrobial peptides with potential therapeutic applications, The International Journal of Molecular Sciences 20 (2019). Available from: https://doi.org/10.3390/ijms20235862.

[88] P. Kumar, J.N. Kizhakkedathu, S.K. Straus, Antimicrobial peptides: diversity, mechanism of action and strategies to improve the activity and biocompatibility in vivo, Biomolecules 8 (2018). Available from: https://doi.org/10.3390/biom8010004.

[89] V.S. Skosyrev, E.A. Kulesskiy, A.V. Yakhnin, Y.V. Temirov, L.M. Vinokurov, Expression of the recombinant antibacterial peptide sarcotoxin IA in *Escherichia coli* cells, Protein Expression and Purification 28 (2003) 350−356. Available from: https://doi.org/10.1016/S1046-5928(02)00697-6.

[90] M. Lamberty, D. Zachary, R. Lanot, C. Bordereau, A. Robert, J.A. Hoffmann, et al., Insect immunity. Constitutive expression of a cysteine-rich antifungal and a linear antibacterial peptide in a termite insect, Journal of Biological Chemistry 276 (2001) 4085−4092. Available from: https://doi.org/10.1074/jbc.M002998200.

[91] J. Koehbach, Structure-activity relationships of insect defensins, Frontiers in Chemistry 5 (2017) 45. Available from: https://doi.org/10.3389/fchem.2017.00045.

[92] R.D.O. Dias, O.L. Franco, Cysteine-stabilized $\alpha\beta$ defensins: from a common fold to antibacterial activity, Peptides 72 (2015) 64−72. Available from: https://doi.org/10.1016/j.peptides.2015.04.017.

[93] A.A. Baxter, I.K.H. Poon, M.D. Hulett, The lure of the lipids: how defensins exploit membrane phospholipids to induce cytolysis in target cells, Cell Death and Disease 8 (2017) e2712. Available from: https://doi.org/10.1038/cddis.2017.69.

[94] K. Yamada, S. Natori, Characterization of the antimicrobial peptide derived from sapecin B, an antibacterial protein of *Sarcophaga peregrina* (flesh fly), Biochemistry Journal 298 (1994) 623−628. Available from: https://doi.org/10.1042/bj2980623.

[95] X.X. Xu, X. Zhong, H.Y. Yi, X.Q. Yu, *Manduca sexta* gloverin binds microbial components and is active against bacteria and fungi, Developmental and Comparative Immunology 38 (2012) 275−284. Available from: https://doi.org/10.1016/j.dci.2012.06.012.

[96] J. Nesa, A. Sadat, D.F. Buccini, A. Kati, A.K. Mandal, O.L. Franco, Antimicrobial peptides from: *Bombyx mori*: a splendid immune defense response in silkworms, RSC Advances 10 (2019) 512−523. Available from: https://doi.org/10.1039/c9ra06864c.

[97] S. Islam, S. Bezbaruah, J. Kalita, A review on antimicrobial peptides from *Bombyx mori* L and their application in plant and animal disease control, Journal of Advances in Biology & Biotechnology 9 (2016) 1−15. Available from: https://doi.org/10.9734/jabb/2016/27539.

[98] L.L. Yang, M.Y. Zhan, Y.L. Zhuo, Y.M. Pan, Y. Xu, X.H. Zhou, et al., Antimicrobial activities of a proline-rich proprotein from *Spodoptera litura*, Developmental and Comparative Immunology 87 (2018) 137−146. Available from: https://doi.org/10.1016/j.dci.2018.06.011.

[99] H. Reddy, S. A, A. C, Critical assessment of *Bombyx mori* haemolymph extract on *Staphylococcus aureus* an in vitro and in silico approach, Journal of Proteomics and Bioinformatics 9 (2016) 1−6. Available from: https://doi.org/10.4172/jpb.1000410.

[100] S.E. Brown, A. Howard, A.B. Kasprzak, K.H. Gordon, P.D. East, The discovery and analysis of a diverged family of novel antifungal moricin-like peptides in the wax moth *Galleria mellonella*, Insect Biochem. Molecular Biology 38 (2008) 201−212. Available from: https://doi.org/10.1016/j.ibmb.2007.10.009.

[101] X. Xu, A. Zhong, Y. Wang, B. Lin, P. Li, W. Ju, et al., Molecular identification of a Moricin family antimicrobial peptide (Px-Mor) from *Plutella xylostella* with activities against the opportunistic human pathogen *Aureobasidium pullulans*, Frontiers in Microbiology 10 (2019) 2211. Available from: https://doi.org/10.3389/fmicb.2019.02211.

[102] S. Mukhopadhyay, A.S. Bharath Prasad, C.H. Mehta, U.Y. Nayak, Antimicrobial peptide polymers: no escape to ESKAPE pathogens—a review, World Journal of Microbiology and Biotechnology 36 (2020) 131−132. Available from: https://doi.org/10.1007/s11274-020-02907-1.

[103] N. Raheem, S.K. Straus, Mechanisms of action for antimicrobial peptides with antibacterial and antibiofilm functions, Frontiers in Microbiology 10 (2019) 2866. Available from: https://doi.org/10.3389/fmicb.2019.02866.

[104] A. Hollmann, M. Martinez, P. Maturana, L.C. Semorile, P.C. Maffia, Antimicrobial peptides: interaction with model and biological membranes and synergism with chemical antibiotics, Frontiers in Chemistry 6 (2018) 204. Available from: https://doi.org/10.3389/fchem.2018.00204.

[105] C.F. Le, C.M. Fang, S.D. Sekaran, Intracellular targeting mechanisms by antimicrobial peptides, Antimicrobial Agents and Chemotherapy 61 (2017). Available from: https://doi.org/10.1128/AAC.02340-16.

[106] J.E. Faust, P.Y. Yang, H.W. Huang, Action of antimicrobial peptides on bacterial and lipid membranes: a direct comparison, The Biophysical Journal 112 (2017) 1663−1672. Available from: https://doi.org/10.1016/j.bpj.2017.03.003.

[107] M. Cytryńska, M. Rahnamaeian, A. Zdybicka-Barabas, K. Dobslaff, T. Züchner, G. Sacheau, et al., Proline-rich antimicrobial peptides in medicinal maggots of *Lucilia sericata* interact with bacterial DnaK but do not inhibit protein synthesis, Frontiers in Pharmacology 11 (2020) 1. Available from: https://doi.org/10.3389/fphar.2020.00532.

[108] A. Vilcinskas, M. Wedde, Insect inhibitors of metalloproteinases, IUBMB Life 54 (2002) 339−343. Available from: https://doi.org/10.1080/15216540216040.

# Amphibian host defense peptides

A. Anju Krishnan[1], A.R. Sarika[2], K. Santhosh Kumar[1] and Arunan Chandravarkar[2]
[1]Chemical Biology Lab, Rajiv Gandhi Centre for Biotechnology, Thiruvananthapuram, Kerala, India, [2]Kerala State Council for Science, Technology and Environment, Thiruvananthapuram, Kerala, India

## 8.1 Antimicrobial peptides: critical component of innate immune system

Amphibians and insects have enjoyed remarkable evolutionary success and emerged as the most successful clade of all organisms constituting close to 80% of all animal life forms. The defense systems of most of these organisms have multilayer chemical defense system which consists of many pharmacologically active molecules like amines, alkaloids, peptides in their serous cells. These compounds are considered as the effector molecules of innate immunity, and are secreted to the skin surface upon stimulation and provide protection against potential pathogenic microorganisms [1,2]. Antimicrobial peptides (AMPs) or cationic host defense peptides (CHDPs) are an evolutionarily conserved component of the innate immune response, which act as the principal defense system for the majority of living organisms, and are found among all classes of life ranging from prokaryotes to humans [3,4]. AMPs generally act directly or indirectly to provide protection from microbes, by exhibiting broad range of antimicrobial activity against both gram positive, gram negative as well as other forms of organisms ranging from viruses to parasites or by modulating the innate immune response of the host. Besides their broad spectrum antimicrobial activity, AMPs were reported to be involved in angiogenesis, inflammatory response, and cell signaling [5].

AMPs are generally defined as oligopeptides of less than 50 amino acid residues with an overall positive charge (generally $+2$ to $+9$), with high proportion of cationic amino acids. The presence of a high proportion of cationic and hydrophobic residuals greatly affects their antimicrobial activity. Cationic residues such as Arginine, Lysine, Histidine mediate the interactions with negatively charged bacterial lipids, whereas hydrophobic residues including Tryptophan, Phenylalanine, Leucine mediate amphipathic structure formation and thereby induce membrane damage [6]. They act primarily on bacterial cell membranes, disrupt it and eventually induce bacterial cell death. This unique mode of action reduces the possibility of bacteria to develop resistance against peptide. Thus AMPs can serve as complementary to antibiotics.

## 8.2 Antimicrobial peptide from amphibians

The current understanding of AMPs has been obtained from studies of those isolated from amphibian skin secretions which are a rich source of these peptides [2,7–10], and 98% of known AMPs were identified from the skin secretions of amphibians belonging to various families [11]. Amphibian skin secretion has been known as a rich source of pharmacologically active peptides for a long time [12,13]. Scientific interest in amphibian AMPs began after the landmark studies on purification of bombinin from *Bombina variegata* in 1970 and isolation of magainins from *Xenopus laevis* in 1987. Since then, thousands of AMPs have been isolated either from the frog skin secretion or skin extraction.

Amphibians being the ancient creatures and the first group of organisms that form a connecting link between land and water are forced to adapt and survive in a variety of conditions laden with pathogenic microorganisms and have highly evolved immune system [14]. In amphibian these peptides are stored in their granular glands and are released upon stimulation [15]. These peptides contribute to animals' innate immunity and defend them from the invading pathogenic microorganisms and therefore they represent therapeutically valuable antiinfective agents. Skin secretions of frogs of *Hylidae* family, *Ranidae* family, *Pipidae* family, and *Hyperoliidae* family are rich source of AMPs [16–21]. Amphibians synthesize multiple structurally related peptides with antimicrobial activity specific to each species such that no two species harbor identical peptides. AMPs isolated from amphibian skin secretions are 10–46 residues long, cationic, have a number of hydrophobic residues, and are amphipathic in nature [22,23]. Majority of these peptides adopt α-helical conformation or β-sheeted structures. AMPs possess high hydrophobic moment and a wide nonpolar face, in contrast to narrow polar face, that enables them to effectively interact with negatively charged bacterial membrane [24]. Majority of the studies reported that AMPs depolarize and disrupt the membrane integrity by pore or channel formation, thereby leading to cell death [25]. In addition, they effectively cross the outer and inner membrane to target intracellular components to mediate its action. The effectiveness of these peptides to induce membrane destabilization has made them potential candidates to be developed as therapeutic agents. Conventional antibiotics have specific targets, based on enzyme inhibition and takes days to exert its action while AMPs can quickly promote lyses on bacteria, fungi, protozoa, etc. As many as 1093 AMPs derived from amphibians are listed in latest antimicrobial peptide database (APD-3) (https://aps.unmc.edu/).

### 8.2.1 Antimicrobial peptides isolated from African frogs

The frogs in the family *Pipidae* are found mainly in African continent and northern part of South America. This family contains 33 species in 5 genres; that is, *Hymenochirus, Pipa, Pseudhymenochirus, Silurana,* and *Xenopus* [26]. The first AMP magainin was discovered in *X. laevis*—African clawed frog [27]. Magainin showed broad spectrum antimicrobial activity, and it consists of two closely related peptides having 23 amino acids, and differs by two amino acids substitutions at 10 and 22 positions. Later, Zasloff et al. showed that two other peptides—xenopsin-precursor fragment (XPF) and Peptidyl-glycylleucine-carboxyamide also known as Peptide beginning with Glycine and ending with Leucine Amide, or PGLa—that are found in the secretion of granular gland of *X. laevis* [28] also showed broad spectrum antimicrobial activity. All these peptides are amphiphilic in nature and are membrane disruptive. However, they exhibit little hemolytic activity against human erythrocytes. Skin secretions *Hymenochirus boettgeri* belonging to two populations from different geographical locations gave five structurally related peptides, termed together as hymenochirins, having broad antimicrobial properties [29]. Conlon and coworkers isolated a broad spectrum AMP, Kassinatuerin-1, from the skin of *Kassina senegalensis* (*Hyperoliidae* family which contains over 250 species in 19 genus). The peptide is active against gram-negative, gram-positive bacterium and yeast *Candida albicans*, at varying minimum inhibitory concentrations [30]. A structurally similar peptide Kassinatuerin-2 was also found in high quantity along with Kassinatuerin-1, but was devoid of any antimicrobial activity due to the absence of its cationic nature and unlike magainin C-termini of these peptides are amidated.

## 8.2.2 Antimicrobial peptide isolated from amphibians in North America

Though amphibians in North America belong to the families of *Bufonidae, Eleutherodactylidae, Hylidae, Leiopelmatidae, Ranidae,* and *Scaphiopodidae,* AMPs were observed only in *Leiopelmatidae* ("tailed frogs") and *Ranidae* ("true frogs") families. AMP such as Ascaphins is separated from *Ascaphus truei* (*Leiopelmatidae*), and peptides belonging to the brevinin-1, esculentin-1, esculentin-2, palustrin-1, palustrin-2, ranacyclin, ranatuerin-1, ranatuerin-2, and temporin families have been isolated from *Ranidae* family. Current taxonomic recommendations divide North American frogs from the family *Ranidae* into two genera: *Lithobates* and *Rana* [31]. Ranalexin was separated from the skin of bullfrog (*Rana catesbeiana*). This 20-amino acid long peptide contains a single intramolecular disulfide bond which forms a heptapeptide ring within the molecule making it structurally similar to the bacterial antibiotic polymyxin which also contains a similar heptapeptide [32]. Brevenin and esculentin are similar AMPs isolated from *Rana brevipoda porsa* and *Rana esculenta*, but these species are present in Asia and Europe, respectively.

Another class of nine AMPs were isolated from the skin of American bullfrog, *Rana catesbeiana* termed as ranatuerins 1–9 which show activity toward *Staphylococcus aureus*. Ranatuerins 1 and 4 contain an intramolecular disulfide bridge forming a heptapeptide but ranatuerins 2 and 3 form a hexapeptide ring by intramolecular disulfide bridge. Other peptides in this series ranatuerins 5–9 having 12–14 amino acids long show sequence similarity toward hemolytic peptides A1 and B9 isolated from the skin of *R. esculenta* [33]. From the North American pickerel frog *Rana palustris*, 22 peptides having growth-inhibitory activity against bacteria and yeast were isolated [34]. Among them 13 AMPs are similar to peptides isolated from Ranid frogs (brevinine, esculentin-1, esculentin-2, ranatuerin-2, and temporin) and 9 peptides show little structural similarity with other known AMPs hence they are classified in a different family, termed as palustrin. Four peptides with 27 to 28 amino acid long belong to palustrin-1 which have intramolecular disulfide forming a heptapeptide ring, 3 peptides (palustrin-2) having 31 amino acid residues have a cyclic heptapeptide ring, and 2 peptides (palustrin-3) with 48 amino acids have a cyclic hexapeptide ring.

Ascaphins are a group of AMPs isolated from *Ascaphus truei* (family: *Leiopelmatidae*), the most primitive extant frog. Eight structurally related peptides termed ascaphins 1–8 are extracted from the skin secretions of *A. truei* [35], they all contain 23 amino acids in free form except ascaphin-1 and ascaphin-8 whose carboxy terminals are amidated. These eight peptides show little structural similarity to other AMPs isolated from frog skin of other species. Though ascaphin-8 shows highest action against pathogen, it has the greatest hemolytic activity thus its use as therapeutic agent is limited. Ten AMPs with differential activity toward gram-positive and gram-negative bacteria were isolated from *Rana clamitans*. They belong to ranatuerin, ranalexin, and temporin families [36].

### 8.2.3 Antimicrobial peptides isolated from amphibians in South America

South America is home to 2623 anuran species from 163 genera and 24 families [37]. Many AMPs were isolated from the families of *Phyllomedusidae*, *Hylidae*, and *Leptodactylidae*. Dermaseptin-I, a 34-residue lysine-rich AMP peptide with nonhemolytic activity was isolated from South American arboreal frog *Phyllomedusa sauuvgii* (family: *Hylidae*) [38]. The peptide is strongly basic, because of the presence of lysine residues, toward the amino part and contains long alternating hydrophobic moieties. The C-terminal region is hydrophilic and negatively charged. Four AMPs which are similar to dermaseptin-I were isolated from *Phyllomedusa dacnicolor* (family: *Phyllomedusidae*) [39].

Another class of dermaseptin AMPs was isolated from South American frogs *Agalychnis annae* (family: *Phyllomedusidae*) and *Pachymedusa dacnicolor* (family: *Phyllomedusidae*, genus: *Agalychnis*) [40]. Five of the isolated peptides are functional homologs to dermaseptin-I isolated from *P. sauuvgii* (family: *Hylidae*). From the skin secretions of *Phyllomedusa bicolor* (*Phyllomedusidae*), a South American frog, six polycationic (lysine-rich), helical peptides having 24–33 amino acid residues long were isolated. These peptides belong to dermaseptin family and are termed as dermaseptin B1–B6. Phylloxin is another peptide which was isolated from frog *P. bicolor*. (family: *Phyllomedusidae*) [41]. Though phylloxin is isolated from *Phyllomedusidae* family, it is not structurally homologous to either the dermaseptins or the dermorphin/deltorphins peptides, but exhibits structural similarities with levitide precursor fragment and XPF which are AMPs isolated from the skin of an evolutionary distant frog species, *X. laevis*. Phylloxin, unlike dermaseptins, show very narrow AMPs. The first heterodimeric AMP, distinctin, was isolated from a tree frog *Phyllomedusa distincta* (family: *Hylidae*). This peptide contains two different polypeptide chains connected by an intermolecular disulfide bridge [42]. In water, distinctin forms a noncovalent four α-helix bundle [43], and this kind of feature gives the peptide a high proteolytic stability. Pseudins are four AMPs that are extracted from the skin of the paradoxical frog, *Pseudis paradoxa* (family: *Hylidae*). All the four peptides are structurally related but do not have any sequence similarity with other AMPs reported from same family. Of these four peptides, pseudin-2 shows greater antimicrobial activity with low hemolytic activity [44]. Sixteen AMPs were isolated from different species of *Leptodactylidae* family. Four AMPs termed as ocellatins 1—4 were isolated from the skin secretion of the South American frog *Leptodactylus ocellatus*. All are C-terminal-amidated peptides and varies in amino acid residues from 21 to 25. All these peptides have low net positive charge, and their hemolytic properties are higher [45,46].

Fallaxin, a 25 amino acid residue and C-terminally amidated peptide, was isolated from the skin secretions of the mountain chicken frog *Leptodactylus fallax* [47]. This peptide has structural similarity with members of the ranatuerin-2 family isolated from the skins of frogs of the genus *Rana* that are only distantly related to the *Leptodactylidae*. Fallaxin has low action against gram-negative bacteria and no activity toward gram-positive bacteria and the yeast *C. albicans* [48]. The skin extraction of Caribbean frog, *Leptodactylus validus*, gave three AMPs termed as ocellatin-V1, ocellatin-V2, and ocellatin-V3. These peptides have low cationic properties and low amphipathicity, hence exhibit low AMP potency. Eight new peptides, ocellatins-PT1 to PT8, were isolated from the skin secretion of the frog *Leptodactylus pustulatus*, and possess structural similarities with other AMPs from the skin secretion of *Leptodactylus* genus frogs. Ocellatins-PT1 to PT5 are 25 amino acid residues long and are amidated at the C-terminus like other peptides in this family, and ocellatins-PT6 to PT8 (32 amino acid residues) have free carboxylates. All peptides, except for ocellatin-PT2, have antimicrobial activity against at least one gram-negative strain [49].

Phylloseptins (PS) are another class of antibacterial and antiprotozoal peptide isolated from the skin secretion of frogs from *Phyllomedusidae* family [50,51]. They are having length of 19—21 amino acid residues with a highly conserved N-terminal domain, FLSLIP, and C-terminal amidation. PS-1 to PS-6 were isolated from the skin secretions of frogs *Pithecopus hypochondrialis* and *P. oreades*. PS-1 showed great activity against gram-positive and gram-negative bacteria without much hemolytic activity in mammalian cells. One of the members in the phylloseptins, phylloseptin-PBa, which was isolated from the skin secretion of *P. baltea*, showed cancer cell cytotoxicity [47]. Another peptide, phylloseptin-PV1 (PPV1), isolated from *P. vaillantii* (family: *Hydlidae*) is found very potential against gram-positive but weaker in the case of gram-negative bacteria [52]. Hyposin (HA) is similar AMP isolated from the skin of *P. hypochondrialis azurea*. From the skin extracts, 22 peptides were isolated and a set of new AMPs family termed as hyposins were identified [53]. They are 11–15 residue long and are rich in Lys and Arg residues make it strong cationic in nature. Except HA-5 all hyposin peptides are amidated at C-terminus.

Hylins are a group of three peptides isolated from South American frogs. Hylins b1 and b2 are 19 amino acid long peptides isolated from *Hyla biobeba* (synonym *Hyla lundii*, family: *Hylidae*, genus: *Boana*), and are structurally very similar (amino acids at position 14 is Gly in Hy-b2 where as it is His in Hy-b1) [54]. Hylin a1, isolated from *Hypsiboas albopunctatus* (synonym: *Boana albopunctata*; family: *Hylidae*, genus: *Boana*) is 18 amino acid residues long AMP [55].

Three structurally related host defense peptides (frenatin 2.1S, 2.2S, 2.3S) were isolated from the skin secretions of *Sphaenorhynchus lacteus* (family: *Hylidae*). These peptides have limited sequence similarities with other frenatin series that were isolated from Australian frog species [56]. Frenatin 2.1S and 2.2S are C-terminal amidated and 2.3S is the free form of 2.1S. The C-terminal-amidated frenatins have more activity against gram-negative bacteria, gram-positive bacteria, yeast pathogen *C. albicans* and *Acinetobacter baumannii*, and *Stenotrophomonas maltophilia* than the free frenatin (2.3S). However, frenatin 2.3S protect cells from yellow fever viral infection [57]. Another peptide found in *S. lacteus* was BF2 which has same primary sequence of buforin II (BF2) isolated from *Bufo bufo gargarizans* stomach tissue [58].

## 8.2.4 Antimicrobial peptide isolated from amphibians in Australia

Australia is home to 248 species within 6 families: *Hylidae, Limnodynastidae, Microhylidae, Myobatrachidae, Ranidae*, and *Bufonidae*. Many of the AMPs that are isolated belong to the family *Hylidae* and genus *Litoria*. About 80 different peptides are isolated from the genus *Litoria*. Caerins (five peptides) and caeridin-1 were isolated from *Litoria splendida* [59], two peptides belonging to caerin family, caerin 1.6 and 1.7, from the skin of *Litoria xanthomera* [60] and caerin 1.8 and 1.9 from *Litoria chloris* [61] were also isolated. Other species from which AMPs were extracted are *Genimaculata, Eucnemis, Citropa, Aurea, Dahlii*, and *Fallax*. Five AMPs were extracted from the skin glands of the tree frog *Litoria genimaculata*, termed as maculatins. Maculatin 1.1 shows greater activity especially against gram-positive organisms [62]. From the granular dorsal glands of Blue Mountains tree frog *Litoria citropa*, 19 peptides termed as citropins were isolated. In the citropin family of peptides, three peptides citropin 1.1–1.3 show strong wide spectrum antibacterial properties, and they are very similar in their primary structure and their C-termini are amidated [63]. Thirteen aurein peptides were isolated from the granular dorsal glands of the Green and Golden Bell Frog *Litoria aurea*. These peptides are named in three groups (aureins 1–3). Aurein 1.2, aurein 2.2, and aurein 3.1 exhibit pronounced antimicrobial activity among the series [64].

Eleven peptides, termed as dahleins, with varying amino acid sequences (13 residues to 21 residues long) were isolated from the skin secretions of *Litoria dahlia*.

However, most peptides are not active against microbes, and only dahleins 1.1 and 1.2 show some wide-spectrum antimicrobial activity [65]. Three peptides, fallaxidins, from the skin secretion of *Litoria fallax*, show weak to moderate antimicrobial activity, though none of them shows broad antimicrobial activity [66].

AMPs aurein 1.2, citropin 1.1, maculatin 1.1, caerin 1.1, kalata B1, and ChexArg20 which are isolated from various *Litoria* species were studied for their membrane action, using nuclear magnetic resonance (NMR) techniques and various biophysical methods [9,67]. These peptides show similarity in their primary and secondary structures. However, their mode of action on bacterial membrane is different.

### 8.2.5 Antimicrobial peptide isolated from amphibians in Europe

There are seven families of anuran species spread over European continent. AMPs were reported from the family *Ranidae*. The skin secretion of European frog *R. esculenta* gave three AMPs and two of them are structurally similar to brevinin-1 and brevinin-2, and the third peptide is termed as esculentin. Esculentin is a very long peptide, containing 46 amino acids and like other two peptides it has an intramolecular disulfide bridge near to C-terminus. All the three peptides show activity against gram-negative and gram-positive microorganisms [68]. Using cDNA cloning method, several structurally related peptides were isolated from the skin secretion of *Rana temporaria*. These peptides, termed as temporins, have a length of 10–13 residues. All these peptides show varying antimicrobial activities [69]. Simmaco and coworkers reported two cyclic 17-residue peptides termed as ranacyclins E and T [70]. Ranacyclin E was isolated from *R. esculenta* frog skin secretions and ranacyclin T was discovered by screening a cDNA library from *R. temporaria*. Ranacyclins have antimicrobial and antifungal activities.

### 8.2.6 Antimicrobial peptide isolated from amphibians in Asia

Many AMPs are isolated from the genus *Rana* (family: *Ranidae*) which constitute a diverse group with an estimated 250 species spread over all continents except Antarctica [71]. Brevinin-1 and brevinin-2, two AMPs were isolated from the skin of *Rana porosa brevipoda*. Brevinin-2 is active against gram-negative and gram-positive bacteria while brevinin-1 has weak activity against *Escherichia coli*. Both these peptides have a disulfide bridge at the C-termini [72]. From the skin of frog *Rana rugosa*, three AMPs termed as rugosins A, B, and C were isolated which showed structural homology (45%) with brevinin-2, and have intramolecular disulfide bridge at the C-terminus as seen in brevinin peptides. Rugosins A and C show antimicrobial activity against gram-positive while rugosin B shows activity against gram-negative and gram-positive bacteria [73]. Interestingly, from the *Rana* species that is present in North American and Eurasian regions, *Rana okinavana*, a family of acyclic brevinin-1 peptides were isolated. This 24-mer peptides lack the disulfide bridge at the C-terminus as seen in the case of brevinin peptides, because of the absence of Cys at 18 and 24 [74]. Six AMPs in the brevinin and temporin peptide families were also isolated from the skin of the Tsushima brown frog *Rana tsushimensis*. Brevinin-1, which lacks the disulfide bridge at the C-terminus, is structurally similar to peptide that was isolated from *R. okinavana*. Brevinin-2, isolated from *R. tsushimensis*, showed broad spectrum activity against a range of gram-negative and gram-positive bacteria [75].

Tigerinins 1−4 are four AMPs isolated from Indian frog *Rana tigerina*. These are very small (tigerin-1 and tigerin-4 are 11 residues long and tigerin-2 and tigerin-3 are 12 residues long), nonhelical, cationic AMPs. All of these peptides show an intramolecular disulfide bridge at the C-terminus [76]. Two AMPs termed as japonicin-1 and japonicin-2, showing differential growth-inhibitory activity against the gram-negative bacterium, *E. coli* and the gram-positive bacterium *S. aureus*, were isolated from an extract of the skin of the Japanese brown frog *Rana japonica* [77]. Like other Rana peptides, these too contain C-terminus disulfide bridge. Four structurally similar peptides belonging to the family japonicin-2 was isolated from the skin of *Rana chaochiaoensis* [78]. Six AMPs, dybowskins, were isolated from the skin secretions of *Rana dybowskii*. All these peptides have the C-terminal disulfide bridge which is conserved in other AMPs derived from the genus *Rana*. Dybowskin peptides show broad spectrum antimicrobial activity against gram-positive and gram-negative bacteria [79]. From the skin secretions of *Rana pleuraden*, four AMPs designated as pleurain A were isolated. These peptides contain 26 amino acids. Pleurains also contain a disulfide bridge at the C-terminus region [80]. Lai and coworkers reported 34 AMPs which belong to 9 families, rugosin, gaegurin, temporin, and six niagroains (niagroain B, niagroain C, niagroain D, niagroain E, niagroain I, and niagroain K) from the skin secretions of *Rana nigrovittata* [81]. Two other peptides similar to brevinin-2 family (brevinin-2 RN1 and RN2) were also isolated from this frog. Unlike brevinin-2 which is a 33 amino acid-containing peptide, brevinin-2RN contains only 30 amino acids [82]. From the skin secretions of *Rana shuchinae*, five AMPs termed as shuchin (shuchin 1−5) were isolated. The peptides show broad spectrum antibacterial activities against gram-positive, gram-negative bacteria and yeast [83,84].

The first anionic AMP was isolated from the tree frog *Polypedates puerensis* (family: *Rhacophoridae*), termed as PopuDef. PopuDef showed moderate antimicrobial activities against *Pseudomonas aeruginosa* and *S. aureus* and relatively weak activities against *E. coli* and *Bacillus subtilis* [85]. Five AMPs belonging to temporin, rugosin, and gaegurin families were isolated from the skin secretions of *Limnonectes kuhlii* (family: *Dicroglossidae*) [86]. Most of them show broad spectrum antimicrobial activity and antifungal activity.

### 8.2.6.1 Western Ghats: the treasure house for antimicrobial peptides

The Western Ghats (Green shade in Fig. 8.1), also known as the *Sahyadri Hills*, is a chain of hills of varied width and height running parallel to the western coast of India. It extends from the Satpura Range in the north, to south past Goa, through Karnataka into Kerala and Tamil Nadu and ends at Kanyakumari. Owing to the

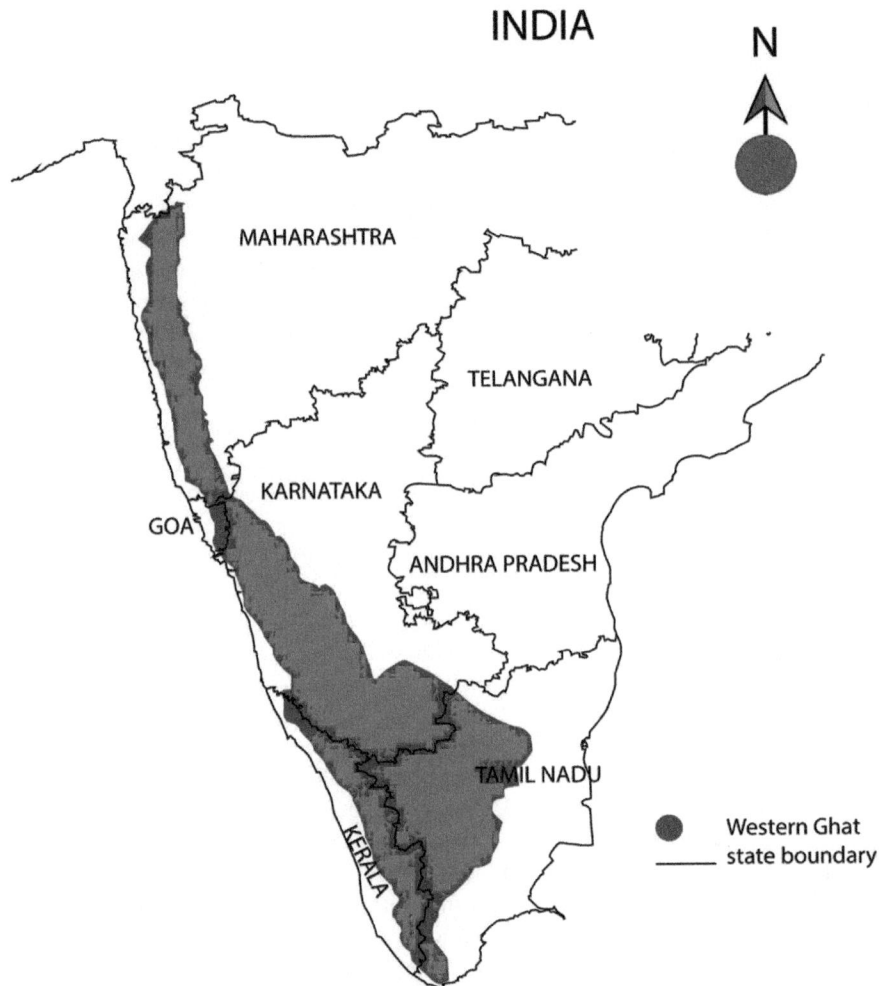

**Figure 8.1** Map of southern part of India showing Western Ghats region spread across various states.

biological diversity, this mountain range is considered as 1 among the 36 biodiversity hot spots identified throughout the world and has the highest proportion of endemic species, especially the lower vertebrates such as amphibians and reptiles [87–89]. It is rich in flora consisting of 7402 flowering plants, out of which 5588 are endemic to this region. It has a wide range of faunal diversity—nearly 508 avian species, 288 fish species, 203 reptilian species, 139 mammal species, and 181 amphibian species [90].

In addition to the species richness, Western Ghats offers an array of bioactive substance, and among these bioactive substances, AMPs got significant attention. The amphibians of the Western Ghats are diverse and unique, with more than 80%

endemic to the region. Most of the endemic species have their distribution in the rainforests of these mountains (https://news.mongabay.com/2013/09/photo-essay-indias-western-ghats-is-a-haven-for-endemic-amphibians). Though extensive studies were carried out on skin secretion of frogs belonging to the temperate region to identify biologically active peptides, limited studies were carried out to identify skin-secreted peptides of the habitants of tropical climate, especially those confined to the Western Ghats. In the Western Ghats, there are 121 species of frogs from 23 genera which fall under 7 families. The frogs of the genus *Rana* (family: *Ranidae*) are extremely diverse group and are widely distributed throughout the world and have the ability to synthesize wide variety of peptides with potent antimicrobial activity [91−93]. Except in polar regions, southern parts of South America, and in major part of Australia, they are distributed worldwide and reportedly comprise of 379 species [94].

Many genera that are initially included in *Rana* are now placed as separate genus. The *Hylarana*, that was categorized under the subgenus of the genus *Rana*, is now recognized as a distinct genus [95]. The bronzed frog (*Hylarana temporalis* also called as *Rana temporalis*, *Sylvirana temporalis*, or golden-backed frog) is a unique species of true frog usually found in the riparian evergreen forests of the Western Ghats and the highlands of southwestern Sri Lanka [96]. Santhosh and coworkers isolated AMPs from the skin secretion of various species of *Ranidae* family of frogs. Fifteen AMPs were isolated from *H. temporalis*, found in the Western Ghats. The peptides belong to esculentin-2 family (four peptides: E2HLte1−E2HLte4), esculentin-1 family (three peptides: B1HLte1−B1HLte3), brevinin-2 family (five peptides: B2HLte1−B2HLte5), and ranacyclin family (three peptides: RHLte1−RHLte3) [97]. From *H. temporalis*, three peptides under brevinin family [S(1TEa, 2Tea, 2TEb)] also were isolated *and* all the peptides showed higher antimicrobial activity against gram-negative than against gram-positive bacteria [98]. Besides these peptides, bradykinin-related peptides were also isolated from the secretion of *H. temporalis* [99]. Santhosh and coworkers identified and categorized five novel brevinin family cationic antibacterial peptides by peptidomic approach that were named brevinin-1 family (B1CTcu1−B1CTcu5), that all showed effectiveness against various gram-positive and gram-negative bacteria. They were isolated from skin secretion of *Clinotarsus curtipes* of the Western Ghats, a member of the *Ranidae* family of frogs which had a toad-like disposition [100]. Two novel peptides of brevinin-1 family, HYba1 and HYba2, were isolated from the skin secretion of *Hydrophylax bahuvistara* (family: *Ranidae*), an endemic frog species of the Western Ghats, India [101].

Apart from the antimicrobial activity against gram-negative and gram-positive bacteria, these peptides were shown to have low hemolytic activity. Lectin-like peptides were first reported from *H. bahuvistara*. This is the first report regarding the identification of lectin-like peptides from an Indian frog. The peptide did not show antimicrobial activity against tested gram-negative and gram-positive bacteria, whereas the specificity of the peptide can be exploited for drug targeting. The synergistic effect of HYba1 and HYba2 with traditional antibiotics provides the advantage of combinatorial therapy, which can be an effective strategy to combat antibiotic-resistant pathogens. The peptide named, Urumin, isolated from *H. bahuvistara* showed activity against drug-resistant H1 influenza strains [102]. Two AMPs, B1 and B2 belonging to brevinin family of AMPs, were isolated from the skin secretion of *Indosylvirana aurantiaca* (family: *Ranidae*), an endemic frog of Western Ghats. The amidated form of both B1 and B2 showed broad spectrum activity against gram-positive and gram-negative bacteria [8]. A recent study showed that novel peptide (belonging to brevinin family) isolated from *I. aurantiaca* showed antiviral activity against Asian, African, and South American Zika virus strains and four serotypes of dengue virus [103].

## 8.3 Conclusion

Amphibian skin secretion has been known as a rich source of pharmacologically active peptides for a long time. The nature and antimicrobial activity of peptides vary from species to species across the family. Amphibians form a connecting link between land-living and water-living species and so are forced to adapt and survive in a variety of conditions laden with pathogenic microorganisms. The peptides isolated from amphibians and their modified versions show powerful activity against pathogens and these can eventually replace antibiotics.

## References

[1] R.E. Hancock, D.S. Chapple, Peptide antibiotics, Antimicrobial Agents and Chemotherapy 43 (1999) 1317–1323.
[2] R.C. Stebbins, N.W. Cohen (Eds.), A Natural History of Amphibians, Princeton University Press, Princeton, NJ, 1995.
[3] M. Osusky, G. Zhou, L. Osuska, R.E. Hancock, W.K. Kay, S. Misra, Transgenic plants expressing cationic peptide chimeras exhibit broad-spectrum resistance to phytopathogen, Nature Biotechnology 18 (2000) 1162–1166.
[4] M. Zasloff, Antimicrobial peptides of multicellular organism, Nature 415 (2002) 389–395.
[5] M. Zaiou, Multifunctional antimicrobial peptides: therapeutic targets in several human diseases, Journal of Molecular Medicine 85 (2007) 317–329.
[6] L.M. Yin, M.A. Edwards, J. Li, C.M. Yip, C.M. Deber, Roles of hydrophobicity and charge distribution of cationic antimicrobial peptides in peptide-membrane interaction, Journal of Biological Chemistry 287 (2012) 7738–7745.
[7] M. Amiche, C. Galanth, Dermaseptins as models for the elucidation of membrane-acting helical amphipathic antimicrobial peptides, Current Pharmaceutical Biotechnology 12 (2011) 1184–1193.
[8] P. Thomas, T.V. Vineeth Kumar, V. Reshmy, K.S. Kumar, S. George, A mini review on the antimicrobial peptides isolated from the genus *Hylarana* (Amphibia: Anura) with a proposed nomenclature for amphibian skin peptides, Molecular Biology Reports 39 (2012) 6943–6947.

[9] D.I. Fernandez, J.D. Gehman, F. Separovic, Membrane interactions of antimicrobial peptides from Australian frogs, Biochimica et Biophysica Acta (BBA)-Biomembranes 1788 (8) (2009) 1630–1638.
[10] P. Nicolas, C. El Amri, The dermaseptin superfamily: a gene-based combinatorial library of antimicrobial peptides, Biochimica et Biophysica Acta 1788 (8) (2008) 1537–1550.
[11] C.H. Chen, T.K. Lu, Development and challenges of antimicrobial peptides for therapeutic applications, Antibiotics 9 (1) (2020) 1–20.
[12] G. Bertaccini, G.D. Caro, The effect of physalaemin and related polypeptides on salivary secretion, Journal of Physiology 181 (1965) 68–81.
[13] A. Anastasi, V. Erspamer, J.M. Cei, Isolation and amino acid sequence of Physalaemin, the main active polypeptide of the skin of Physalaemus fuscumacula, Archives of Biochemistry and Biophysics 108 (1964) 341–348.
[14] L.A. Rollins-Smith, L.A.K. Reinert, C.J. O'leary, L.E. Houston, D.C. Woodhams, Antimicrobial peptide defenses in amphibian skin, Integrative and Comparative Biology 45 (2005) 137–142.
[15] B.J. Benson, M.E. Hadley, In vitro characterization of adrenergic receptors controlling skin gland secretion in two anurans *Rana pipiens* and *Xenopus laevis*, Comparative Biochemistry and Physiology 30 (1969) 857–864.
[16] A. Siano, M.V. Húmpola, E. de Oliveira, F. Albericio, A.C. Simonetta, R. Lajmanovich, et al., Antimicrobial peptides from skin secretions of *Hypsiboas pulchellus* (Anura: *Hylidae*), Journal of Natural Products 77 (4) (2014) 831–841.
[17] M. Zhou, T. Chen, B. Walker, C. Shaw, Lividins: novel antimicrobial peptide homologs from the skin secretion of the Chinese Large Odorous frog, Rana (Odorrana) livida: identification by "shotgun" cDNA cloning and sequence analysis, Peptides 27 (9) (2006) 2118–2123.
[18] J.M. Conlon, M. Mechkarska, Host-defense peptides with therapeutic potential from skin secretions of frogs from the family pipidae, Pharmaceuticals 7 (2014) 58–77.
[19] Y. Lu, Y. Ma, X. Wang, J. Liang, C. Zhang, K. Zhang, et al., The first antimicrobial peptide from sea amphibian, Molecular Immunology 45 (2008) 678–681.
[20] E. Königa, O.R.P. Bininda-Emonds, Evidence for convergent evolution in the antimicrobial peptide system in anuran amphibians, Peptides 32 (2011) 20–25.
[21] Q. Che, Y. Zhou, H. Yang, J. Li, X. Xu, R. Lai, A novel antimicrobial peptide from amphibian skin secretions of *Odorrana grahami*, Peptides 29 (2008) 529–535.
[22] M. Simmaco, G. Mignogna, D. Barra, Antimicrobial peptides from amphibian skin: what do they tell us? Biopolymers 47 (1998) 435–450.
[23] D. Barra, M. Simmaco, Amphibian skin: a promising resource for antimicrobial peptides, Trends in Biotechnology 13 (1995) 205–209.
[24] Z. Oren, Y. Shai, Mode of action of linear amphipathic alpha-helical antimicrobial peptides, Biopolymers 47 (1998) 451–463.
[25] F. Guilhelmelli, N. Vilela, P. Albuquerque, L.S. Derengowski, I. Silva-Pereira, C.M. Kyaw, Antibiotic development challenges: the various mechanisms of action of antimicrobial peptides and of bacterial resistance, Frontiers in Microbiology 4 (2013) 1–12.
[26] L.S. Ford, D.C. Cannatella, The major clades of frogs, Herpetological Monographs 7 (1993) 94–117.
[27] M. Zasloff, Magainins, a class of antimicrobial peptides from Xenopus skin: isolation, characterization of two active forms, and partial cDNA sequence of a precursor, Proceedings of National Academy of Sciences 84 (1987) 5449–5453.

[28] E. Soravia, G. Martini, M. Zasloff, Antimicrobial properties of peptides from Xenopus granular gland secretions, FEBS Letters 228 (1988) 337−340.

[29] M. Mechkarska, M. Prajeep, L. Coquet, J. Leprince, T. Jouenne, H. Vaudry, et al., The hymenochirins: a family of host-defense peptides from the Congo dwarf clawed frog *Hymenochirus boettgeri* (Pipidae), Peptides 35 (2012) 269−275.

[30] B. Mattute, F.C. Knoop, J.M. Conlon, Kassinatuerin-1: a peptide with broad-spectrum antimicrobial activity isolated from the skin of the hyperoliid frog, *Kassina senegalensis*, Biochemistry and Biophysical Research Communications 268 (2000) 433−436.

[31] J.M. Conlon, J. Kolodziejek, N. Nowotny, Antimicrobial peptides from the skins of North American frogs, Biochimica et Biophysica Acta (BBA)-Biomembranes 1788 (2009) 1556−1563.

[32] D.P. Clark, S. Durell, W.L. Maloy, M. Zasloff, Ranalexin. A novel antimicrobial peptide from bullfrog (*Rana catesbeiana*) skin, structurally related to the bacterial antibiotic, polymyxin, Journal of Biological Chemistry 269 (1994) 10849−10855.

[33] J. Goraya, F.C. Knoop, J.M. Conlon, Ranatuerins: antimicrobial peptides isolated from the skin of the American bullfrog, *Rana catesbeiana*, Biochemistry and Biophysical Research Communications 250 (1998) 589−592.

[34] Y.J. Basir, F.C. Knoop, J. Dulka, J.M. Conlon, Multiple antimicrobial peptides and peptides related to bradykinin and neuromedin N isolated from skin secretions of the pickerel frog, *Rana palustris*, Biochimica et Biophysica Acta (BBA) 1543 (2000) 95−105.

[35] J.M. Conlon, A. Sonnevend, C. Davidson, D.D. Smith, P.F. Nielsen, The ascaphins: a family of antimicrobial peptides from the skin secretions of the most primitive extant frog, *Ascaphus truei*, Biochemistry and Biophysical Research Communications 320 (2004) 170−175.

[36] T. Halverson, Y.J. Basir, F.C. Knoop, J.M. Conlon, Purification and characterization of antimicrobial peptides from the skin of the North American green frog *Rana clamitans*, Peptides 21 (4) (2000) 469−476.

[37] T.S. Vasconcelos, F.R. Silva, T.G. Santos, V.H.M. Prado, D.B. Provete, Biogeographic Patterns of South American Anuran, Springer Nature Switzerland AG 2019.

[38] A. Mor, V.H. Nguyen, A. Delfour, D. Migliore-Samour, P. Nicolas, Isolation, amino acid sequence, and synthesis of dermaseptin, a novel antimicrobial peptide of amphibian skin, Biochemistry 30 (36) (1991) 8824−8830.

[39] A. Mor, P. Nicolas, Isolation and structure of novel defensive peptides from frog skin, European Journal of Biochemistry 219 (1994) 145−154.

[40] C. Wechselberger, Cloning of cDNAs encoding new peptides of the dermaseptin-family, Biochimica et Biophysica Acta 1388 (1) (1998) 279−283.

[41] T.N. Pierre, A.A. Seon, M. Amiche, P. Nicolas, Phylloxin, a novel peptide antibiotic of the dermaseptin family of antimicrobial/opioid peptide precursors, European Journal of Biochemistry 267 (2) (2000) 370−378.

[42] C.V. Batista, A. Scaloni, D.J. Rigden, L.R. Silva, A.R. Romero, R. Dukor, et al., A novel heterodimeric antimicrobial peptide from the tree-frog *Phyllomedusa distincta*, FEBS Letters 494 (2001) 85−89.

[43] D. Raimondo, G. Andreotti, N. Saint, P. Amodeo, G. Renzone, M. Sanseverino, et al., A folding-dependent mechanism of antimicrobial peptide resistance degradation unveiled by solution structure of distinctin, Proceedings of National Academy of Sciences USA 102 (2005) 6309−6314.

[44] L. Olson 3rd, A.M. Soto, F.C. Knoop, J.M. Conlon, Pseudin-2: an antimicrobial peptide with low haemolytic activity from the skin of the paradoxical frog, Biochemistry and Biophysical Research Communications 288 (2001) 1001−1005.

[45] A.C.C. Nascimento, L.C. Zanotta, C.M. Kyaw, E.N.F. Schwartz, C.A. Schwartz, A. Sebben, et al., Ocellatins: new antimicrobial peptides from the skin secretion of the South American frog *Leptodactylus ocellatus* (Anura: Leptodactylidae), The Protein Journal 23 (2004) 501–508.

[46] A.C.C. Nascimentoa, A. Chapeaurougec, J. Peralesc, A. Sebbenb, M.V. Sousaa, W. Fontesa, et al., Purification, characterization and homology analysis of ocellatin 4, a cytolytic peptide from the skin secretion of the frog *Leptodactylus ocellatus*, Toxicon 50 (2007) 1095–1104.

[47] L.A. Rollins-Smith, J.D. King, P.F. Nielsen, A. Sonnevend, J.M. Conlon, An antimicrobial peptide from the skin secretions of the mountain chicken frog *Leptodactylus fallax* (Anura: Leptodactylidae), Regulatory Peptides 124 (2005) 173–178.

[48] J.D. King, J. Leprince, H. Vaudry, L. Coquet, T. Jouenne, J.M. Conlon, Purification and characterization of antimicrobial peptides from the Caribbean frog, *Leptodactylus validus* (Anura: Leptodactylidae), Peptides 29 (2008) 1287–1292.

[49] M.M. Marani, F.S. Dourado, P.V. Quelemes, A.R. de Araujo, M.L.G. Perfeito, E.A. Barbosa, et al., Characterization and biological activities of ocellatin peptides from the skin secretion of the frog *Leptodactylus pustulatus*, Journal of Natural Products 78 (7) (2015) 1495–1504.

[50] J.R.S.A. Leite, L.P. Silva, M.I.S. Rodrigues, M.V. Prates, G.D. Brand, B.M. Lacava, et al., Phylloseptins: a novel class of anti-bacterial and anti-protozoan peptides from the *Phyllomedusa* genus, Peptides 26 (4) (2005) 565–573.

[51] Y. Wan, C. Ma, M. Zhou, X. Xi, L. Li, D. Wu, et al., Phylloseptin-PBa—a novel broad-spectrum antimicrobial peptide from the skin secretion of the peruvian purple-sided leaf frog (*Phyllomedusa baltea*) which exhibits cancer cell cytotoxicity, Toxins 7 (2015) 5182–5193.

[52] Y. Liu, D. Shi, J. Wang, X. Chen, M. Zhou, X. Xi, et al., A novel amphibian antimicrobial peptide, phylloseptin-PV1, exhibits effective anti-staphylococcal activity without inducing either hepatic or renal toxicity in mice, Frontiers in Microbiology 11 (2020) 1–14.

[53] A.H. Thompson, A.J. Bjourson, D.F. Orr, C. Shaw, S. McClean, Amphibian skin secretomics: application of parallel quadrupole time-of-flight mass spectrometry and peptide precursor cDNA cloning to rapidly characterize the skin secretory peptidome of *Phyllomedusa hypochondrialis azurea*: discovery of a novel peptide family, the hyposins, Journal of Proteome Research 6 (2007) 3604–3613.

[54] M.S. Castro, R.H. Matsushita, A. Sebben, M.V. Sousa, W. Fontes, Hylins: bombinins H structurally related peptides from the skin secretion of the Brazilian tree-frog *Hyla biobeba*, Protein and Peptide Letters 12 (1) (2005) 89–93.

[55] M.S. Castro, T.C.G. Ferreira, E.M. Cilli, E. Crusca Jr, M.J.S. Mendes-Giannini, A. Sebben, et al., Hylina1, the first cytolytic peptide isolated from the arboreal South American frog *Hypsiboas albopunctatus* ("spotted treefrog"), Peptides 30 (2) (2009) 291–296.

[56] J.M. Conlon, M. Mechkarska, G. Radosavljevic, S. Attoub, J.D. King, M.L. Lukic, et al., A family of antimicrobial and immunomodulatory peptides related to the frenatins from skin secretions of the Orinoco lime frog *Sphaenorhynchus lacteus* (Hylidae), Peptides 56 (2014) 132–140.

[57] C. Muñoz-Camargo, M.C. Méndez, V. Salazar, J. Moscoso, D. Narváez, M.M. Torres, et al., Frog skin cultures secrete anti-yellow fever compounds, Journal of Antibiotics 69 (2016) 783–790.

[58] C. Muñoz-Camargo, V.A. Salazar, L. Barrero-Guevara, S. Camargo, A. Mosquera, H. Groot, et al., Unveiling the multifaceted mechanisms of antibacterial activity of buforin ii and frenatin 2.3s peptides from skin micro-organs of the orinoco lime tree-frog (*Sphaenorhynchus lacteus*), International Journal of Molecular Sciences 19 (8) (2018) 2170.
[59] D.J.M. Stone, R.J. Waugh, J.H. Bowie, J.C. Wallace, M.J. Tyler, Peptides from Australian frogs. Structures of the caerins and caeridin-1 from *Litoria splendida*, Journal of the Chemical Society, Perkin Transactions I 23 (1992) 3173–3178.
[60] S.T. Steinborner, R.J. Waugh, J.H. Bowie, J.C. Wallace, M.J. Tyler, S.L. Ramsay, New caerin antimicrobial peptides from the skin glands of the Australian tree frog *Litoria xanthomera*, Journal of Peptide Sciences 3 (1997) 181–185.
[61] S.T. Steinborner, G.J. Currie, J.H. Bowie, J.C. Wallace, M.J. Tyler, New antibiotic caerin 1 peptids from the skin secretion of the Australian tree frog *Litoria chloris*. Comparison of the activities of caerin 1 peptides from the genus *Litoria*, Journal of Peptide Research 51 (1998) 121–126.
[62] T. Rozek, R.J. Waugh, S.T. Steinborner, J.H. Bowie, M.J. Tyler, J.C. Wallace, The maculatin peptides from the skin glands of the tree frog *Litoria genimaculata*: a comparison of the structures and antimicrobial activities of maculatin 1.1 and caerin 1.1, Journal of Peptide Sciences 4 (1998) 111–115.
[63] K.L. Wegener, P.A. Wabnitz, J.A. Carver, J.H. Bowie, B.C. Chia, J.C. Wallace, et al., Host defence peptides from the skin glands of the Australian blue mountains tree-frog *Litoria citropa*. Solution structure of the antibacterial peptide citropin 1.1, European Journal of Biochemistry 265 (1999) 627–637.
[64] T. Rozek, K.L. Wegener, J.H. Bowie, I.N. Olver, J.A. Carver, J.C. Wallace, et al., The antibiotic and anticancer active aurein peptides from the Australian Bell rogs *Litoria aurea* and *Litoria raniformis* the solution structure of aurein 1.2, European Journal of Biochemistry 267 (2000) 5330–5341.
[65] K.L. Wegener, C.S. Brinkworth, J.H. Bowie, J.C. Wallace, M.J. Tyler, Bioactive dahlein peptides from the skin secretions of the Australian aquatic frog *Litoria dahlii*: sequence determination by electrospray mass spectrometry, Rapid Communications in Mass Spectrometry 15 (2001) 1726–1734.
[66] R.J. Jackway, J.H. Bowie, D. Bilusich, I.F. Musgrave, K.H. Surinya-Johnson, M.J. Tyler, et al., The fallaxidin peptides from the skin secretion of the Eastern Dwarf Tree Frog *Litoria fallax*. Sequence determination by positive and negative ion electrospray mass spectrometry: antimicrobial activity and cDNA cloning of the fallaxidins, Rapid Communications in Mass Spectrometry 22 (2008) 3207–3216.
[67] S. Zhu, M.-A. Sani, F. Separovic, Interaction of cationic antimicrobial peptides from Australian frogs with lipid membranes, Peptide Science 110 (3) (2018) 1–10.
[68] M. Simmaco, G. Mignogna, D. Barra, F. Bossa, Novel antimicrobial peptides from skin secretion of the European frog *Rana esculenta*, FEBS Letters 324 (2) (1993) 159–161.
[69] M. Simmaco, G. Mignogna, S. Canofeni, R. Miele, M.L. Mangoni, D. Barra, Temporins, antimicrobial peptides from the European red frog *Rana temporaria*, European Journal of Biochemistry 242 (1996) 788–792.
[70] M.L. Mangoni, N. Papo, G. Mignogna, D. Andreu, Y. Shai, D. Barra, et al., Ranacyclins, a new family of short cyclic antimicrobial peptides: biological function, mode of action, and parameters involved in target specificity, Biochemistry 42 (2003) 14023–14035.
[71] W.E. Duellman, L. Trueb, Biology of Amphibians, The Johns Hopkins University Press, Baltimore and London, 1994.

[72] N. Morikawa, K. Hagiwara, T. Nakajima, Brevinin-1 and -2, unique antimicrobial peptides from the skin of the frog *Rana brevipoda porsa*, Biochemistry and Biophysical Research Communications 189 (1992) 184—190.
[73] S. Suzuki, Y. Ohe, T. Okubo, T. Kakegawa, K. Tatemoto, Isolation and characterization of novel antimicrobial peptides, rugosins A, B and C, from the skin of the frog, *Rana rugosa*, Biochemistry and Biophysical Research Communications 189212 (1995) 249—254.
[74] J.M. Conlon, A. Sonnevend, T. Jouenne, L. Coquet, D. Cosquer, H. Vaudry, et al., A family of acyclic brevinin-1 peptides from the skin of the Ryukyu brown frog *Rana okinavana*, Peptides 26 (2005) 185—190.
[75] J.M. Conlon, N. Al-Ghaferi, B. Abraham, A. Sonnevend, L. Coquet, J. Leprince, et al., Antimicrobial peptides from the skin of the Tsushima brown frog *Rana tsushimensis*, Comparative Biochemistry and Physiology Part C: Toxicology & Pharmacology 143 (2006) 42—49.
[76] K.P. Sai, M.V. Jagannadham, M. Vairamani, N.P. Raju, A.S. Devi, R. Nagaraj, et al., Tigerinins: novel antimicrobial peptides from the Indian frog *Rana tigerina*, Journal of Biological Chemistry 276 (2001) 2701—2707.
[77] T. Isaacson, A. Soto, S. Iwamuro, F.C. Knoop, J.M. Conlon, Antimicrobial peptides with atypical structural features from the skin of the Japanese brown frog *Rana japonica*, Peptides 23 (2002) 419—425.
[78] J.M. Conlon, J. Leprince, H. Vaudry, H. Jiansheng, P.F. Nielsen, A family of antimicrobial peptides related to japonicin-2 isolated from the skin of the chaochiao brown frog *Rana chaochiaoensis*, Comparative Biochemistry and Physiology Part C: Toxicology & Pharmacology 144 (2006) 101—105.
[79] S.S. Kim, M.S. Shim, J. Chung, D.-Y. Lim, B.J. Lee, Purification and characterization of antimicrobial peptides from the skin secretion of *Rana dybowskii*, Peptides 28 (2007) 1532—1539.
[80] X. Wang, Y. Song, J. Li, H. Liu, X. Xu, R. Lai, et al., A new family of antimicrobial peptides from skin secretions of *Rana pleuraden*, Peptides 28 (2007) 2069—2074.
[81] Y. Ma, C. Liu, X. Liu, J. Wu, H. Yang, Y. Wang, et al., Peptidomics and genomics analysis of novel antimicrobial peptides from the frog, *Rana nigrovittata*, Genomics 95 (2010) 66—71.
[82] X. Liu, R. Liu, L. Wei, H. Yang, K. Zhang, J. Liu, et al., Two novel antimicrobial peptides from skin secretions of the frog, *Rana nigrovittata*, Journal of Peptide Sciences 17 (1) (2011) 68—72.
[83] R. Zheng, B. Yao, H. Yu, H. Wang, J. Bian, F. Feng, Novel family of antimicrobial peptides from the skin of *Rana shuchinae*, Peptides 31 (2010) 1674—1677.
[84] J. Pei, G. Zhao, B. Wang, H. Wang, Three novel antimicrobial peptides from the skin of *Rana shuchina*, Gene 521 (2013) 234—237.
[85] L. Wei, H. Che, Y. Han, J. Lv, L. Mu, L. Lv, et al., The first anionic defensin from amphibians, Amino Acids 47 (2015) 1301—1308.
[86] G. Wang, Y. Wang, D. Ma, H. Liu, J. Li, K. Zhang, et al., Molecular Biology Reports 40 (2013) 1097—1102.
[87] N. Myers, R.A. Mittermeier, C.G. Mittermeier, G.A.B. da Fonseca, J. Kent, Biodiversity hotspots for conservation priorities, Nature 403 (2000) 853—858.
[88] R.J.R. Daniels, Endemic fishes of the Western Ghats and the Satpura hypothesis, Current Science 81 (3) (2001) 240—244.
[89] N. Dahanukar, R. Raut, A. Bhat, Distribution, endemism and threat status of freshwater fishes in the Western Ghats of India, Journal of Biogeography 31 (2004) 123—136.

[90] WorldAtlas. <https://www.worldatlas.com/articles/western-ghats-biodiversity-hotspot.html>.

[91] Y.J. Basir, F.C. Knoop, J. Dulka, J.M. Conlon, Multiple antimicrobial peptides and peptides related to bradykinin and neuromedin N isolated from skin secretions of the pickerel frog, *Rana palustris*, Biochimica et Biophysica Acta (BBA)-Protein Structure and Molecular Enzymology 1543 (2000) 95−105.

[92] C. Wang, L. Qian, C. Zhang, W. Guo, T. Pan, J. Wu, et al., A new species of *Rana* from the Dabie Mountains in eastern China (Anura, Ranidae), Zookeys 724 (2017) 135−153.

[93] Z. Huang, C. Yang, D. Ke, DNA barcoding and molecular phylogeny in Ranidae, Mitochondrial DNA 27 (2016) 4003−4007. Available from: https://doi.org/10.3109/19401736.2014.989522.

[94] D. Frost, Amphibian Species of the World: An Online Reference (Version 6), American Museum of Natural History, NY, 2016. Available from: http://research.amnh.org/vz/herpetology/amphibia123/index.php//Amphibia/Anura.

[95] D. Gawor, R. Hendrix, M. Vences, W. Böhme, T. Ziegler, Larval morphology in four species of Hylarana from Vietnam and Thailand with comments on the taxonomy of *H. nigrovittata sensu latu* (Anura: Ranidae), Zootaxa 2051 (2009) 1−25.

[96] S.D. Biju, S. Garg, S. Mahony, N. Wijayathilaka, G. Senevirathne, M. Meegaskumbura, DNA barcoding, phylogeny and systematics of Golden-backed frogs (Hylarana, Ranidae) of the Western Ghats-Sri Lanka biodiversity hotspot, with the description of seven new species, Contributions to Zoology 83 (2014) 269−335.

[97] V. Reshmy, K.S. Kumar, S. Ganga, Full length cDNA desired novel peptides belonging to Esculentin family from skin of Indian Bronzed Frog *Hylarana temporalis*, Research Journal of Biochemistry 6 (2011) 71−74.

[98] V. Reshmy, V. Preeji, A. Parvin, K.S. Kumar, S. George, Three novel antimicrobial peptides from the skin of the Indian bronzed frog *Hylarana temporalis* (Anura: Ranidae), Journal of Peptide Science 17 (2011) 342−347.

[99] V. Reshmy, V. Preeji, A. Parvin, K.S. Kumar, S. George, Molecular cloning of a novel bradykinin-related peptide from the skin of Indian bronzed frog *Hylarana temporalis*, Genomic Insights 3 (2010) 23−28.

[100] P. Abraham, S. George, K.S. Kumar, Novel antibacterial peptides from the skin secretion of the Indian bicoloured frog *Clinotarsus curtipes*, Biochimie 97 (2014) 144−151.

[101] T.V.V. Kumar, R. Asha, G. Shyla, S. George, Identification and characterization of novel host defense peptides from the skin secretion of the fungoid frog, *Hydrophylax bahuvistara* (Anura: Ranidae), Chemical Biology & Drug Design 92 (2018) 1409−1418. Available from: https://doi.org/10.1111/cbdd.12937.

[102] D.J. Holthausen, S.H. Lee, V.T.V. Kumar, N.M. Bouvier, F. Krammer, A.H. Ellebedy, et al., An amphibian host defense peptide is virucidal for human h1 hemagglutinin-bearing influenza viruses, Immunity 46 (4) (2017) 587−595.

[103] S.H. Lee, E.H. Kim, J.T. O'neal, G. Dale, D.J. Holthausen, J.R. Bowen, et al., The amphibian peptide Yodha is virucidal for Zika and dengue viruses, Scientific Reports 11 (2021) 1−12.

# Plant-derived antimicrobial peptides

**9**

Jane Mary Lafayette Neves Gelinski[1], Bernadette Dora Gombossy de Melo Franco[2] and Gustavo Graciano Fonseca[3]

[1]Laboratory of Protein Chemistry and Biochemistry, Department of Cell Biology, University of Brasilia-Federal District, Brasilia, Federal District, Brazil, [2]Food Research Center — FoRC, Sao Paulo University, USP, Sao Paulo, SP, Brazil, [3]Faculty of Natural Resource Sciences, School of Business and Science, University of Akureyri, Akureyri, Iceland

Antimicrobial peptides (AMPs) are small peptides ($\leq 100$ amino acid residues) with antibacterial, antiviral, antifungal, antiparasitic, and antitumor activities [1,2]. They constitute an important part of the innate nonspecific host defense system of different organisms [3,4], presenting characteristics that vary according to source, activity, structure, and amino acid composition [2]. These compounds derived from enzymatic hydrolysis of plant proteins are biologically active and have great potential for biotechnological applications.

AMPs are categorized into different families mainly based on the amino acid sequence, electrical charge, and number of cysteine residues [5]. There are more than 2500 natural AMPs registered in the antimicrobial peptide database [6]. In general, they are amphipathic peptides, but also cationic, mainly due to arginine (Arg) and lysine (Lys) residues in the structure [1].

The multifunctional properties of AMPs have directed studies aimed at health promotion, including the development of new ingredients for foods and cosmetics [7] or for therapeutic and immunobiological use [8,9], since they may be less immunogenic than recombinant proteins and antibodies [10]. Some obstacles to the use of AMPs in medicine have been reported, related to toxicity, hemolytic activity, loss of activity due to degradation or even bacterial resistance [8,9,11]. However, the broad antimicrobial activity of AMPs makes them superior to synthetic or chemical antibiotics [12]. Among the well-known bioactive peptides, those derived from plants are the most promising for technological application, in view of the existing plant biodiversity and the ease of analytical processes in relation to those obtained from animals. Here, we present a review on plant-derived AMPs, focusing on their origin, mode of action, structure, biochemical determinants, extraction, and identification methods, bioactivity and functionality, technological and therapeutic applications, and future perspectives.

## 9.1 General characteristics of bioactive peptides derived from plants

As reviewed [13], plant AMPs act in the response of plants to stress and infection by pathogens and constitute a potential of great applicability in the agrochemical and pharmaceutical industries [14]. AMPs have been isolated from various parts of plants, such as roots, seeds, leaves, stems, and flowers [12]. Despite the great diversity of plant species, the discovery of new plant AMPs was slow. Additionally, due to the rapid increase in the therapeutic use of antibiotics AMPs were temporarily overlooked [15]. However, the advent of microbial multidrug resistance stimulated the investigation of bioactive compounds in plants, including some secondary metabolites [5], and the peptides from their defense systems [16].

Plant AMPs or plant host defense peptides (HDPs) have long been known for their potent and broad spectrum of antimicrobial activity against bacteria, fungi, viruses, protozoa, and helminths [12,17]. Studies indicate the positive action of plant AMPs in infectious processes by in vivo induction in animal models. In the review on AMPs of plant origin carried out by Tang et al. [12] was verified that, among other studies, the use of raw extract of Indian herbs from *Piper longum* (fruit) has counteracted infection caused by *Entamoeba*.

The mechanisms of action of AMPs are presented in Fig. 9.1. The carpet model, the barrel stave model, and the toroidal pore model explain that AMPs destabilize the microbial cell membrane, forming pores and causing its rupture, resulting in ions and metabolites leakage and eventually cell death [4,5,18]. Plant HDPs are rich in cysteine, which forms multiple disulfide bridges, responsible for the thermal, chemical, and proteolytic stability, acting as a barrier against pests and pathogens [17,19].

## 9.2 Antimicrobial peptides derived from different plant families

### 9.2.1 Cyclotides

Macrocyclic peptides are present in plants from the *Violaceae*, *Rubiaceae*, and *Cucurbitaceae* families, formed by six residues of cystine, and active against pathologies such as cancer, pain, obesity, and cardiovascular disease [20]. Kalata B1 was the first cyclotide identified and it corresponds to the prototypic peptide of the group with around 30 amino acid residues and a unique cyclic cystine knot (CCK) arrangement of three conserved disulfide bonds [21]. The rigid and compact structure provided by the CCK is responsible for the resistance of cyclotides to physical, chemical, and biological degradation, which are properties that confer a great potential for drug development [22].

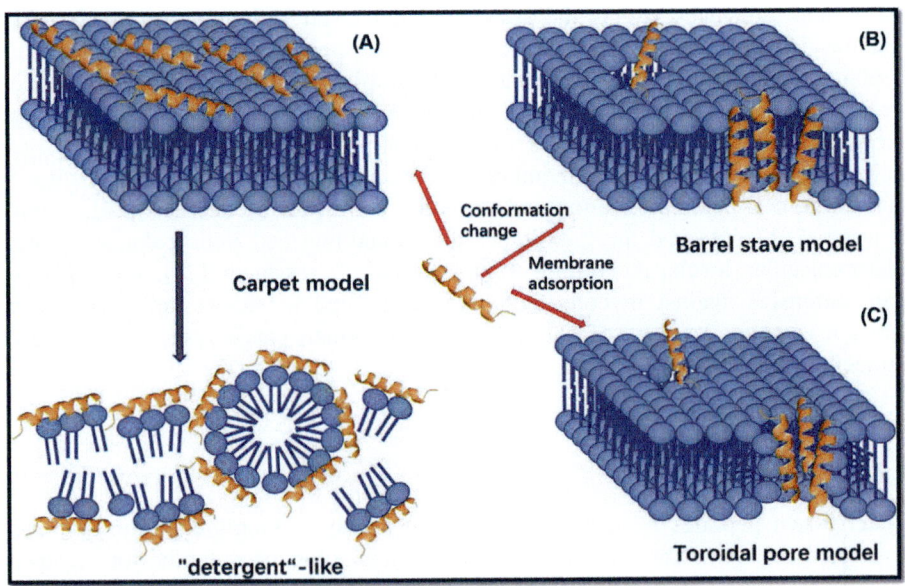

**Figure 9.1** Models of action of antimicrobial activity of antimicrobial peptides (AMPs) on the cell membrane. (A) Carpet model—detergent action; (B) barrel stave model—AMPs paralleling the phospholipid bilayer, forming channels. (C) Toroidal pore model: AMPs vertically embedded in the cell membrane forming a ring-like pore.
*Source*: Adapted and updated with permission from Y. Huan, Q. Kong, H. Mou, H. Yi, Antimicrobial peptides: classification, design, application and research progress in multiple fields, Frontiers in Microbiology 16 (2020) 582779, https://doi.org/10.3389/fmicb.2020.582779.

### 9.2.2 Thionins

Small peptides (5 kDa), with 47 amino acid residues, mainly arginine, lysine, and cysteine, are active against yeasts and most bacteria [23]. Their structure houses two antiparallel helices and a double β-sheet also antiparallel linked by disulfide bridges (S-S), and with a tyrosine residue [4].

### 9.2.3 Defensins

Defensins are small 46 kDa cationic peptides with 45−54 amino acid residues, including cysteine residues. [24,25]. Besides plants, they are produced by humans, inferior vertebrates, and insects [23]. These cationic peptides have multiple and intramolecular disulfide bridges between the conserved cysteine residues which maintain the structure, protecting them from chemical attacks and proteolytic degradation [25]. However, antifungal defensins from plants do not present conservative amino acid sequences, except the cysteine residues and a glycine residue [26]. The cysteine-rich peptides, defensins and defensin-like peptides are the most abundant in plants, for example, in legume-rhizobial symbiosis which can inhibit pathogen growth and control rhizobial differentiation in legume nodules [27]. These are likely to occur by the interaction of defensin peptides with negatively charged molecules in the cytoplasmic membrane of microorganisms (Fig. 9.1), leading to increase in cell permeabilization and depletion and death by necrosis [24]. Studies reported [28] reinforced the importance of maintaining cell ionic balance, even at micromolecular levels. According to the author's assessment [28], the action of plant defensins against mycobacteria may be related to electrostatic interactions between arginine residues of cationic defensins which cause lysis by leakage of intracellular metabolites.

## 9.2.4 Snakins

Cysteine-rich peptides from potatoes, specific to the snaking/gibberellic transcript stimulated acid in Arabidopsis family, with biotechnological potential for therapeutic and agricultural applications [14,29]. There is still no evidence that this characteristic is crucial for the antimicrobial activity of snakins, as they occur in other peptides rich in disulfide bridges. Anyway, Garcia et al. [30] reported that the peptide MsSN1 presented in vitro and in vivo antibacterial and antifungal activities on alfalfa pathogens, observing also that MsSN1-overexpressing alfalfa transgenic plants increased antimicrobial activity against virulent fungi strains. Studies on the potential application of molecules of the snakins family, whether in agriculture or clinic, still lack data regarding the mode of action, and molecular and biological properties, among others [17].

## 9.2.5 Heveins and hevein-like peptides

These AMPs comprise three subfamilies: prototypic 8C and 6C, and 10C-hevein-like peptides, that present conservative chitin-binding cysteine-rich peptides [31,32]. Hevein is found in the latex of the rubber tree *Hevea brasiliensis*, family: Euphorbiaceae. The amino acid sequence contains eight cysteine residues and resembles the nettle's chitin-binding agglutinin (from urtica dioica agglutinin) [32,33]. The in vitro inhibitory activity was early observed in fungi containing chitin [33]. Based on the capability of plant defense proteins system to bind and degrade chitin of fungal cell walls, Slavokhotova et al. [32] observed that recombinant hevein-like AMPs (wheat antimicrobial peptides) inhibited fungalysin in in vitro assays, suppressing hyphal elongation. As fungalysin plays an important role in fungal development, the authors highlighted that heveins and hevein-like peptides have a great potential for the control of plant pathogenic fungi. Table 9.1 shows examples of plant AMPs and the respective plant organ of origin.

Table 9.1 Some plant antimicrobial peptides (AMPs) and the respective plant organ of origin.

| Family/peptide | Structure [protein data bank[a] (PDB)] | Plant species/family | Plant organ | Antimicrobial activity to |
|---|---|---|---|---|
| **Cyclotides**—Kalata B1 and 2, Cycloviolacins I1–6 (cyI1-cyI6) | PDB: 1K48. Kalata B1 peptide [34] | *Violaceae, Rubiaceae, Cucurbitaceae* | Leaves and flowers | Insects, bacteria |
| **Thionins** (Types I–IV) | PDB: 2V9B. Viscotoxin B2 [35] | *Poaceae* | Endosperm of grains, seeds (wheat) | Bacteria, fungi, yeast |
| **Defensins** Peptide PvD1 γ-Hordothionin Peptide NaD1 PhD1, PhD2 ZmESR6 | PDB: 6B55 Flower-specific defensin (protein) [36] | *Triticum turgidum Phaseolus vulgaris Hordeum vulgare Nicotiana alata Petunia hybrida* | Seeds Maize Kernels | Fungi, bacteria, insect |

(*Continued*)

Table 9.1 (Continued)

| Family/peptide | Structure [protein data bank[a] (PDB)] | Plant species/family | Plant organ | Antimicrobial activity to |
|---|---|---|---|---|
| **Snakins** StSN1 and StSN2 | PDB: 5E5Y D-snakin-1 (protein) [37] | *Solanum tuberosum* | Flowers's bulbs and tubers | Plant pathogens (bacteria, fungi) |
| **Hevein**-like AMPs: PMAP1 | PDB: 6LNR Class I chitinase protein [38] | *Broussonetia papyrifera* syn. *Morus papyrifera* L. | Leaves | Plant pathogenic fungi |
| Others: CWE1 peptide extracts Protease inhibitors rich extracts *Lipid transfer proteins (LTPs)* | | *Capsicum annuum* L. *Solanum stramonifolium* Jacq. [39] | Leaves Seeds Seeds | Bacteria, fungi |

[a] According to Refs. [6,17,24,28,39,40]. All 3D view structure from RCSB Protein Data Bank—RSCB/PDB [41].

## 9.3 Extraction and identification of plant antimicrobial peptides

There are many procedures for extraction of AMPs from different plants and plant organs, and for identification and evaluation of their antiviral, antibacterial, anthelmintic, and antifungal activities. For extraction, the conditions for protein hydrolysis are essential, and pH and hydrolysis time are fundamental in obtaining compounds with specific functionalities [42–45]. For characterization, purification, and identification of the peptides, a combination of methods is used, including chromatographic column purification, salt precipitation, ion exchange and C18 reverse phase extraction in solid phase, high-performance liquid chromatography (HPLC, RP-HPLC), and sodium dodecyl sulfate gel electrophoresis—SDS/PAGE method [12,46–48]. The three-dimensional structure and the amino acid sequences of these antimicrobial peptides can be assessed by comparisons with existing databases such as RCSB/PDB-Protein Data Bank[1] [6,41] or others as described [49]. Fig. 9.2 presents a summary of these procedures.

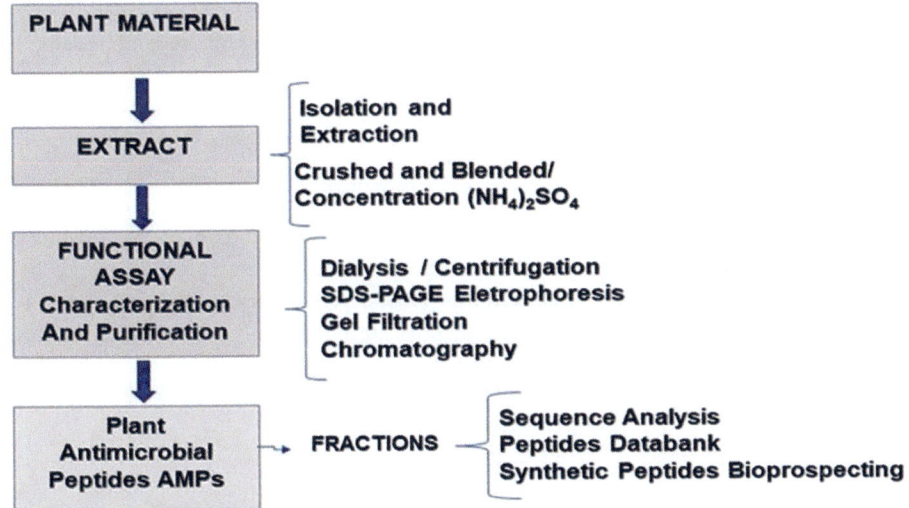

**Figure 9.2** General steps for extraction, purification, sequencing, and identification of antimicrobial peptides (AMPs).
*Source*: Based on S.S. Tang, Z.H. Prodhan, S.K. Biswas, C.F. Le, S.D. Sekaran, Antimicrobial peptides from different plant sources: isolation, characterisation, and purification, Phytochemistry 154 (2018) 94–105, https://doi.org/10.1016/j.phytochem.2018.07.002.

---

[1] Protein Data Bank (https://www.rcsb.org/) includes 3D shapes of proteins, nucleic acids and complex assemblies that help students and researchers understand various aspects of biomedicine and agriculture.

**Table 9.2** Some analytical methods used to assess the antimicrobial activity of substances including plant peptides.

| Organism | Analytical method | Standard reference/ other | Advantages/ disadvantages |
|---|---|---|---|
| Bacteria | • Disk diffusion assay<br>• Agar well diffusion assay<br>• Microtiter plate assay<br>• MICs | [52,53] | Low cost<br>Difficult to determine the concentration inhibitory minimum (MIC) |
| Fungi | • Germinated fungal spore count/broth dilution assays and MICs | [54] | Rapid, low cost |
| Virus | • Plaque reduction assays<br>• Titer reduction assays<br>• Inhibition assays of virus-induced cytopathic effect<br>• End-point titer determination tests | [12,55] | • In general, complex analytical procedure<br>• Require high technical qualification |
| Parasites (protozoa and helminths) | • Antiparasitic activity is assessed by using cell culture containing the inhibiting agent and observation by microscopy | [12,56] | More refined biosafety issues |

The antimicrobial activity of plant AMPs can be evaluated using pure substances or crude extracts, tested against different organisms, including fungi, bacteria, helminths, protozoa, and viruses [12]. The most used protocols are the same used for other types of conventional compounds, such as antibiotics or nonprotein substances [46,50,51]. Table 9.2 presents a summary of the most common assays to test for antimicrobial activity.

## 9.4 Perspectives in technological and therapeutic applications

The multifunctional properties of AMPs have directed studies aimed at health promotion, including the development of new ingredients for foods, biocosmetics, and biological inputs for therapeutic use, where these AMPs can assist in the treatment of many metabolic syndromes [55], such as abnormalities in the metabolism of glucose, lipids, and others that in turn increase the risk of hypertension and heart attack. Peptides that inhibit angiotensin-converting enzyme I (ACE) can be a promising prospect for the treatment of hypertension since side effects are minor compared to current drugs [45,57]. However, plant peptides are probably the most suitable for use in food and nutrition, as they are diverse and present applications beyond the antimicrobial activity, such as antioxidants and antihypertensives [7,45].

The design of synthetic peptides for technological and therapeutic applications depends on proper knowledge on the relationship between the structure and activity [18,58,59]. It is important to consider that biochemical determinants, such as primary sequence, peptide size, hydrophobicity and amphipathicity, among others, can influence the biofunctionality of these molecules [1]. The immense diversity of peptides already identified and countless others that are yet to be discovered or effectively used in therapeutics and other purposes [60], suggests the need for strategic studies based on global research consortia to meet the increasingly urgent demands.

## 9.5 Concluding remarks

Among the main families of plant-derived AMPs are cyclotides, thionines, defensins, hevein, and hevein-like peptides. In general, the multifunctional properties of AMPs open a range of possibilities for the development of new ingredients, mainly for food and nutrition. However, given that some bioactive plant-derived peptides are resistant to physical, chemical, and biological degradation, such properties also confer potential for drug development given that global microbial drug resistance is one of the world's most serious concerns.

## References

[1] N. Shagaghi, E.A. Palombo, A.H.A. Clayton, M. Bhave, Antimicrobial peptides: biochemical determinants of activity and biophysical techniques of elucidating their functionality, World Journal of Microbiology and Biotechnology 34 (2018) 62. Available from: https://doi.org/10.1007/s11274-018-2444-5.
[2] Y. Huan, Q. Kong, H. Mou, H. Yi, Antimicrobial peptides: classification, design, application and research progress in multiple fields, Frontiers in Microbiology 16 (2020) 582779. Available from: https://doi.org/10.3389/fmicb.2020.582779.
[3] A.O. Carvalho, V.M. Gomes, Plant defensins and defensin-like peptides - biological activities and biotechnological applications, Current Pharmaceutical Design 17 (2011) 4270–4293. Available from: https://doi.org/10.2174/138161211798999447.
[4] R. Nawrot, J. Barylski, G. Nowicki, J. Broniarczyk, W. Buchwald, A. Goździcka-Józefiak, Plant antimicrobial peptides, Folia Microbiologica 59 (2014) 181–196. Available from: https://doi.org/10.1007/s12223-013-0280-4.
[5] P.B. Pelegrini, R.P. Del Sarto, O.N. Silva, O.L. Franco, M.F. Grossi-de-Sa, Antibacterial peptides from plants: what they are and how they probably work, Biochemistry Research International (2011) 250349. Available from: https://doi.org/10.1155/2011/250349.
[6] S.K. Burley, C. Bhikadiya, C. Bi, S. Bittrich, L. Chen, G.V. Crichlow, et al., RCSB Protein Data Bank: powerful new tools for exploring 3D structures of biological macromolecules for basic and applied research and education in fundamental biology, biomedicine, biotechnology, bioengineering and energy sciences, Nucleic Acids Research 49 (2021) D437–D451. Available from: https://doi.org/10.1093/nar/gkaa1038.

[7] M. Hajfathalian, S. Ghelichi, P.J. García-Moreno, A.M. Sørensen, C. Jacobsen, Peptides: production, bioactivity, functionality, and applications, Critical Reviews in Food Science and Nutrition (2017). Available from: https://doi.org/10.1080/10408398.2017.1352564.
[8] A.M. Carmona-Ribeiro, L.D.M. Carrasco, Novel formulations for antimicrobial peptides, International Journal of Molecular Sciences 15 (10) (2014) 18040−18083. Available from: https://doi.org/10.3390/ijms151018040.
[9] V. Machado, J.M.L.N. Gelinski, C.M. Baratto, E.M. Borges, V. Vicente, M.M.F. Nascimento, et al., Technological potential of antimicrobial peptides: a systematic review, Indian Journal of Pharmaceutical Sciences 81 (5) (2019) 807−814. Available from: https://doi.org/10.36468/pharmaceutical-sciences.574.
[10] P. Vlieghe, V. Lisowski, J. Martinez, M. Khrestchatisky, Synthetic therapeutic peptides: science and market, Drug Discovery Today O15 (2010) 40−56.
[11] R. Nuti, N.S. Goud, A.P. Saraswati, R. Alvala, M. Alvala, Antimicrobial peptides: a promising therapeutic strategy in tackling antimicrobial resistance, Current Medicinal Chemistry 24 (38) (2017) 4303−4314. Available from: https://doi.org/10.2174/0929867324666170815102441.
[12] S.S. Tang, Z.H. Prodhan, S.K. Biswas, C.F. Le, S.D. Sekaran, Antimicrobial peptides from different plant sources: isolation, characterisation, and purification, Phytochemistry 154 (2018) 94−105. Available from: https://doi.org/10.1016/j.phytochem.2018.07.002.
[13] I.P. Sarethy, Plant peptides: bioactivity, opportunities and challenges, Protein and Peptide Letters 4 (2017) 102−108.
[14] M. Oliveira-Lima, A.M. Benko-Iseppon, J.R.C. Ferreira Neto, S. Rodriguez- Decuadro, E.A. Kido, S. Crovella, et al., Snakin: structure, roles and applications of a plant antimicrobial peptide, Current Protein & Peptide Science 18 (4) (2017) 368−374. Available from: https://doi.org/10.2174/1389203717666160619183140.
[15] L.J. Zhang, R.L. Gallo, Antimicrobial peptides, Current Biology 26 (1) (2016) R14−R19. Available from: https://doi.org/10.1016/j.cub.2015.11.017.
[16] J. Lei, L. Sun, S. Huang, C.Z.P. Li, J. He, V. Mackey, et al., The antimicrobial peptides and their potential clinical applications, The American Journal of Translational Research 11 (2019) 3919−3931.
[17] T. Su, M. Han, D. Cao, M. Xu, Molecular and biological properties of snakins: the foremost cysteine-rich plant host defense peptides, Journal of Fungi 6 (2020) 220. Available from: https://doi.org/10.3390/jof6040220.
[18] M.D.T. Torres, S. Sothiselvam, T.K. Lu, C. Fuente-Nunez, Peptide design principles for antimicrobial applications, Journal of Molecular Biology 431 (2019) 3547−3567. Available from: https://doi.org/10.1016/j.jmb.2018.12.015.
[19] J.P. Tam, S. Wang, K.H. Wong, W.L. Tan, Antimicrobial peptides from plants, Pharmaceuticals 8 (2015) 711−757. Available from: https://doi.org/10.3390/ph8040711.
[20] S.J. de Veer, M.W. Kan, D.J. Craik, Cyclotides: from structure to function, Chemical Reviews 119 (2019) 12375−12421. Available from: https://doi.org/10.1021/acs.chemrev.9b00402.
[21] D.J. Craik, Host-defense activities of cyclotides, Toxins (Basel) 4 (2012) 139−156. Available from: https://doi.org/10.3390/toxins4020139.
[22] J.A. Camarero, M.J. Campbell, The potential of the cyclotide scaffold for drug development, Biomedicines 7 (2019) 31. Available from: https://doi.org/10.3390/biomedicines7020031.

[23] M. Suarez-Carmona, P. Hubert, P. Delvenne, M. Herfs, Defensins: "simple" antimicrobial peptides or broad-spectrum molecules? Cytokine & Growth Factor Reviews 26 (2015) 361–370. Available from: https://doi.org/10.1016/j.cytogfr.2014.12.005.
[24] A. Lacerda, É. Vasconcelos, P. Pelegrini, M.F. Grossi-de-Sá, Antifungal defensins and their role in plant defense, Frontiers in Microbiology 5 (2014). Available from: https://doi.org/10.3389/fmicb.2014.00116.
[25] S.C. Lo, Z.R. Xie, K.Y. Chang, Structural and functional enrichment analyses for antimicrobial peptides, International Journal of Molecular Sciences 21 (2020) 8783.
[26] N. Van der Weerden, M.A. Anderson, Plant defensins: commons fold, multiple functions, Fungal Biology Reviews 26 (2013) 121–131. Available from: https://doi.org/10.1016/j.fbr.2012.08.004.
[27] G. Maróti, J.A. Downie, É. Kondorosi, Plant cysteine-rich peptides that inhibit pathogen growth and control rhizobial differentiation in legume nodules, Current Opinion in Plant Biology 26 (2015) 57–63. Available from: https://doi.org/10.1016/j.pbi.2015.05.031.
[28] P. Méndez-Samperio, Expression and regulation of chemokines in mycobacterial infection, The Journal of Infection 57 (2008) 374–384.
[29] B. Shahin-Kaleybar, A. Niazi, A. Afsharifar, G. Nematzadeh, R. Yousefi, B. Retzl, et al., Isolation of cysteine-rich peptides from *Citrullus colocynthis*, Biomolecules 10 (2020) 1326. Available from: https://doi.org/10.3390/biom10091326.
[30] A.N. García, N.D. Ayub, A.R. Fox, M.C. Gómez, M.J. Diéguez, E.M. Pagano, et al., Alfalfa snakin-1 prevents fungal colonization and probably coevolved with rhizobia, BMC Plant Biology 17 (2014) 248. Available from: https://doi.org/10.1186/s12870-014-0248-9.
[31] K.H. Wong, W.L. Tan, A. Serra, T. Xiao, S.K. Sze, D. Yang, et al., Ginkgotides: proline-rich hevein-like peptides from gymnosperm ginkgo biloba, Frontiers in Plant Science 3 (2016) 1639. Available from: https://doi.org/10.3389/fpls.2016.01639.
[32] A.A. Slavokhotova, T.A. Naumann, N.P. Price, E.A. Rogozhin, Y.A. Andreev, A.A. Vassilevski, et al., Novel mode of action of plant defense peptides - hevein-like antimicrobial peptides from wheat inhibit fungal metalloproteases, The FEBS Journal 281 (2014) 4754–4764. Available from: https://doi.org/10.1111/febs.13015.
[33] M.P. Chapot, W.J. Peumans, A.D. Strosberg, Extensive homologies between lectins from non-leguminous plants, FEBS Letters 195 (1986) 231–234.
[34] L. Skjeldal, L. Gran, K. Sletten, B.F. Volkman, Refined structure and metal binding site of the kalata B1 peptide, Archives of Biochemistry and Biophysics 399 (2002) 142–148. Available from: https://doi.org/10.1006/abbi.2002.2769.
[35] A. Pal, J.E. Debreczeni, M. Sevvana, T. Gruene, B. Kahle, A. Zeeck, et al., Structures of viscotoxins A1 and B2 from European mistletoe solved using native data alone, Acta Crystallographica Section D Biological Crystallography 64 (2008) 985–992. Available from: https://doi.org/10.1107/S0907444908022646.
[36] M. Jarva, K. Phan, C. Humble, F.T. Lay, M. Hulett, M. Kvansakul, Crystal structure of the plant defensin NaD1 complexed with phosphatidic acid, Nature Communications 9 (2018) 1962.
[37] H. Yeung, C.J. Squire, Y. Yosaatmadja, S. Panjikar, E.N. Baker, P.W.R. Harris, et al., Quasi-racemic snakin-1 in P1 before radiation damage, Angewandte Chemie International Edition 55 (2016) 7930–7933.
[38] K.E. Balu, K.S. Ramya, T. Ankur, A. Radha, K. Gunasekaran, Structure of intact chitinase with hevein domain from the plant *Simarouba glauca*, known for its traditional anti-inflammatory efficacy, The International Journal of Biological Macromolecules 161 (2020) 1381–1392. Available from: https://doi.org/10.1016/j.ijbiomac.2020.07.284.

[39] M. Afroz, S. Akter, A. Ahmed, R. Rouf, J.A. Shilpi, E. Tiralongo, et al., Ethnobotany and antimicrobial peptides from plants of the solanaceae family: an update and future prospects, Frontiers in Pharmacology 7 (2020) 565. Available from: https://doi.org/10.3389/fphar.2020.00565.
[40] N.C. Parsley, P.W. Sadecki, C.J. Hartmann, L.M. Hicks, Viola "inconspicua" no more: an analysis of antibacterial cyclotides, Journal of Natural Products 82 (2019) 2537−2543. Available from: https://doi.org/10.1021/acs.jnatprod.9b00359.
[41] RCSB PDB. Protein Data Bank. <https://www.rcsb.org>, 2021 (accessed 20.05.21).
[42] J.S. Moreira, R.G. Almeida, L.S. Tavares, M.O. Santos, L.F. Viccini, I.M. Vasconcelos, et al., Identification of botryticidal proteins with similarity to NBS−LRR proteins in rosemary pepper (Lippia sidoides Cham.) flowers, Protein Journal 30 (2011) 32−38.
[43] E.V. Pingitore, E. Salvucci, F. Sesma, M.E. Nader-Macias, Different strategies for purification of antimicrobial peptides from lactic acid bacteria (LAB), Communicating Current Research and Educational Topics and Trends in Applied Microbiology 1 (2007) 557−568.
[44] S.M. Muhammad, I.A. Sabo, A.M. Gumel, I. Fatima, Extraction and purification of antimicrobial proteins from *Datura Stramonium* seed, Journal of Advances in Biotechnology 18 (2019) 1073−1077. Available from: https://doi.org/10.24297/jbt.v8i0.822.
[45] A. Jakubczyk, M. Karaś, K. Rybczyńska-Tkaczyk, E. Zielińska, D. Zieliński, Current trends of bioactive peptides-new sources and therapeutic effect, Foods 9 (7) (2020) 846. Available from: https://doi.org/10.3390/foods9070846.
[46] Y.Q. Li, Q. Han, J.L. Feng, W.L. Tian, H.Z. Mo, Antibacterial characteristics and mechanisms of ε-poly-lysine against *Escherichia coli* and *Staphylococcus aureus*, Food Control 43 (2014) 22−27.
[47] Y. Shioi, T.H.P. Brotosudarmo, L. Limantara, Separation of photosynthetic pigments by high performance liquid chromatography: comparison of column performance, mobile phase, and temperature, Procedia Chemistry 14 (2015) 202−210.
[48] M. Sakthivel, P. Palani, Isolation, purification and characterization of antimicrobial protein from seedlings of *Bauhinia purpurea* L, The International Journal of Biological Macromolecules 86 (2016) 390−401. Available from: https://doi.org/10.1016/j.ijbiomac.2015.11.086.
[49] D. Das, M. Jaiswal, F.N. Khan, S. Ahamad, S. Kumar, Plant Pep DB: a manually curated plant peptide database, Scientific Reports 10 (2020) 194. Available from: https://doi.org/10.1038/s41598-020-59165-2.
[50] G. Liu, G. Ren, L. Zhao, C.W. Cheng, B. Sun, Antibacterial activity and mechanism of bifidocin A against *Listeria monocytogenes*, Food Control 73 (2016) 854−861.
[51] D. Zhang, Y. He, Y. Ye, Y. Ma, P. Zhang, H. Zhu, et al., Little antimicrobial peptides with big therapeutic roles, Protein and Peptide Letters 26 (8) (2019) 564−578. Available from: https://doi.org/10.2174/1573406415666190222141905.
[52] A.W. Bauer, W.M. Kirby, J.C. Sherris, M. Turck, Antibiotic Susceptibility testing by a standardized single disk method, American Journal of Clinical Pathology 45 (1966) 493−496.
[53] EUCAST 2000, Determination of minimum inhibitory concentrations (mics) of antibacterial agents by agar dilution, Clinical Microbiology and Infection 6 (2000) 509−515.
[54] N. Kovalskaya, Y. Zhao, R.W. Hammond, Antibacterial and antifungal activity of a snakin-defensin hybrid protein expressed in tobacco and potato plant, The Open Plant Science Journal 5 (2011) 29−42. Available from: https://doi.org/10.2174/1874294701105010029.

[55] P. Ranasinghe, Y. Mathangasinghe, R. Jayawardena, A.P. Hills, A. Misra, Prevalence and trends of metabolic syndrome among adults in the Asia-pacific region: a systematic review, BMC Public Health 17 (2017) 101. Available from: https://doi.org/10.1186/s12889-017-4041-1.

[56] F. Freiburghaus, R. Kaminsky, M. Nkunya, R. Brun, Evaluation of African medicinal plants for their in vitro trypanocidal activity, Journal of Ethnopharmacology 55 (1996) 1–11.

[57] G.W. Chen, J.S. Tsai, B.S. Pan, Purification of angiotensin I converting enzyme inhibitory peptides and antihypertensive effect of milk produced by protease-facilitated lactic fermentation, The International Dairy Journal 17 (2007) 641–647.

[58] L.T. Nguyen, E.F. Haney, H.J. Vogel, The expanding scope of antimicrobial peptide structures and their modes of action, Trends in Biotechnology 29 (2011) 464–472. Available from: https://doi.org/10.1016/j.tibtech.2011.05.001.

[59] M.S. Zharkova, D.S. Orlov, O.Y. Golubeva, O.B. Chakchir, I.E. Eliseev, T.M. Grinchuk, et al., Application of antimicrobial peptides of the innate immune system in combination with conventional antibiotics-a novel way to combat antibiotic resistance? Frontiers in Cellular and Infection Microbiology 30 (2019) 128. Available from: https://doi.org/10.3389/fcimb.2019.00128.

[60] T.T. Chai, K.Y Ee, D.T. Kumar, F.A. Manan, F.C. Wong, Plant bioactive peptides: current status and prospects towards use on human health, Protein & Peptide Letters 28 (6) (2021) 623–642. Available from: https://doi.org/10.2174/0929866527999201211195936.

# Mammalian antimicrobial peptides 10

M. Divya Lakshmanan, Swapna M. Nair and B.R. Swathi Prabhu
Yenepoya Research Centre, Yenepoya (Deemed to be University), Mangalore, Karnataka, India

## 10.1 Introduction

Antimicrobial peptides (AMPs) are host defense peptides (HDPs) that exhibit antagonistic effects against broad spectrum microorganisms. AMPs have been identified in prokaryotes, fungi, insects, amphibians, and mammals. Around 2300 AMPs have been recognized and characterized so far [1]. AMPs are generally encoded in the genome of the organism and expressed and secreted by the host defense system, in response to any invading microbes.

Most of the AMPs are cationic and amphiphilic (hydrophilic and hydrophobic) in nature. These cationic AMPs can easily bind with the negatively charged plasma membrane and hoard inside the bacteria to block bacterial functions inducing cell death [1–8]. However, the clinical applications of natural AMPs are restricted based on their relatively short half-life, toxicity, and hemolytic activity in the mammalian system.

## 10.2 History of antimicrobial peptides

Peptides with antimicrobial activity have been first identified in prokaryotic cells. It has been almost 100 years now since Alexander Fleming discovered the antimicrobial activity of lysozyme from nasal mucus, in 1922, followed by the discovery of "gramicidins," an AMP from *Bacillus brevis* in 1939, which was the primary AMPs to be commercialized and marketed as antibiotics [9]. With the discovery of penicillin, an antibiotic by Fleming, and the emergence of diverse forms of natural and synthetic antibiotics, the AMPs went out of the limelight. In the 1960s the evolution of antidrug-resistant pathogens evoked the interest in AMP's importance in host defense mechanism [10].

The first evidence of AMP generation in higher eukaryotes came from the experiment of Hans Boman in 1981 who identified "cecropins," a potent antibacterial, α-helical AMP from Cecropia silk moth [11]. Subsequently, in 1987, Zasloff and his colleagues discovered some cationic AMPs named magainins from the African clawed frog *Xenopus laevis* [12]. AMPs exist across different species, starting from prokaryotes to higher multicellular organisms, as an important peptide crucial for defense mechanism [13]. The discovery of AMPs "lactoferrin," from bovine milk, in 1960s, basic AMPs, from human leukocytes lysosomes, in 1963, followed by a mound of reported AMPs such as cathelicidins, defensins, and their variants, from mammalian systems thereafter. Histatins were discovered in the salivary gland in 1988 [14]. The discovery of the first β-defensin was from the bovine tracheal airway in 1991 [15] followed by the isolation of human β-defensin, HBD1 and 2, in 1995 and 1997, respectively [16]. At present, thousands of AMPs' sequences and structures are present across several AMP databases (Table 10.1).

## 10.3 Mammalian antimicrobial peptides as first-line defense against invading microbes

AMPs have been identified at several sites of mammalian system exposed to microbes, such as the skin, mouth, ears, eyes, mucosa of airways and lung, intestinal linings and inner lining of urinary tract [17,18]. They are produced by different types of immune cells, including neutrophils, eosinophil, and platelets [19—21]. AMPs are now recognized as an essential peptide crucial for our innate immune system and act as first-line defense against pathogens.

**Table 10.1** Antimicrobial peptide databases.

| Database | Type | Source | Year | Database URL |
|---|---|---|---|---|
| ANTIMIC | Sequence Structure | Natural AMPs | 2004 | http://research.i2r.a-star.edu.sg/Templar/DB/ANTIMIC/ |
| DAMPD | Sequence Structure | Synthetic AMPs | 2012 | http://apps.sanbi.ac.za/dampd/BioTools.php |
| APD | Sequence Structure | Synthetic AMPs | 2004 | http://aps.unmc.edu/AP/main.html |
| AMPer | Sequence Structure | Animal/plant AMPs | 2007 | http://www.cnbi2.com/cgi-bin/amp.pl |
| AMDD | Sequence Structure | Synthetic AMPs | 2012 | http://www.amddatabase.info |
| ADAPTABLE | Sequence Structure | Synthetic AMPs | 2019 | http://gec.u-picardie.fr/adaptable |
| Ocins database | Sequence Structure | Natural AMPs | 2019 | https://doi.org/10.1099/acmi.0.000034 |
| APD3 | Sequence Structure | Natural AMPs | 2015 | http://aps.unmc.edu/AP/ |
| DRAMP | Sequence Structure | Natural and synthetic AMPs | 2016 | http://dramp.cpu-bioinfor.org/ |

Evidences suggest that expression and types of AMPs may vary depending on (1) the age of the individual; (2) prevailing physiological conditions of the individual; and (3) may be tissue/site specific as well. AMPs show differential expression in different phases of human growth and development. While cathelicidin LL-37 is predominantly seen in the skin of neonates [22], β-defensin-2 (hBD2) is found in older individuals. Fetal keratinocytes have significantly higher levels of human S 100 protein, human β-defensin-3 (hBD3), and cathelicidin expression rate than postnatal skin cells [23]. Apart from Psoriasin (s100A7), an abundant antimicrobial protein, RNase7 and hBD3 are expressed differentially in different healthy human skin [24]. Dermcidin, the major AMPs in human sweat, is also expressed differentially based on physiological condition of a given individual, for instance dermcidin are differentially expressed in response to exercise [25]. AMPs found in human tears are lysozyme and lactoferrin [26,27]. Oral epithelial cells primarily express defensins (29–35 amino acids) and histatins against microbial and fungal pathogens and play important role in preventing oral cavity [28–30].

AMPs secreted either constitutively or can be induced infection, injury or by inflammatory response [29]. For instance, α-defensins and dermcidin tend to be produced constitutively in the skin, whereas the majority of other identified AMPs are inducible [31–33]. Cathelicidins and defensins are important in immunomodulation, apoptosis, and angiogenesis [34] and play a vital part in the general health of the organism. These peptides link innate immune responses to the adaptive immune system through their functional molecules.

## 10.4 Classification of mammalian antimicrobial peptides

Mammalian AMPs are reported from the body fluids and tissues of humans and cattle, most of them falling under cathelicidins and defensins families. Mammalian AMPs are categorized according to their amino acid sequence, structure, and activity [35].

### 10.4.1 Classification of antimicrobial peptides based on amino acid sequence

AMPs are short peptides of almost 55 amino acids with net charge +5.6. Human AMPs have higher average length and net charge when compared with those from other sources (32.4 amino acids and pH +3.2). Human AMPs are diverse with sequence lengths varying from 10 amino acids as in the case of neurokinin A to 149 amino acid long, for instance, Reg IIIα with broad pH values, for example, −3 (β-amyloid peptide); +20 (CXCL9). Diversity in the sequence of mammalian AMPs is a direct reflection of their structural and functional diversity [36].

#### 10.4.1.1 Proline-rich peptides

Proline-rich AMP (PrAMP) generally targets protein synthesis and gains entry into the bacterial cytoplasm with the help of SbmA, a transporter protein present in the inner membrane. Their mechanism of action is through inhibiting the binding of aminoacyl-tRNA with the peptidyl transferase center. Alternatively, certain other PrAMPs may interfere with the decoding of stop signals by trapping protein release factors, at the active centers of the ribosome during the translation termination. On the other hand, Tur1A, from *Tursiops truncates* (dolphin) which is orthologous to the Bac7, a bovine PrAMP, interferes with the translation elongation by binding within the ribosomal tunnel. This impedes ribosome function thus hindering protein synthesis. PrAMPs preserve a short motif with proline (P) and arginine (R) repeats across species through other sequences that may be divergent depending on species or tissue origin. The well-known examples are the bovine PrAMPs—Bac5 and Bac7 with conserved sequences "PPXR" and "PRPX," respectively. Most of the mammalian PrAMPs are specific toward Gram-positive bacteria but crustacean PrAMP1 is reported to show antimicrobial activity against both Gram-positive and Gram-negative bacteria [37].

### 10.4.1.2 Tryptophan and arginine-rich antimicrobial peptides

In AMPs rich in tryptophan and arginine, tryptophan preferably act on the interface area of the lipid bilayer, while arginine, with its basic nature, exhibits preference for peptides having cationic charges and can contribute toward hydrogen bond interactions, which make it an ideal molecule for the bacterial membrane interaction. The tryptophan residues present play an important role in promoting ion-pair-$\pi$ interactions, thereby enhancing the AMP—membrane interactions. Indolicidin and triptrpticin are examples of the well-characterized peptides, rich in Arg and Trp residues showing antimicrobial activity. Synthetic Trp-rich and Arg-rich AMPs constructed with the sequence of mammalian lactoferricin have also proved to be highly effective against bacteria [38].

### 10.4.1.3 Histidine-rich peptides

Histidine-rich AMPs display membrane permeation activity and exhibit its antimicrobial potential through cell membrane break and destruction. For instance, HV2, a histidine-rich AMP, inhibits bacterial swarming and exhibits a robust anti-inflammatory effect by lowering the expression of tumor necrosis factor $\alpha$ (TNF-$\alpha$). The LAH4 family of amphipathic peptides designed by inserting four histidine residues in leucine and alanine based on the sequence of linear cationic peptide magainin also exhibits well-characterized antibacterial activity with bacterial cell-penetrating properties [39].

### 10.4.1.4 Glycine-rich antimicrobial peptides

Attacins and diptericins are glycine-rich AMPs produced by many organisms. These nonpolar residues which accounts for about 14%—22% of the peptide has an important effect on the protein folding and functions of the peptide chain. Glycine-rich AMPs derived from salmonid cathelicidins, unlike the conventional AMPs, activate phagocyte-mediated microbicidal mechanisms. The glycine-rich central—symmetrical GG3 is a proven drug acting against clinically important Gram-negative bacteria [40].

## 10.4.2 Classification of antimicrobial peptides based on the structure

Widely accepted classification for AMPs is based on their structure. They are classified into four groups such as (1) linear α-helical peptides, (2) β-sheet peptides (with two or more disulfide bonds), (3) linear extension structure, and (4) peptides with both α-helix and β-sheet. Apart from these common naturally occurring structures of AMPs, cyclic peptides and AMP with more complex topologies are also reported. Many peptide structures get activated upon interactions with the membranes of their target cells. Indolicidin is the perfect example for this and changes its structure along with DNA [21]. Most of the AMPs come under these four classes of categories, with few exceptions. Few of them even contain two different structural components.

α-Helix and β-sheet are the most commonly studied structures of AMPs. In α-helix structure, the two adjacent amino acids are placed at a distance of 0.15 nm apart, with an angle of 100 degrees with respect to the center from a top view. AMPs such as protegrin, magainin, indolicidins both cyclic and coiled, falls under this class. β-Sheets are formed with at least two beta-strands with cysteine residues forming a bisulfide bridge between these strands. Some well-characterized AMPs include (1) α-defensins and β-defensins, (2) cathelicidin LL37, (3) histatin family, and (4) thrombocidine [41].

### 10.4.2.1 Defensins

Defensins are cysteine-rich cationic peptides, identified from higher organisms and functions as both cell-signaling molecules and antimicrobials. The peptides connected by six cysteine with three disulfide bridge giving them their topology. Based on the position of bisulfide bond, defensins can be divided into alpha, beta, and theta (Fig. 10.1).

α-Defensins, with six members, are peptides of approximately 2–6 kDa, are generally found in neutrophils, macrophages, and Paneth cells of the small intestinal crypts of Lieberkühn. Of the six types of human α-defensins identified, 1–4 are myeloid α-defensins recognized as human neutrophil peptides 1–4 (HNP1–4). All enteric defensins known so far are α-defensins, and hence 5, 6 expressed in the Paneth cell of the small intestine is designated as human defensin (HD5, 6). All of them share a typical structure consisting of a triple-stranded β-sheet, stabilized by three paired cysteine disulfide bonds. Some of the mammalian α-defensins are produced from β-defensins and the disulfide linkage pattern between the three pairs of conserved cysteine residues structurally separates both classes [42]. α-Defensins have been identified from several primates and rodents.

β-Defensin, with two members, are peptides of 6–8 kDa, produced by various epithelial cells, macrophages, neutrophils, granulocytes, and NK cells. For instance, β-defensin-1 is abundantly found in neutrophil cells, and β-defensin-2 is found in psoriatic lesional skin [32–35]. The structure of β-defensins comprises of triple-stranded β-pleated structures with an N-terminal α-helix (Fig. 10.1). Three types of β-defensins from humans, designated as human β-defensin-1 (hBD1), -2 (hBD2), and -3 (hBD3) [43], have been structurally elucidated to date. The β-defensins

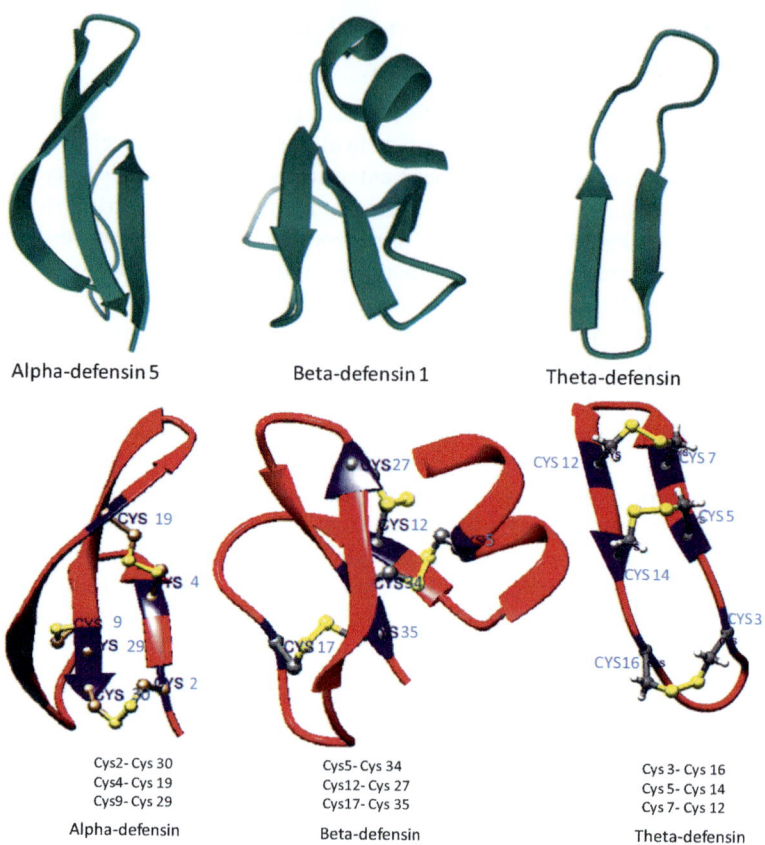

**Figure 10.1** Structure of alpha, beta, and theta defensins. Cys—Cys linkage is displayed. Cysteine residues are in *blue* and disulfide bond are displayed in *yellow*.

family of AMPs are encoded by more than one gene and have been reported for all mammalian species explored so far. hBD1 are antagonists for adenovirus, Gram-positive, and Gram-negative bacteria, but is salt-sensitive and its activity is inhibited by NaCl. On the other hand, hBD2 exhibit the least specificity toward Gram-positive bacteria, but acts against Gram-negative bacteria, with 10 times higher potency than hBD1 against yeast. hBD3 acts against Gram-positive and Gram-negative bacteria and is resistant to salt, but its expression is inhibited by glucocorticoids and their analogs [44]. In general, hBDs safeguard the oral, respiratory, reproductive, and enteric tissues from the invading pathogens and associated pathophysiology.

Theta-defensins are cyclic peptides of eight amino acids ($\sim 2$ kDa), produced by binary ligation of truncated α-defensins (Fig. 10.1). Theta-defensins are so far reported to be present only in nonhuman primates. The antiparallel β-sheets of theta-defensins are linked by three disulfide bonds to give the appearance of a ladder, which confers their structural stability [45].

Both α-defensins and β-defensins can assemble on the bacterial cell wall and block cell wall biosynthesis by interacting with lipid II precursors involved in the biosynthesis of the bacterial cell wall.

## 10.4.2.2 Cathelicidins

Cathelicidins are proteins produced in the lysosomes of macrophages, dendritic cells, and polymorphonuclear leukocytes (PMNs), with broad spectrum of microbicidal activity against bacteria, fungi, and enveloped viruses. In addition, keratinocytes and epithelial cells of the respiratory and gastrointestinal tracts express these AMPs. As these peptides are initially observed in the mammalian bone marrow myeloid cells, they are also designated as myeloid AMPs (Fig. 10.2). All the 30 members of this family identified across different mammalian species show a common modular structure comprising of multiple domains. Cathelicidins share highly conserved N-terminal domain coding for the signal sequence and pro-region (cathelin). On the other hand, almost all AMPs of cathelicidin family showed substantial heterogeneity for C-terminal domain, that encode the mature peptide [46]. C-terminal domain may be α-helical or β-hairpin or maybe tryptophan or proline/arginine rich depending on the species. Five distinct groups of peptides are identified within this family depending on the difference in their C-terminal domain such as: (1) cyclic dodecapeptides with one disulfide bond, (2) porcine protegrins with two disulfide bonds, (3) peptides with α-helical structure, (4) tryptophan-enriched peptides like indolicidin and other peptides enriched with arginine and proline residues, and (5) tandem repeats of organized short molecules like bactenecins. Some of the well-identified cathelicidins includes LL-37, rCRAMP, RL-37, CAP11, and mCRAMP from primates and rodents [47].

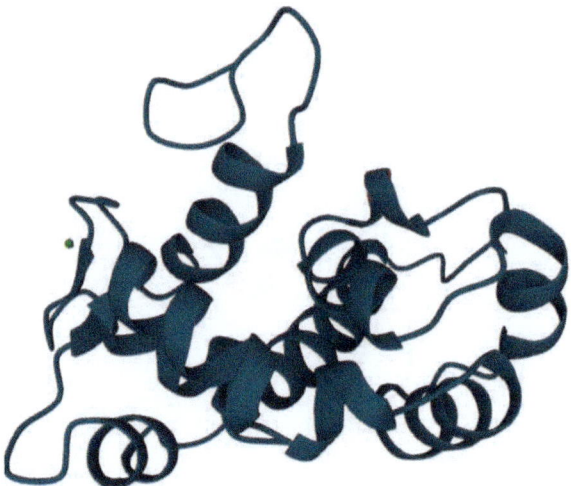

**Figure 10.2** Structure of cathelicidins.

## 10.4.2.3 Histatins

Histatins, with three main members, are histone rich, low molecular weight cationic peptides (Fig. 10.3), secreted from parotid and submandibular salivary glands in human, and play an important role in wound closure. Histatin 1 and histatin 3 are encoded by two separate genes HTN1 and HTN3, respectively. Other histatins are derived from the proteolytic cleavage of histatin 1 and 3. Histatin 5, a major histatin, is derived from histatin 3. Histatin 1 and 3 are the main histatin subfamily comprising more than 80% of histatin in the saliva. Histatin 5 comprises 24 amino acids sequence, that are identical with N-terminal sequence of histatin 3. Histatin 3 have additional eight amino acid in their C-terminal end (Fig. 10.3) [48]. Other degraded products such as histatin 2 and histatin 4 are the end products of autoproteolytic degradation. On the other hand, histatin 2 is the slow degradative product of histatin 1 and histatin 4 mainly arises as rapid breakdown product of histatin 3.

Histatin mainly functions against *Candida albicans* and maintains oral hemostasis, regulates tooth pellicle formation, and helps metal ion interactions, especially copper and nickel, in the saliva, generating ROS in the infected cells, ultimately leading to the death of the pathogens [49,50].

## 10.4.2.4 Thrombocidin

These are cationic, amphipathic AMPs secreted from mammalian blood platelet α-granules, in response to thrombin treatment under in vitro conditions and hence designates as thrombocidins. However, in in vivo condition platelets are activated and aggregated upon direct contact with bacteria causing the release of micobicidins that eliminate the bacteria [51]. Two types of thrombocidins, thrombocidin-1 (TC-1) and TC-2 are reported, and are truncated products of the CXC chemokines neutrophil-activating peptide-2 (NAP-2/CXC) and connective tissue-activating peptide-III (CTAP-III), respectively. For instance, TC-1 is derived from the chemokine NAP-2

| Sequence | Name |
|---|---|
| DSHEKRHHGYRRKFHEKHHSHREFPFYGDYGSNYLYDN | Histatin 1 |
| RKFHEKHHSHREFPFYGDYGSNYLYDN | Histatin 2 |
| DSHAKRHHGYKRKFHEKHHSHRGYRSNYLYDN | Histatin 3 |
| RKFHEKHHSHRGYRSNYLYDN | Histatin 4 |
| DSHAKRHHGYKRKFHEKHHSHRGY | Histatin 5 |

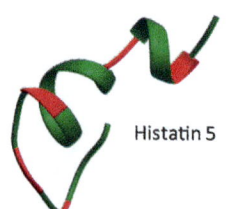

Histatin 5

**Figure 10.3** Sequence and structure of histatins. Histatin 2 is the degradative product of histatin 1. Histatins 4 and 5 are the degradative products of histatin 3. In the cartoon representation of the structure, histidine residues are displayed in *red* color.

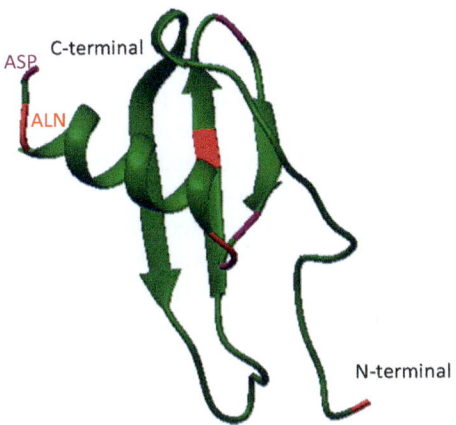

Neutrophil-Activating Peptide-2

**Figure 10.4** Structure of neutrophil-activating peptide-2. TC-1 is derived by truncation of ALA (shown as *red* colored) and ASP (shown as *pink* colored) residue of NAP-2 at the C-terminal.

(Fig. 10.4) by terminal deletion of ALA and ASP residues through proteolysis at the C-terminal. On the other hand, TC-2 is the truncated product of CTAP-III produced by the deletion of two C-terminal amino acids (Ala-Asp), as similar to TC-1.

Though TC-1 and TC-2 are potent AMPs, their parent peptides lack bactericidal and fungicidal activities. TC-1 is more potent than TC-2 against bacterial and fungal pathogens [52].

### 10.4.3 Classification of antimicrobial peptides based on the activity

According to the ADP3 database statistics, mammalian AMPs can be categorized into several groups such as antibacterial, antiviral, antifungal, antiparasitic, antihuman immunodeficiency virus (HIV), and antitumor peptides [53].

#### 10.4.3.1 Antibacterial peptides

Most of the natural and synthetic AMPs such as nisin, defensins, and cecropins have antibacterial property and are antagonistic to most of the clinically relevant pathogens and food-borne organisms [20]. HNP1−4 and HD-5 display antibacterial properties against common Gram-positive and Gram-negative bacteria such as *Staphylococcus aureus*, *Escherichia coli*, and *Enterobacter aerogenes*. In contrast, no such activity is reported for HD-6 [54]. On the other hand, HNP-1 and HNP-3 alleviate the cytopathogenic effects induced by *Clostridium difficile* toxin B [55]. A recombinant peptide human β-defensin (HBD)-2 is effective against infections developing from the use of prosthesis implantation [56]. The peptide PAC-113 or P-113 which is an analog of naturally occurring molecule histatin-5 secreted from the salivary gland [57] displays broad spectrum antibacterial activity in vitro.

## 10.4.3.2 Antifungal peptides

Most of the antifungal peptides (AFPs) show broad spectrum antifungal activities against clinically important pathogenic fungi, food-borne fungus, and plant pathogens. AMPs isolated from *Lactobacillus plantarum* TE10 which is a mixture of 37 antifungal peptides was found effective against the common maize pathogen *Aspergillus flavus* [58]. Human defensins have potential antagonistic activity against chronic treatment-resistant fungal infections of toenails Human, $\alpha$-melanocyte-stimulating hormone, derived peptide is known to be effective against vaginal candidiasis [59]. The peptide PAC-113 or P-113 is effective against *C. albicans* and is used as a constituent of mouthwash against oral candidiasis commonly found in HIV patients [60].

## 10.4.3.3 Antiviral peptides

Antiviral peptides (AVPs) exert their effect either by preventing attachment of the virus and cell membrane fusion or by destructing the virus envelope and inhibiting replication of the virus. Human $\alpha$-defensins and cathelicidin-derived AMP LL-37 inhibits herpes simplex virus and human immunodeficiency virus in vitro by preventing the binding of the viral envelope protein gp120 of HIV with the host immune cell receptor, CD4 in vitro [61,62]. It can also repress the replication of HIV-1 [63,64]. A synthetic antiviral peptide Fuzeon (enfuvirtide) is one of the AVPs that is commercialized as an anti-HIV drug [33]. Another example of the AVP is Epi-1 (nonmammalian source), which is effective against hand-foot-and-mouth diseases commonly seen in young children. Swine intestinal AMP (SIAMP) shows good inhibitory activity on IBV (infectious bronchitis virus).

## 10.4.3.4 Antiparasitic peptides

Antiparasitic peptides against malaria and leishmaniasis include cathelicidin, temporins-SHd. Epi-1, a marine AMP from bony fish is a nonmammalian AMP, has been effective against *Trichomonas vaginalis* by destroying its membrane [65].

## 10.4.3.5 Anticancer peptides

Anticancer peptides (ACPs) regulate cancer initiation and progression through inhibiting the proliferation and invasion of cancer cells through different mechanisms as described below:

1. facilitate the recruitment of immune cells such as dendritic cells and cytotoxic T-cells to kill tumor cells
2. act as signaling molecules inducing the cancer cells to undergo necrosis or apoptosis
3. act as antiangiogenesis factors preventing tumor nutrition and inhibit metastasis
4. triggering the activation of regulatory proteins responsible for inhibiting the gene transcription and translation of tumor cells

Main factors influencing the anticancer activity of ACPs are net charge and hydrophobicity [65]. The role of ACPs in cancer regulation was first defined in cancers of the urogenital tract, where human-β-defensin-1 expressed was associated with epithelial cell cancer initiation. Similarly, downregulation of hBD-1 was corelated with renal carcinoma progression and prostate cancer malignancy, whereas its expression remained unaltered in benign conditions [66]. Anticancer functions for α-defensins were reported from in vitro and in vivo studies, where it kills tumor cells through membrane disruption. For example, HNP1−3 treatment was found cytotoxic to multiple myeloma cells [67] and treatment of hBD-1, hBD-2, and hBD-3 inhibited oral squamous carcinoma. On the other hand, hBD-2 and hBD-3 induced cell proliferation in osteosarcoma [68]. Hence, hBD may be considered a biomarker in early cancer detection and in determining the risk associated with the malignant transformation of benign lesions [68]. Likewise, human α-defensin-3 expression was altered in lymphomas and may be considered as a biomarker for lymphocyte-associated cancers [69]. Similarly, α-defensin-6 may be a potential biomarker for deciding the malignant transformation of formerly benign epithelial colon cells [70]. The LL37 is effective against patients with metastatic melanoma when administered directly via intratumor route.

### 10.4.3.6 Immunomodulatory and chemotactic peptides

AMPs can act as chemotactic-signaling molecules to attract lymphocytes. The α-defensins, that is, HNP1−3 and β-defensins such as human β-defensin-1, -3, and -4 have the ability to recruit neutrophils, monocytes, immature dendritic cells (iDC), and memory T-cells to the site of inflammation. On the other hand, human cathelicidin LL-37 has a key role in the chemotaxis of most of the immune cells except dendritic cells which is facilitated by the G protein-coupled receptor, FPRL1. The AMP that has gained clinical importance is hLF1−11, comprising of the first 11 residues in the N-terminal of human lactoferricin, used for microbial infections commonly occurring in the transplant recipients receiving immunosuppressants [59]. In humans, it was found that the β-defensin is responsible for the chemotaxis of the memory T-cells as well as iDC via GPCR CC-chemokine receptor 6 which is different from the activity carried out by LL-37 [71]. LL-37 functions in arthritic lesions as a chemotactic factor, mainly via its interaction with monocytic formyl-peptide receptors, inducing arterial monocyte influx. The reduction in lesion sizes with lower macrophage numbers with reduced atherosclerosis was found in CRAMP knockout mice [72]. A study on a pristine-induced arthritic rat model has emphasized the role of rCRAMP, the rat ortholog of LL-37 in arthritis [73]. Chronic periodontitis and overgrowth of *Actinobacillus actinomycetemcomitans* was found in congenital neutropenia patients who lacked LL-37 in plasma or saliva and reduced levels of HNP-(1−3) in their neutrophils [74].

### 10.4.3.7 Antimicrobial peptides in tissue regeneration and wound healing

The role played by AMPs in wound healing and tissue regeneration is well established. HNP-1 and HNP-2 regulates β-catenin signaling and promotes proliferation and collagen synthesis of lung fibroblasts. This treatment strategy is considered effective against fibroproliferative lesions as lung fibrosis [75]. All known members of HBDs promote wound closure by inducing the migration of keratinocytes [67,68,75,76]. HBD-2 was found to positively influence the vascularization process, and promote migration HBD-2 along with hBD-3 might also be effective in bone tissue regeneration [67,77]. AMP PX450 has been effective in curing the burn wounds in mice. AMP D2A21 has entered phase III trials for alleviating burn wound infection.

### 10.4.3.8 Antimicrobial peptides in ophthalmology

AMPs have proven ophthalmological application. A study has reported the application of rabbit α-defensin (NP-1) for ophthalmic infections [56]. Lactoferrin B, because of its potential antimicrobial activity, is considered as a promising drug candidate for ophthalmological applications.

### 10.4.3.9 Antimicrobial peptides in fertility

Recent studies show that defensins also play an important role in fertilization [78]. Human β-defensin 1 (hBD-1) has important role in the innate immunity as it is expressed in female genital tracts, fallopian tube and endometrium and placenta of pregnant uterus. The defensins sustain the genitourinary health, and fertility status, and endow the capability to conceive and implant the zygote, mainly by controlling microbes associated with inflammation, inflammatory disorders, and anatomical anomalies, influencing fertility. Generally, in females, hBD-1 is said to be differentially expressed during the menstrual cycle and plays a prominent role during pregnancy in protecting the uterus and the fetus. In males, the expression of β-defensins has been reported in the prostate, testis, and sperm where these molecules may enable protection against infections and other functions influencing sperm mortality and activity [79,80].

## 10.5 Common mechanism of action of mammalian antimicrobial peptides

Human AMPs are the host defense peptides with antagonistic actions against Gram-positive and Gram-negative bacteria, fungi, enveloped viruses, parasite, and altered self-cells and cancer cells. The most accepted modes of targeting by AMPs involve: (1) membrane-targeting mechanism, (2) cell wall-targeting mechanism, (3) intracellular-targeting mechanism, and (4) immunomodulatory mechanism.

### 10.5.1 Membrane-targeting mechanism

Most mammalian AMPs possess a net positive charge and facilitate binding with the negatively charged bacterial membrane. Once inside the bacterial cell membrane, peptides initiate the disruption of pathogens by using three mechanisms (Fig. 10.5) described below [18,79].

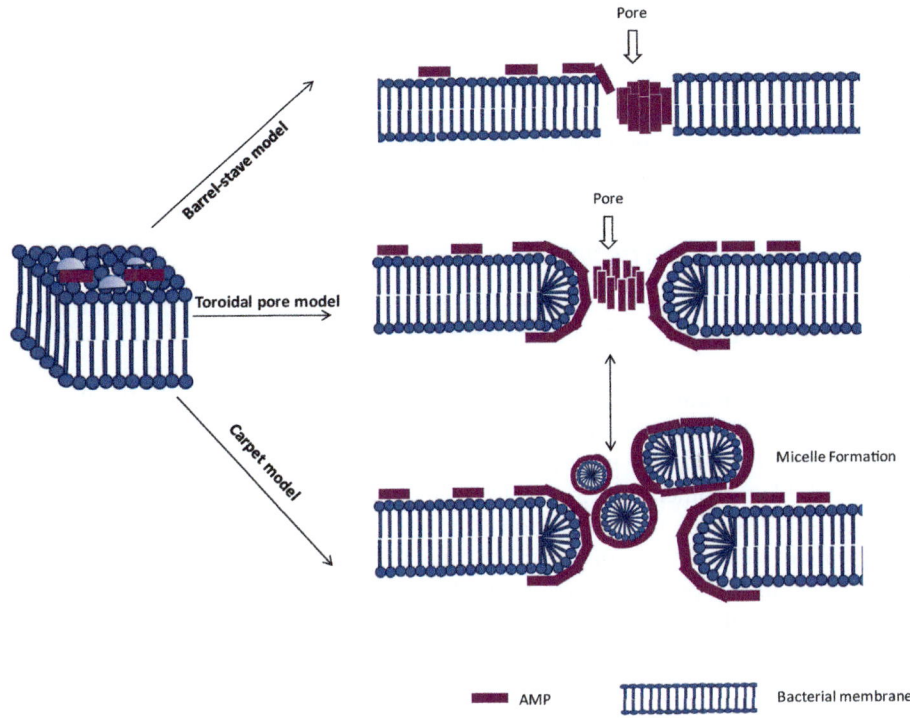

**Figure 10.5** Membrane-targeting mechanism of action of mammalian antimicrobial peptides.

AMPs acting through *toroidal pore mechanism* will absorb on a membrane surface, then aggregate and force the membrane to thin, which will expand the head region. Here peptides that are inserted get arranged vertically in the lipid bilayer; but lack specific peptide–peptide interactions. Due to this action, lipid bilayer bends and makes the upper and lower leaflet to meet together disrupting the hydrophobic and hydrophilic arrangement of the bilayer. This gives a toroidal appearance to the pore, which has been formed. This will disrupt the regular seclusion of polar and nonpolar parts of the bacterial membrane by providing an alternative surface for lipid hydrocarbon and nonpolar heads to interact with [80,81]. In *barrel stave model* of mechanism of action, peptides place in an upright direction to the surface of the plasma membrane, forming the "staves" in a barrel-fashioned bundle. In contrast to toroidal pore formation, hydrophobic and hydrophilic organization of the lipids in the bilayer is well sustained in the barrel stave model. Here, peptides inserted into the membrane are arranged parallel to the lipid bilayer with hydrophilic regions facing the lumen of the pore and the hydrophobic regions interacting with the lipid moieties of the microbial membrane. *Carpet formation* is the most commonly studied mechanism describing the action of peptides [18,82]. AMPs adsorb parallel to the phospholipids until it reaches a threshold level that covers the membrane, forming a "carpet." This induces the membrane to lose its integrity mimicking the detergent-like effect, leading to the formation of micelles. This micelle formation arising due to a complete loss of membrane integrity is sometimes referred to as the detergent-like model (Fig. 10.5) The carpet model is devoid of any specific interactions between the membrane peptides and the bound peptide monomers; also, the peptide need not be inserted into the hydrophobic core to induce any transmembrane pores or specific peptide structure formation.

### 10.5.2 Cell wall-targeting mechanism

In the majority of the Gram-positive bacteria, AMPs cross the cytoplasmic membrane only on traversing cell walls composed of lipoteichoic acids (LTAs) and peptidoglycan. AMPs do not interfere with peptidoglycan synthesis through peptide binding, but in contrast, make a complex with the precursor molecule like lipid 11, that is required for the peptidoglycan cell wall synthesis, facilitating membrane pore formation and loss of membrane integrity. Peptidoglycan synthesis is initiated from lipid 11 monomer [81,83]. Lipid 11 monomer consists of 11 subunits long, polyisoprenoid anchor of C55 carbon (C55-PP) chain linked to one disaccharide-pentapeptide subunit through pyrophosphate linkage. The lipid II monomer gets translocated to the periplasm of bacterial membrane where it integrates into the developing peptidoglycan network. In mammals human β-defensin 3 binds to the pyrophosphate sugar moiety of the lipid II, thus interfering with peptidoglycan synthesis [76,82−84]. Thus the versatile mechanisms in combating the bacteria make AMPs superior to antibiotics. AMPs that target peptidoglycan facilitate penetration of the membrane resulting in its disruption.

Many cocci and bacilli have LTA as a constituent of the bacterial cell wall component. Many AMPs, such as LL-37, melittin, and cecropin, facilitate membrane disruption through *LTA* binding. The binding of positively charged AMPs to anionic teichoic acids may result in entrapment of AMPs or facilitates its entry to the cytoplasmic membrane via polyanionic ladder building. These interactions affect membrane permeability; lead to cell division inhibition or result in the delocalization of peripheral membrane proteins. LTA binding is also believed to induce conformational changes in the peptide resulting in active site inaccessibility to AMPs, which attenuates its membrane-disintegrating activity and abrogates its bactericidal activity toward *S. aureus* [76,85].

### 10.5.3 Targeting intracellular processes

The APMs such as indolicidin, human α-defensin-4, and human α-defensin-5 interact with plasma membrane and gain entry into the cytoplasm where they block critical cellular process, such as inhibition of nucleic acid or protein synthesis, acid synthesis. Human α-defensin-5 is found to translocate into the cytoplasm of *E. coli* where they accumulate at the cell division plate showcasing antimicrobial activity due to the presence of the target in the cytoplasm [76,82−84,86]. Indolicidin can induce remarkable depolarization of *E. coli* cell membrane and upon entry to cytoplasm can block DNA synthesis and subsequently promotes filamentation of cells [85,87]. PR-39 can gain entry into the bacterial membrane without compromising the membrane integrity, and inhibits DNA and protein synthesis leading to bacterial cell death.

### 10.5.4 Immunomodulatory mechanism

AMPs activate the immune responses such as attraction, activation, and differentiation of lymphocytes, bringing down inflammation by downregulating the proinflammatory cytokines expression [86–89]. The AMPs like LL-37 and β-defensin act as chemoattractant promoting the recruitment of mast cells, dendrites, leukocytes [88–92].

Cathecidin LL-37, which is expressed in several types of immune cells, facilitates the recruitment of other immune cells to the place of microbial attack through binding and activating the peptide receptor-like 1 [91,93]. In addition, it also interacts and activates primary receptor integrin $\alpha_M\beta_2$ (Mac-1) on the bone marrow cell surface thereby promoting the migration of mononuclear cells to the site of infection and enhancing the phagocytosis against Gram-positive and Gram-negative bacteria. Furthermore, LL-37 has the ability to bind with mast cell surface receptor, GPCR Mas-related gene X2 (MrgX2) thereby recruiting it to the site of inflammatory lesions. LL-37 also interacts with the P2X7 receptor in LPS-primed macrophages thereby promoting caspase-1 activation and subsequent processing and release of IL-1β. It also enhances ROS generation in neutrophils [91–97].

Mammalian defensins exhibit a proinflammatory function and are expressed in most of the immune cells such as: lymphocytes, neutrophils, NK cells, monocytes, and macrophages. HNP1–3 induces the secretion of TNF-α and IFN-γ from macrophages which acts in an autocrine manner upregulating the expression of CD32 (FcγRIIB) and CD64 (FcγRI), thereby increasing its phagocytosis [95,98]. In addition, HBD1 can stimulate the expression of other cytokines and chemokines such as IL-6, IL-10, IFN-γ-inducible protein IP-10, MCP-1, and macrophage inflammatory protein-3α (MIP-3α) and RANTES/CCL5 in human keratinocytes. In addition, HBD2, HBD3, and HNP1 function as proinflammatory molecules stimulating the mononuclear and polymorphonuclear cells to secrete TNF-α, IL-10, and IL-6 in an inflammatory condition [94–100].

## 10.6 Clinical applications of antimicrobial peptides

AMPs have rapid bactericidal activity and is developing as a novel therapeutic approach with potential to overcome certain disadvantages associated with antibodies. AMPs is evolving as a promising antibacterial agent for dental, surgical, ophthalmological, and wound healing applications, to list a few. AMPs are intrinsically involved in angiogenesis, wound closures and tissue regeneration, intracellular signaling, etc., which increased their attention in the field of medicine. However, only a few AMPs are approved for clinical use and certain AMPs are currently in the last phases of clinical development. Table 10.2 lists the details of AMPs approved for clinical application and in different phases of clinical trials.

Table 10.2 Selected antimicrobial peptides in the clinical phase of development.

| Peptide name | Source | Phase | Application | Administration | Clinical trial identifier | Sponsor/collaborator | References |
|---|---|---|---|---|---|---|---|
| Natural interferon-α or multiferon | Leukocyte fraction of human blood | Approved by Swedish Medical Products Agency — 2006 | Effective against metastatic renal cell carcinoma Improves cellular function/ growth and immune defense | Administration through subcutaneous route | | Intron/ Roferon-A/ Roche | |
| Omiganan | Derivative of bovine indolicidin | Phase II/III | Effective against catheter infections, rosacea, acne vulgaris | Topical gel | NCT02576847 NCT00231153 | Cutanea Life Sciences Mallinacckrodt | [101–103] |
| CZEN002 | Dimeric octamer derived from human α-MSH | Phase II | Effective against vaginal candidiasis | Vaginal gel | | Cutanea Life Sciences | [100] |
| Novexatin (NP-213) | Human defensins | Phase II | Recalcitrant nail infections | Topical brush-on-treatment | | | [100] |
| PXL01 | Human lacto ferricin | Phase II | Used as antiadhesion barriers in hand surgery | Hyaluronic acid-based hydrogel | NCT01022242 | | [100] |
| PAC-113 | Derivative of histatin 3 (human saliva) | Phase II | Oral candidiasis in HIV seropositive patients | Mouthwash | NCT00659971 | Pacgen Biopharmaceuticals Corporation/ Quintiles, Inc. | [73,104] |

| | | | | | | |
|---|---|---|---|---|---|---|
| OP-145 (AMP60.4Ac) | Cathelicidin family (derived from LL37) | Phase II | Treatment of chronic suppurative otitis media | | OctoPlus BV | [105] |
| Lactoferrin | Histatin analog | Phase II | HIV infection | | Jason Baker/Ventri Bioscience/ Minneapolis Medical Research Foundation | [106] |
| DPK-060 | Derived from human kininogen | Phase II | Bacteremia of atopic dermatitis and acute external otitis | Topical ointment | NCT01522391 and NCT 01447017 | [97–99] |
| LL-37 | Human LL-37 | Phase I/II | Recalcitrant venous leg ulcers | Polyvinyl alcohol-based solution | | [100] |
| hLF1–11 | Human lacto ferricin | Phase I/II | Effective against pathogenic infections in hematopoietic stem cell transplant recipients with immunosuppressant treatment | Administration through intravenous route | NCT00509938h | AM-Pharma | [105,106] |
| | | I/II (withdrawn) | Treatment of bacteremia by *Staphylococcus epidermidis* | | | |

The structural derivative of a human AMP, DPK-060, shows a wide spectrum of antimicrobial action. The ointment of it has efficacy to treat the infection in atopic dermatitis and acute external otitis and has reached clinical phase II trials [98–100,105–107]. The molecule CZEN-002, a peptide dimer successively derived from α-melanocyte-stimulating hormone, is effective against vaginal candidiasis [105,108]. hLF1−1, a cationic fragment comprising of the first 11 amino acid residues of human lactoferricin N-terminal, when administered intravenously is effective against broad spectrum fungal and bacterial infections found in stem cell transplant recipients administered with immunosuppressants. The phase I/II clinical trials for both the applications with hLF1−11 were withdrawn prior to the enrollment [106,107,109,110]. The human cathelicidin LL-37 has a key role in the chemotaxis of monocytes, T-cells, and neutrophils, but not in dendritic cells which are mediated by the G protein-coupled receptor, FPRL1. In humans, it was found that the β-defensin is responsible for the chemotaxis of the memory T-cells as well as iDC via GPCR CC-chemokine receptor 6, which is different from the activity carried out by LL-37 as it is not involved in chemotaxis of iDCs [71,108]. A clinical trial for optimizing the dose of LL37 with its therapeutic activity developed against metastatic melanoma administered intratumorally is being evaluated in a phase I stage. LL37 may contribute significantly to arthritis development as from a study in a rat model, arthritis was induced with pristine which was found to show upregulation of rCRAMP, the rat ortholog of LL-37 [73,109]. The wound healing mechanism of LL-37 is not yet understood completely but is known to involve several cellular repair mechanisms such as angiogenesis, inflammation, and epithelialization. The chemoattractant effects of LL-37 help in the stimulation of reepithelialization on epithelial cells [71,104,110,111]. The derivative of LL-37, that is, OP-145 (AMP60−4Ac) has come up as a successful candidate drug in phase II clinical trials against chronic suppurative otitis media [38].

The peptide PAC-113 or P-113 is used in the form of a mouth rinse for combating candidiasis developed in the oral cavity of HIV patients [57,111].

Omiganan, an indolicin derivative belonging to cathelicidin family, was one among the initial AMPs that successfully completed phase III clinical trials against several pathogens causing rosacea, catheter infections, atopic dermatitis, genital warts, and acne vulgaris [57,60,101,104,111]. Apart from direct administration of AMPs, a strategy to enhance the endogenous production of AMP thereby enhancing innate response to combat infection is underway.

## 10.7 Current and future prospects and challenges in developing antimicrobial peptides

The prevalence of increasing high-level multidrug resistance (MDR) has become the urge for the discovery of mammalian AMPs. AMPs like lactoferrin, cathelicidin, defensins, dermaseptins have been now studied in the reproductive tract of mammals which will regulate fertility and prevent sexually transmitted diseases, and this make them a good choice of candidates in contraceptive agents for vaginal prophylaxis [102,103,112].

Many challenges are associated with the development and clinical use of AMPs. Studies with AMPs are working toward overcoming the complications restricting its utility in clinical applications and patient care. Stability is a major concern as the majority of the AMPs discovered so far are highly sensitive to atmospheric conditions and exhibit low metabolic stability. Metabolic stability is influenced by the physiological level of relevant ions and physiological barriers in the cellular environment. High ionic strength and the presence of protease in the physiological environment lower the stability of the AMPs. Apart from the stability issue, many discrepancies have been reported between in vitro versus in vivo efficacy, making it difficult in accurate prediction of the results. Another major setback in the development of AMPs is their low oral availability.

In conclusion, mammalian AMPs are structurally and functionally variable potent therapeutic candidates, showing promising antimicrobial and immune-boosting effects against many microbial drug-resistant strains. Apart from the antimicrobial actions, AMPs contribute to cellular signaling, immune modulation, antitumor functions, and drug transportation. So it is very important to have proper knowledge of the versatile biological characteristics of AMPs for the clinical development of peptide-based therapies.

# References

[1] H.G. Boman, Antibacterial peptides: basic facts and emerging concepts, Journal of Internal Medicine 254 (2003) 197–215.
[2] T. Ganz, R.I. Lehrer, Defensins, Current Opinion in Immunology 6 (1994) 584–589.
[3] M. Zasloff, Antimicrobial peptides of multicellular organisms, Nature 415 (2002) 359–365.
[4] R.E.W. Hancock, H.G. Sahl, Antimicrobial and host-defense peptides as new anti-infective therapeutic strategies, Nature Biotechnology 24 (2006) 1551–1557.
[5] Y. Lai, R.L. Gallo, AMPed up immunity: how antimicrobial peptides have multiple roles in immune defense, Trends in Immunology 30 (2009) 131–141.
[6] N.Y. Yount, M.R. Yeaman, Emerging themes and therapeutic prospects for anti-infective peptides, Annual Review of Pharmacology and Toxicology 52 (2012) 337–360.
[7] Z. Wang, G. Wang, APD: the antimicrobial peptide database, Nucleic Acids Research 32 (2004) D590–D592.
[8] G. Wang, X. Li, Z. Wang, APD2: the updated antimicrobial peptide database and its application in peptide design, Nucleic Acids Research 37 (2009) D933–D937.
[9] E.P. Abraham, E. Chain, An enzyme from bacteria able to destroy penicillin, Nature 146 (1940) 837.
[10] D.G. White, M.N. Alekshun, P.F. McDermott, Frontiers in Antimicrobial Resistance: A Tribute to Stuart B. Levy, ASM Press, Washington, DC, 2005.
[11] H. Steiner, D. Hultmark, A. Engstrom, et al., Sequence and specificity of two antibacterial proteins involved in insect immunity, Nature 292 (1981) 246–248.
[12] M. Zasloff, Magainins, a class of antimicrobial peptides from Xenopus skin: isolation, characterization of two active forms, and partial cDNA sequence of a precursor, Journal of Ethnopharmacology 23 (1988) 360.

[13] K.L. Brown, R.E. Hancock, Cationic host defense (antimicrobial) peptides, Current Opinion in Immunology 18 (2006) 24–30.
[14] A.A. Bahar, D. Ren, Antimicrobial peptides, , Pharmaceuticals (Basel) 6 (2013) 1543–1575.
[15] A.J. Ouellette, D.P. Satchell, M.M. Hsieh, S.J. Hagen, M.E. Selsted, Characterization of luminal Paneth cell α-defensins in mouse small intestine. Attenuated antimicrobial activities of peptides with truncated amino termini, Journal of Biological Chemistry 275 (2000) 33969–33973.
[16] R.P. Darveau, A. Tanner, R.C. Page, The microbial challenge in periodontitis, Periodontology 2000 (14) (1997) 12–32.
[17] B.R. Da Silva, V.A. de Freitas, L.G. Nascimento-Neto, V.A. Carneiro, F.V. Arruda, A. S. de Aguiar, et al., Antimicrobial peptide control of pathogenic microorganisms of the oral cavity: a review of the literature, Peptides 36 (2012) 315–321.
[18] T.L. Tollner, C.L. Bevins, G.N. Cherr, Multifunctional glycoprotein DEFB126—a curious story of defensin-clad spermatozoa, Nature Reviews Urology 9 (2012) 365–375.
[19] B.J. Muhialdin, H.L. Algboory, H. Kadum, N.K. Mohammed, N. Saari, Z. Hassan, Antifungal activity determination for the peptides generated by *Lactobacillus plantarum* TE10 against *Aspergillus flavus* in maize seeds, Food Control 109 (2020) 106898.
[20] F. Zahedifard, H. Lee, J.H. No, M. Salimi, N. Seyed, A. Asoodeh, et al., Comparative study of different forms of Jellein antimicrobial peptide on *Leishmania* parasite, Experimental Parasitology 209 (2020) 107823.
[21] Q. Sun, K. Wang, R. She, W. Ma, F. Peng, H. Jin, Swine intestine antimicrobial peptides inhibit infectious bronchitis virus infectivity in chick embryos, Poultry Science 89 (2010) 464–469.
[22] G. Marchini, V. Lindow, H. Brismar, B. Ståbi, V. Berggren, A.K. Ulfgren, et al., The newborn infant is protected by an innate antimicrobial barrier: peptide antibiotics are present in the skin and vernix caseosa, British Journal of Dermatology 147 (2002) 11271134.
[23] M. Gschwandtner, S. Zhong, A. Tschachler, V. Mlitz, S. Karner, A. Elbe-Bürger, et al., Fetal human keratinocytes produce large amounts of antimicrobial peptides: involvement of histone-methylation processes, Journal of Investigative Dermatology 134 (2014) 2192–2201.
[24] M. Wittersheim, J. Cordes, U. Meyer-Hoffert, J. Harder, J. Hedderich, R. Gläser, Differential expression and in vivo secretion of the antimicrobial peptides psoriasin (S100A7), RNase 7, human beta-defensin-2 and-3 in healthy human skin, Experimental Dermatology 22 (2013) 364–366.
[25] M. Lointier, C. Aisenbrey, A. Marquette, J.H. Tan, A. Kichler, B. Bechinger, Membrane pore formation correlates with the hydrophilic angle of histidine-rich amphipathic peptides with multiple biological activities, Biochimica et Biophysica Acta (BBA)-Biomembranes 1862 (2020) 183212.
[26] H.G. Hanstock, J.P. Edwards, N.P. Walsh, Tear lactoferrin and lysozyme as clinically relevant biomarkers of mucosal immune competence, Frontiers in Immunology 10 (2019) 1178.
[27] R. Gläser, U. Meyer-Hoffert, J. Harder, J. Cordes, M. Wittersheim, J. Kobliakova, et al., The antimicrobial protein psoriasin (S100A7) is upregulated in atopic dermatitis and after experimental skin barrier disruption, Journal of Investigative Dermatology 129 (2009) 641–649.
[28] A.M. McDermott, Antimicrobial compounds in tears, Experimental Eye Research 117 (2013) 53–61.

[29] M. Underwood, L. Bakaletz, Innate immunity and the role of defensins in otitis media, Current Allergy and Asthma Reports 11 (2011) 499−507.
[30] J. Sato, M. Nishimura, M. Yamazaki, K. Yoshida, Y. Kurashige, M. Saitoh, et al., Expression profile of drosomycin-like defensin in oral epithelium and oral carcinoma cell lines, Archives of Oral Biology 58 (2013) 279−285.
[31] M. Arias, E.F. Haney, A.L. Hilchie, J.A. Corcoran, M.E. Hyndman, R.E. Hancock, et al., Selective anticancer activity of synthetic peptides derived from the host defence peptide tritrpticin, Biochimica et Biophysica Acta (BBA)-Biomembranes 1862 (2020) 183228.
[32] A. Walrant, A. Bauzá, C. Girardet, I.D. Alves, S. Lecomte, F. Illien, et al., Ionpair-π interactions favor cell penetration of arginine/tryptophan-rich cell-penetrating peptides, Biochimica et Biophysica Acta (BBA)-Biomembranes 1862 (2020) 183098.
[33] C. Imjongjirak, P. Amphaiphan, W. Charoensapsri, P. Amparyup, Characterization and antimicrobial evaluation of SpPR-AMP1, a proline-rich antimicrobial peptide from the mud crab *Scylla paramamosain*, Developmental & Comparative Immunology 74 (2017) 209−216.
[34] R.M. Van Harten, E. Van Woudenbergh, A. Van Dijk, H.P. Haagsman, Cathelicidins: immunomodulatory antimicrobials, Vaccines 6 (2018) 63.
[35] C. Li, C. Zhu, B. Ren, X. Yin, S.H. Shim, Y. Gao, et al., Two optimized antimicrobial peptides with therapeutic potential for clinical antibiotic-resistant *Staphylococcus aureus*, European Journal of Medicinal Chemistry 183 (2019) 111686.
[36] J.D. Wade, F. Lin, M.A. Hossain, R.M. Dawson, Chemical synthesis and biological evaluation of an antimicrobial peptide gonococcal growth inhibitor, Amino Acids 43 (2012) 2279−2283.
[37] J.H. Lee, K.S. Cho, J. Lee, J. Yoo, J. Lee, J. Chung, Diptericin-like protein: an immune response gene regulated by the anti-bacterial gene induction pathway in *Drosophila*, Gene 271 (2001) 233−238.
[38] J. Wang, S. Chou, L. Xu, X. Zhu, N. Dong, A. Shan, et al., High specific selectivity and membrane-active mechanism of the synthetic centrosymmetric α-helical peptides with Gly-Gly pairs, Scientific Reports 5 (2015) 1−19.
[39] F. D'Este, M. Benincasa, G. Cannone, M. Furlan, M. Scarsini, D. Volpatti, et al., Antimicrobial and host cell-directed activities of Gly/Ser-rich peptides from salmonid cathelicidins, Fish and Shellfish Immunology 59 (2016) 456−468.
[40] G. Wang, Human antimicrobial peptides and proteins, Pharmaceuticals (Basel) 7 (2014) 545−594.
[41] C.G. Wilde, J.E. Griffith, M.N. Marra, J.L. Snable, R.W. Scott, Purification and characterization of human neutrophil peptide 4, a novel member of the defensin family, Journal of Biological Chemistry 264 (1989) 11200−11203.
[42] R.I. Lehrer, T. Ganz, Defensins of vertebrate animals, Current Opinion in Immunology 14 (2002) 96−102.
[43] J. Zou, C. Mercier, A. Koussounadis, C. Secombes, Discovery of multiple beta-defensin like homologues in teleost fish, Molecular Immunology 44 (2007) 638−647.
[44] S. Joly, C. Maze, P.B. McCray Jr, J.M. Guthmiller, Human β-defensins 2 and 3 demonstrate strain-selective activity against oral microorganisms, Journal of Clinical Microbiology 42 (2004) 1024−1029.
[45] M.E. Selsted, θ-defensins: cyclic antimicrobial peptides produced by binary ligation of truncated α-defensins, Current Protein and Peptide Science 5 (2004) 365−371.
[46] M. Zanetti, R. Gennaro, D. Romeo, Cathelicidins: a novel protein family with a common proregion and a variable C-terminal antimicrobial domain, FEBS Letters 374 (1995) 1−5.

[47] M. Zanetti, The role of cathelicidins in the innate host defenses of mammals, Current Issues in Molecular Biology 7 (2005) 179–196.
[48] P. de Sousa-Pereira, F. Amado, J. Abrantes, R. Ferreira, P.J. Esteves, R. Vitorino, An evolutionary perspective of mammal salivary peptide families: cystatins, histatins, statherin and PRPs, Archives of Oral Biology 58 (2013) 451–458.
[49] H. Nikawa, C. Jin, S. Makihira, T. Hamada, L.P. Samaranayake, Susceptibility of *Candida albicans* isolates from the oral cavities of HIV-positive patients to histatin-5, The Journal of Prosthetic Dentistry 88 (2002) 263–267.
[50] T. Cabras, M. Patamia, S. Melino, R. Inzitari, I. Messana, M. Castagnola, et al., Pro-oxidant activity of histatin 5 related Cu (II)-model peptide probed by mass spectrometry, Biochemical and Biophysical Research Communications 358 (2007) 277–284.
[51] M. Riool, A. de Breij, P.H. Kwakman, E. Schonkeren-Ravensbergen, L. de Boer, R.A. Cordfunke, et al., Thrombocidin-1-derived antimicrobial peptide TC19 combats superficial multi-drug resistant bacterial wound infections, Biochimica et Biophysica Acta (BBA)-Biomembranes 1862 (2020) 183282.
[52] J. Krijgsveld, S.A. Zaat, J. Meeldijk, P.A. van Veelen, G. Fang, B. Poolman, et al., Thrombocidins, microbicidal proteins from human blood platelets, are C-terminal deletion products of CXC chemokines, Journal of Biological Chemistry 275 (2000) 20374–20381.
[53] Y. Huan, Q. Kong, H. Mou, H. Yi, Antimicrobial peptides: classification, design, application and research progress in multiple fields, Frontiers in Microbiology (2020) 2559.
[54] B. Ericksen, Z. Wu, W. Lu, R.I. Lehrer, Antibacterial activity and specificity of the six human α-defensins, Antimicrobial Agents and Chemotherapy 49 (2005) 269–275.
[55] C. Verma, S. Seebah, S.M. Low, L. Zhou, S.P. Liu, J. Li, et al., Defensins: antimicrobial peptides for therapeutic development, Biotechnology Journal: Healthcare Nutrition Technology 2 (2007) 1353–1359.
[56] S.H. Shin, Y.S. Lee, Y.P. Shin, B. Kim, M.H. Kim, H.R. Chang, et al., Therapeutic efficacy of halocidin-derived peptide HG1 in a mouse model of *Candida albicans* oral infection, Journal of Antimicrobial Chemotherapy 68 (2013) 1152–1160.
[57] E.F. Haney, D. Pletzer, R.E. Hancock, Impact of host defense peptides on chronic wounds and infectionsIn: Chronic Wounds, Wound Dressings and Wound Healing, Springer, Cham, 2018pp. 3–19.
[58] M. Arias, E.F. Haney, A.L. Hilchie, J.A. Corcoran, M.E. Hyndman, R.E. Hancock, et al., Selective anticancer activity of synthetic peptides derived from the host defence peptide tritrpticin, Biochimica et Biophysica Acta (BBA)-Biomembranes 1862 (2020) 183228.
[59] C.D. Fjell, J.A. Hiss, R.E. Hancock, G. Schneider, Designing antimicrobial peptides: form follows function, Nature Reviews. Drug Discovery 11 (2012) 37–51.
[60] K. Yu, J.C. Lo, M. Yan, X. Yang, D.E. Brooks, R.E. Hancock, et al., Anti-adhesive antimicrobial peptide coating prevents catheter associated infection in a mouse urinary infection model, Biomaterials 116 (2017) 69–81.
[61] E. Hazrati, B. Galen, W. Lu, W. Wang, Y. Ouyang, M.J. Keller, et al., Human α-and β-defensins block multiple steps in herpes simplex virus infection, The Journal of Immunology 177 (2006) 8658–8666.
[62] W. Wang, S.M. Owen, D.L. Rudolph, A.M. Cole, T. Hong, A.J. Waring, et al., Activity of α-and θ-defensins against primary isolates of HIV-1, The Journal of Immunology 173 (2004) 515–520.
[63] T.L. Chang, J. Vargas Jr., A. DelPortillo, M.E. Klotman, Dual role of α-defensin-1 in anti−HIV-1 innate immunity, The Journal of Clinical Investigation 115 (2005) 765–773.

[64] C.E. Mackewicz, J. Yuan, P. Tran, L. Diaz, E. Mack, M.E. Selsted, et al., α-Defensins can have anti-HIV activity but are not CD8 cell anti-HIV factors, AIDS (London, England) 17 (2003) F23−F32.
[65] M. Lointier, C. Aisenbrey, A. Marquette, J.H. Tan, A. Kichler, B. Bechinger, Membrane pore-formation correlates with the hydrophilic angle of histidine-rich amphipathic peptides with multiple biological activities, Biochimica et Biophysica Acta (BBA)-Biomembranes 1862 (2020) 183212.
[66] A. Lichtenstein, T. Ganz, M.E. Selsted, R.I. Lehrer, In vitro tumor cell cytolysis mediated by peptide defensins of human and rabbit granulocytes, Blood 68 (1986) 1407−1410.
[67] D. Kraus, J. Deschner, A. Jäger, M. Wenghoefer, S. Bayer, S. Jepsen, et al., Human β-defensins differently affect proliferation, differentiation, and mineralization of osteoblast-like MG63 cells, Journal of Cellular Physiology 227 (3) (2012) 994−1003.
[68] J. Winter, A. Pantelis, R. Reich, M. Martini, D. Kraus, S. Jepsen, et al., Human beta-defensin-1, -2, and -3 exhibit opposite effects on oral squamous cell carcinoma cell proliferation, Cancer Investigation 29 (2011) 196−201.
[69] N. Escher, B. Spies-Weisshart, M. Kaatz, C. Melle, A. Bleul, D. Driesch, et al., Identification of HNP3 as a tumour marker in CD4+ and CD4− lymphocytes of patients with cutaneous T-cell lymphoma, European Journal of Cancer 42 (2006) 249−255.
[70] M.Y. Radeva, F. Jahns, A. Wilhelm, M. Glei, U. Settmacher, K.O. Greulich, et al., Defensin α-6 (DEFA 6) overexpression threshold of over 60-fold can distinguish between adenoma and fully blown colon carcinoma in individual patients, BMC Cancer 10 (2010) 1−6.
[71] D. Yang, Q. Chen, O. Chertov, J.J. Oppenheim, Human neutrophil defensins selectively chemoattract naive T and immature dendritic cells, Journal of Leukocyte Biology 68 (2000) 9−14.
[72] Y. Döring, M. Drechsler, S. Wantha, K. Kemmerich, D. Lievens, S. Vijayan, et al., Lack of neutrophil-derived CRAMP reduces atherosclerosis in mice, Circulation Research 110 (2012) 1052−1056.
[73] P. Chotjumlong, J.G. Bolscher, K. Nazmi, V. Reutrakul, C. Supanchart, W. Buranaphatthana, et al., Involvement of the P2X7 purinergic receptor and c-Jun N-terminal and extracellular signal-regulated kinases in cyclooxygenase-2 and prostaglandin E2 induction by LL-37, Journal of Innate Immunity 5 (2013) 72−83.
[74] E.V. Valore, D.J. Wiley, T. Ganz, Reversible deficiency of antimicrobial polypeptides in bacterial vaginosis, Infection and Immunity 74 (2006) 5693−5702.
[75] W. Han, W. Wang, K.A. Mohammed, Y. Su, α-Defensins increase lung fibroblast proliferation and collagen synthesis via the β-catenin signaling pathway, The FEBS Journal 276 (2009) 6603−6614.
[76] F. Niyonsaba, H. Ushio, N. Nakano, W. Ng, K. Sayama, K. Hashimoto, et al., Antimicrobial peptides human β-defensins stimulate epidermal keratinocyte migration, proliferation and production of proinflammatory cytokines and chemokines, Journal of Investigative Dermatology 127 (2007) 594−604.
[77] A. Baroni, G. Donnarumma, I. Paoletti, I. Longanesi-Cattani, K. Bifulco, M.A. Tufano, et al., Antimicrobial human beta-defensin-2 stimulates migration, proliferation and tube formation of human umbilical vein endothelial cells, Peptides 30 (2009) 267−272.
[78] A.E. King, R.W. Kelly, J.-M. Sallenave, A.D. Bocking, J.R.G. Challis, Innate immune defences in the human uterus during pregnancy, Placenta 28 (2007) 1099−1106.
[79] A.W. Horne, S.J. Stock, A.E. King, Innate immunity and disorders of the female reproductive tract, Reproduction (Cambridge, England) 135 (2008) 739.

[80] E. Com, F. Bourgeon, B. Evrard, T. Ganz, D. Colleu, B. Jégou, et al., Expression of antimicrobial defensins in the male reproductive tract of rats, mice, and humans, Biology of Reproduction 68 (2003) 95–104.
[81] L. Yang, T.M. Weiss, R.I. Lehrer, H.W. Huang, Crystallization of antimicrobial pores in membranes: magainin and protegrin, Biophysical Journal 79 (2000) 2002–2009.
[82] S. Qian, W. Wang, L. Yang, H.W. Huang, Structure of the alamethicin pore reconstructed by x-ray diffraction analysis, Biophysical Journal 94 (2008) 3512–3522.
[83] K. He, S.J. Ludtke, W.T. Heller, H.W. Huang, Mechanism of alamethicin insertion into lipid bilayers, Biophysical Journal 71 (1996) 2669–2679.
[84] D. Münch, H.G. Sahl, Structural variations of the cell wall precursor lipid II in Gram-positive bacteria—impact on binding and efficacy of antimicrobial peptides, Biochimica et Biophysica Acta (BBA)-Biomembranes 1848 (2015) 3062–3071.
[85] N. Malanovic, K. Lohner, Antimicrobial peptides targeting gram-positive bacteria, Pharmaceuticals 9 (2016) 59.
[86] O. Levy, Antimicrobial proteins and peptides: anti-infective molecules of mammalian leukocytes, Journal of Leukocyte Biology 76 (2004) 909–925.
[87] D. Wade, A. Boman, B. Wåhlin, C.M. Drain, D. Andreu, H.G. Boman, et al., All-D amino acid-containing channel-forming antibiotic peptides, Proceedings of the National Academy of Sciences 87 (1990) 4761–4765.
[88] Y.H. Nan, K.H. Park, Y. Park, Y.J. Jeon, Y. Kim, I.S. Park, et al., Investigating the effects of positive charge and hydrophobicity on the cell selectivity, mechanism of action and anti-inflammatory activity of a Trp-rich antimicrobial peptide indolicidin, FEMS Microbiology Letters 292 (2009) 134–140.
[89] A. Nijnik, R. Hancock, Host defence peptides: antimicrobial and immunomodulatory activity and potential applications for tackling antibiotic-resistant infections, Emerging Health Threats Journal 2 (2009) 7078.
[90] A.L. Hilchie, K. Wuerth, R.E.W. Hancock, Immune modulation by multifaceted cationic host defense (antimicrobial) peptides, Nature Chemical Biology 9 (2013) 761–768.
[91] A.M. van der Does, P.S. Hiemstra, N. Mookherjee, Antimicrobial host defence peptides: immunomodulatory functions and translational prospects, Antimicrobial Peptides (2019) 149–171.
[92] R.E. Hancock, E.F. Haney, E.E. Gill, The immunology of host defence peptides: beyond antimicrobial activity, Nature Reviews Immunology 16 (2016) 321–334.
[93] N. Mookherjee, K.L. Brown, D.M.E. Bowdish, S. Doria, R. Falsafi, K. Hokamp, et al., Modulation of the TLR-mediated inflammatory response by the endogenous human host defense peptide LL-37, The Journal of Immunology 176 (2006) 2455–2464.
[94] D. Yang, Q. Chen, A.P. Schmidt, G.M. Anderson, J.M. Wang, J. Wooters, et al., LL-37, LL-37, the neutrophil granule–and epithelial cell–derived cathelicidin, utilizes formyl peptide receptor–like 1 (FPRL1) as a receptor to chemoattract human peripheral blood neutrophils, monocytes, and T cells, The Journal of Experimental Medicine 192 (2000) 1069–1074.
[95] A. Neumann, L. Völlger, E.T. Berends, E.M. Molhoek, D.A. Stapels, M. Midon, et al., Novel role of the antimicrobial peptide LL-37 in the protection of neutrophil extracellular traps against degradation by bacterial nucleases, Journal of Innate Immunity 6 (2014) 860–868.
[96] G.S. Tjabringa, D.K. Ninaber, J.W. Drijfhout, K.F. Rabe, P.S. Hiemstra, Human cathelicidin LL-37 is a chemoattractant for eosinophils and neutrophils that acts via formyl-peptide receptors, International Archives of Allergy and Immunology 140 (2006) 103–112.

[97] V.K. Lishko, B. Moreno, N.P. Podolnikova, T.P. Ugarova, Identification of human cathelicidin peptide LL-37 as a ligand for macrophage integrin αMβ2 (Mac-1, CD11b/CD18) that promotes phagocytosis by opsonizing bacteria, Research and Reports in Biochemistry 2016 (2016) 39−55.
[98] J. Agier, E. Brzezińska-Błaszczyk, J. Agier, E. Brzezińska-Błaszczyk, Cathelicidins and defensins regulate mast cell antimicrobial activity, Postepy higieny i medycyny doswiadczalnej 70 (2016) 618−636.
[99] Y. Cao, F. Chen, Y. Sun, H. Hong, Y. Wen, Y. Lai, et al., LL-37 promotes neutrophil extracellular trap formation in chronic rhinosinusitis with nasal polyps, Clinical & Experimental Allergy 49 (7) (2019) 990−999.
[100] S.H. Yoon, I. Hwang, E. Lee, H.J. Cho, J.H. Ryu, T.G. Kim, et al., Antimicrobial peptide Ll-37 drives rosacea-like skin inflammation in an Nlrp3-dependent manner, Journal of Investigative Dermatology 141 (2021) 2885−2894.
[101] P. Méndez-Samperio, The human cathelicidin hCAP18/LL-37: a multifunctional peptide involved in mycobacterial infections, Peptides 31 (2010) 1791−1798.
[102] S.A. Abdel Monaim, E.J. Ramchuran, A. El-Faham, F. Albericio, B.G. de la Torre, Converting teixobactin into a cationic antimicrobial peptide (AMP), Journal of Medicinal Chemistry 60 (2017) 7476−7482.
[103] A. Giuliani, G. Pirri, S.F. Nicoletto, Antimicrobial peptides: an overview of a promising class of therapeutics, Open Life Sciences 2 (2007) 1−33.
[104] R. Shaykhiev, C. Beisswenger, K. Kändler, J. Senske, A. Püchner, T. Damm, et al., Human endogenous antibiotic LL-37 stimulates airway epithelial cell proliferation and wound closure, American Journal of Physiology-Lung Cellular and Molecular Physiology 289 (2005) L842−L848.
[105] L. Boge, A. Umerska, N. Matougui, H. Bysell, L. Ringstad, M. Davoudi, et al., Cubosomes post-loaded with antimicrobial peptides: characterization, bactericidal effect and proteolytic stability, International Journal of Pharmaceutics 526 (2017) 400−412.
[106] R. Nordström, L. Nyström, O.C. Andrén, M. Malkoch, A. Umerska, M. Davoudi, et al., Membrane interactions of microgels as carriers of antimicrobial peptides, Journal of Colloid and Interface Science 513 (2018) 141−150.
[107] J. Håkansson, L. Ringstad, A. Umerska, J. Johansson, T. Andersson, L. Boge, et al., Characterization of the in vitro, ex vivo, and in vivo efficacy of the antimicrobial peptide DPK-060 used for topical treatment, Frontiers in Cellular and Infection Microbiology 9 (2019) 174.
[108] C.D. Fjell, J.A. Hiss, R.E. Hancock, G. Schneider, Designing antimicrobial peptides: form follows function, Nature Reviews. Drug Discovery 11 (2012) 124.
[109] K.E. Greber, M. Dawgul, Antimicrobial peptides under clinical trials, Current Topics in Medicinal Chemistry 17 (2017) 620−628.
[110] W.J. Velden, T.M. van Iersel, N.M. Blijlevens, J.P. Donnelly, Safety and tolerability of the antimicrobial peptide human lactoferrin 1−11 (hLF1−11), BMC Medicine 7 (2009) 1−8.
[111] S. Tokumaru, K. Sayama, Y. Shirakata, H. Komatsuzawa, K. Ouhara, Y. Hanakawa, et al., Induction of keratinocyte migration via transactivation of the epidermal growth factor receptor by the antimicrobial peptide LL-37, The Journal of Immunology 175 (2005) 4662−4668.
[112] H.S. Sader, K.A. Fedler, R.P. Rennie, S. Stevens, R.N. Jones, Omiganan pentahydrochloride (MBI 226), a topical 12-amino-acid cationic peptide: spectrum of antimicrobial activity and measurements of bactericidal activity, Antimicrobial Agents and Chemotherapy 48 (2004) 3112−3118.

# Antimicrobial peptides from marine environment

M.S. Aishwarya[1], R.S. Rachanamol[2], A.R. Sarika[3], J. Selvin[4] and A.P. Lipton[1]

[1]Centre for Marine Science & Technology (CMST), Manonmaniam Sundaranar University, Rajakkamangalam, Tamil Nadu, India, [2]Department of Microbiology, A.J. College of Science & Technology, Thonnakkal, Kerala, India, [3]Kerala State Council for Science, Technology and Environment, Thiruvananthapuram, Kerala, India, [4]Department of Microbiology, Pondicherry University, Kalapet, Puducherry, India

## 11.1 Introduction

The marine environment is an excellent source of novel antimicrobial compounds. About half of the global biodiversity comprise of marine organisms [1] and hence could be projected as a major reservoir of natural compounds with potent bioactivities. The extreme, competitive, and hostile environment coupled with the astounding biodiversity provides prospects for diverse bioactive substances. Numerous drugs were produced out of marine natural products and some of which are currently under clinical trials [2]. Among the various bioactive compounds produced by the marine organisms, the marine antimicrobial peptides (AMPs) have received widespread interest in the recent times.

AMPs are bioactive peptides which form vital components of the innate immune response. In certain organisms, they act as immunomodulators [3] and are part of the primary defense mechanism against pathogens [4]. Most AMPs exhibit a broad spectrum activity against invading microorganisms; the infectivity varies with the nature of infecting microbes [5–7]. In addition to the broad spectrum activity, the reduced toxicity and lesser chances of resistance development by the target cells [8] also make them ideal drug candidate with potential to replace the antibiotics.

AMPs are mostly cationic, amphipathic, and form variety of secondary structures [9]. These diverse structural forms along with other factors like size, hydrophobic nature, the net charge, amphipathic stereo geometry, and self-association of the peptide to the membrane also contribute to the activity spectrum of AMPs [9,10]. The target for the AMPs is signified by the cell membrane composition of the Gram + ve and Gram − ve bacteria and hence could preferentially target the pathogen without affecting the host cells [11]. Not only against bacteria, but several marine AMPs isolated to date have shown potent activities against wide range of microorganisms including fungi, actinomycetes, viruses, yeasts, protozoa, and parasites including nematodes [12,13].

Several marine AMPs with pronounced biological activities were purified from diverse marine organisms till date as evidenced from numerous studies [14,15]; however, still the vital marine source has not been effusively explored and extensive studies are required on to isolation and developing of novel drug leads. This chapter provides an extensive view of the marine AMPs isolated so far from diverse marine organisms and overview of their structure and activity spectrum.

## 11.2 Antimicrobial peptides from marine invertebrates

Marine invertebrates which include the sponges, tunicates, and molluscs have imparted the largest number of marine-derived secondary compounds.

### 11.2.1 Antimicrobial peptides from marine sponges

Several promising bioactive compounds were elucidated from marine sponge extracts over the years. One of the prominent features of marine sponge bioactive compounds is the ability of a single compound to express several bioactivities. This unique nature of sponge metabolites ranked them top among the natural products [16]. Marine sponges are a major source of amino acid-derived metabolites. Sponge AMPs include cyclic peptides (like depsipeptides, proline-rich cyclopeptides), linear peptides, peptide lactones. There are a major papuamides class of cyclic depsipeptides, Callipeltin A, Neamphamide A, Homophymine A−E, Mirabamides, Theopapuamides, and Celebesides A and B. The first AMP isolated from the marine sponge with bioactivity was Discodermin A with D-amino acids and unusual amino acids like tert-leucine, cysteionic acid, sarcosine [17]. Similarly, *Theonella* sp. synthesize bicyclic peptides with prolific bioactivity which possess unusual amino acids, histidinolamine, 3-methyl-p-bromoohenylalaninme, 2-amino-4-hydroxyadipic acid [18]. Marine sponges are also rich sources of ribosomal and nonribosomal peptides with potential antifungal activity. Fusetani and Matsunaga [19] identified antifungal peptides from marine sponges are found to be nonribosomal peptides with unusual amino acids. The AMPs from marine sponges are detailed in Table 11.1.

### 11.2.2 Antimicrobial peptides from marine molluscs

The phylum Mollusca consists of soft-bodied invertebrates which includes slugs, snails, squid, mussels, octopus, bivalves, and gastropods [46]. The communication and defense of the phyla is through secondary metabolite production and the AMPs from mussels are formed in hemocytes, stored as granules and react to bacterial endotoxins by attachment to cell and degranulation [47]. Marine mussels produce cysteine-rich AMPs relating to the defensin, myticin, mytilin, and mytimycin families [48,49]. A prototypal example of such peptide with unusual cysteine (104-aa long with 10 cysteine residues and mw of 11.28 kDa) derived from the hemolymph

Table 11.1 Antimicrobial peptides from marine sponges and their activity.

| Marine sponge | Peptide | Chemical nature | Biological activity | References |
|---|---|---|---|---|
| Aciculites orientalis | Aciculitins A–C | Bicyclic peptide | Antifungal (against Candida albicans) | [20] |
| Theonella species | Mutremdamide A and Koshikamides C–H | Depsipeptides | Antiviral (HIV) | [21] |
| Geodia sp. | Geodiamolides A and B | Cyclic peptide | C. albicans | [22] |
| Lithistida sp. | Theonegramide | Cyclopeptides | C. albicans ATCC 32354 strain | [23] |
| Discodermia kiiensis | Discodermins | Tetradecapeptides | Antibacterial (G + ve and − ve), antifungal | [24] |
| Theonella sp. | Koshikamides F and H | Depsipeptide | Anti-HIV | [25,26] |
| Theonella swinhoei | Theonellamide G Cyclolithistide A Nagahamide A | Bicyclic glycopeptides Cyclodepsipeptide Cyclodepsipeptide | Antifungal (C. albicans) Antifungal Antibacterial | [27–29] |
| Jaspis (Fiji Islands) | Jaspamide (jasplakinolide) | Cyclic heptapeptide | Candida | [25] |
| Halichondria cylindrata | Halicylindramide A–C, Halicylindramide D, E | Cyclic depsipeptide | Mortierella ramanniana | [30] |
| Siliquariaspongia mirabilis | Mirabamides A–D | Cyclic depsipeptide | Antiviral | [31,32] |
| Stelletta clavosa | Mirabamides E–H | Depsipeptides | Anti-HIV | [33] |
| Homophymia sp. | Homophymine A | Cyclic depsipeptide | Anti-HIV | [34] |
| Neosiphonia superstes | Neosiphoniamolide A | Cyclic depsipeptide | Antifungal | [35] |

(Continued)

Table 11.1 (Continued)

| Marine sponge | Peptide | Chemical nature | Biological activity | References |
|---|---|---|---|---|
| *Siliquariaspongia mirabilis* | Theopapuamides B and C, Celebesides A–C | Depsipeptide | Antifungal, antiviral | [36] |
| *Sidonops microspinosa* | Microsposamide | Cyclic depsipeptide | Anti-HIV | [37] |
| *Neamphius huxleyi* | Neamphamide A | Cyclic depsipeptide | Anti-HIV | [38] |
| *Discodermia* sp. | Discobahamin A and B | Cyclic depsipeptide | Antifungal (*C. albicans*) | [39] |
| *Stelletta clavosa* | Stellettapeptin A and B | Cyclic depsipeptide | Anti-HIV | [40] |
| *Callyspongia aerizusa* | Callyaerins A–F and H | Proline-rich peptides | *B. subtilis*, *C. albicans*, anti tuburcular | [41] |
| *Callipeltin* sp. and *Latrunculia* sp. | Callipeltins | Cyclic depsipeptide | Antifungal, antiviral activity | [42] |
| *Stylissa caribica* | Stylissamide G | Proline-rich peptides | *C. albicans*, *Trichophyton mentagrophytes*, and *Microsporum audouinii* | [43] |
| *Microscleroderma herdmani* | Microsclerodermins J and K | Cyclic peptide | Antifungal | [44] |
| *Stylissa caribica* | Stylissamide G | Proline-rich peptides | *C. albicans*, *T. mentagrophytes* and *M. audouinii* | [45] |

of the mussels *Mytilus coruscus* is Myticusin-1 [46]. This molecule had shown potent activity against Gram + vein comparison to Gram-ve bacteria and fungus. An identical AMP Myticusin-beta (11.18 kDa) was isolated from the mantle of the hard-shelled mussel, *M. coruscus* [50]. Yao et al. [51] characterized a six-cysteine defensin-2 (HdDef) similar to an arthropod defensin, from small abalone *Haliotis diversicolor*. As reported by Yao et al. more than 40 different mollusc AMPs are recorded in the antimicrobial sequences database (http://aps.unmc.edu/AP/main.php).

## 11.2.3 Antimicrobial peptides from ascidians

### 11.2.3.1 Tunicates

Tunicates belong to marine sessile filter-feeding invertebrates and have around 3010 species with the subphylum ascidians, sea squirts, tunicates, sea tulips, sea livers, etc. The AMPs of tunicates defined as evolutionary conserved effector molecules and are produced as a result of the host defense mechanisms and are usually circulated in the hemocytes. Turgencin A and B are novel, potent Cys-rich AMPs isolated from *Synoicum turgens* (Arctic sea squirt) with 36 amino acid residues and 3 disulfide bridges [52]. The authors computationally predicted Turgencin A to be a cationic peptide flanked by two lysine residues on either end displaying an internal stretch (residues 18−27) of an unusual amino acid PGGW as central core, a 10-residue-long sequence that was used to produce a novel-truncated lead AMP. In their extended study, the optimized peptide StAMP-9 displayed bactericidal activity and no cytotoxic effects against mammalian cells [52,53]. Two novel peptides named halocyntin and papillosin comprising 26 and 34-aa residues were isolated and characterized from *Halocynthia papillosa* hemocytes. While both displayed activity against Gram + ve and Gram − ve bacteria, papillosin was more potent [54]. Dicynthaurin, a novel AMP (30 residue monomers) of 6.2 kDa from hemocytes of a tunicate, *Halocynthia aurantium*; the sequence of which was not homologous to any earlier identified peptides. The circular dichroism studies of dicynthaurin in membrane-mimetic environments show mostly alpha-helical conformations. The peptide showed activity against Gram + ve and Gram − ve bacteria [55].

## 11.2.4 Antimicrobial peptides from crustaceans

Crustaceans (such as crab, shrimp, and krill) are less complex organisms that are devoid of an adaptive immune system. These invertebrates lacking an adaptive immune system solely depend on internal innate immune factors like AMPs and lectins formed by circulating immune-competent cells like hemocytes. These peptides are less specific and are fast and efficient against microbes [56] and nontoxic toward eukaryotic host that make them desirable candidates to that of conventional antibiotics. Crustacean AMPs are usually cationic, gene-encoded molecules like hemocytes, which may be derived from unusual proteins [56]. The peptide families like antilipopolysaccharide factor are common in crustaceans [57]. Comprehensively studied in arthropods, the AMPs appear as a component for defense during the course of evolution [47].

AMP studies in crustaceans initiated in mid-1990s [56]. A proline-rich AMP with broad inhibitory spectrum was purified from the hemocyte of shore crab, *Carcinus maenas* that was supposedly one of the first AMP reported and that has 60% identity toward bactenecin 7 [58], and crustins and penaeidins, from the hemolymph of estuarine crab and shrimp in 2010 [56]. Armadillidin, the glycine-rich AMP was purified from the terrestrial isopod *Armadillidium vulgare* [59,60]. Arasin 1, a proline-arginine-rich AMP with 37-aa isolated from the spider crab *Hyas araneus*, possessed antifungal activity [61]. Callinectin, AMP similar to Arasin 1, was purified from the blue crab *Callinectes sapidus*, with three isomers and differs in functional group on the tryptophan residue [62].

Hyastatin (11.7 kDa Gly-rich peptide) from *H. araneus* displayed antimicrobial activity against yeasts, Gram + ve and Gram − ve bacteria [63]. A Gly-rich AMP (Spgly-AMP) gene identified in mud crab, *Scylla paramamosain*, putatively encoded a 26-aa signal peptide and 37-aa mature peptide which displayed broad spectrum antibacterial activity and high thermal stability [64]. A 17-residue cationic β-sheet AMP, Tachyplesin I (mw 2.36 Da) from horseshoe crab *Tachypleus tridentatus* with antiparallel β-sheet connected by a β-turn [65] was highly stable at high temperature and 0.1% trifluroacetic acid and displayed broad spectrum antibacterial activity [66]. Three types of this peptide had so far been identified—Tachyplesin I, II, and III; all having amphipathic β-hairpin structure, are highly positively charged, and vary by only very few amino acid residues with antimicrobial activities. The AMP, scygonadin from the male gonads of mud crab *S. paramamosain*, scygonadin [67] and its homologous protein SCY2 [68] inhibited several pathogens. Another peptide, Scyreprocin and its recombinant product from the same species displayed potent antifungal and antibiofilm activities [69].

Another group of peptides with unique structure are the Penaeidins present in the shrimp *Penaeus vannamei* (Decapoda) with antimicrobial activity mainly against Gram + ve bacteria [70]. AMP from tiger shrimp named Monodoncin (mw 5−8 kDa and 55-aa) composed of two β-sheets in N-terminus and one α-helix in C-terminus showed antimicrobial property against Gram + ve and Gram − ve bacteria and fungi [71]. A group of AMPs related to β-defensins are the big defensins, first isolated from marine chelicerate, the horseshoe crab *T. tridentatus* which inhibited bacteria and fungi. Big defensins are largely described in marine organisms, mainly molluscs and, to a much lesser extent, in ancestral chelicerates (horseshoe crabs) and early-branching chordates (amphioxus) [72]. The diverse peptides from crustaceans so far identified with potent antimicrobial activities are indicative of the fact that they are unparalleled resources of bioactive substances.

## *11.2.5 Antimicrobial peptides from marine worms*

Marine worms included under several different phyla Nematoda, Platyhelminthes, Annelida, Chaetognatha, Hemichordata, and Phoronida. Perinerin is a 51-aa residue cationic, hydrophobic, and linear AMP with arginine and four cysteine residues was isolated from the Asian marine clamworm *Perinereis aibuhitensis*, a marine polychaete, which lives in the sediment of estuaries [73]. Ms-Hemerycin, AMP with 14-aa residues and mw 17.36 kDa from the polychaete *Marphysa sanguinea*, a marine lugworm inhibited Gram + ve and Gram + ve bacteria [74].

### 11.2.6 Antimicrobial peptides from Cnidaria

The phyla Cnidaria comprises more than 10,000 species and have no external defense structures. The AMPs are biophylaxis systems of the species. The antimicrobial secondary metabolites of these organisms are produced from the need for self-protection. The metabolites derived from terpene biosynthesis dominate among the compounds produced by the cnidarians [75]. To date there are only few studies on the AMPs encoded by the coral genomes [76].

Damicornin, a 40 residue AMP is the first reported scleractinian AMP from a scleractinian coral, *Pocillopora damicornis*. The cationic AMP has six cysteine molecules linked by three intramolecular disulfide bridges and C-terminal amidation, which is similar to defensins. It was active against Gram + ve bacteria and the fungus, *Fusarium oxysporum* [77]. Aurelin, AMP from the mesoglea of a scyphoid jellyfish *Aurelia aurita*, is a 40-aa cationic peptide with mw 4.296 kDa, and had modest antibacterial activity against both Gram + ve and Gram − ve bacteria [78].

### 11.2.7 Antimicrobial peptides from Echinodermata

The humoral immunity of invertebrates is characterized by the production of proteins with hemolytic and hemagglutinating properties within the cells [79] and the synthesis of AMPs secreted from the cell [80]. Strongylocins 1 and 2 are cysteine-rich cationic AMPs from coelomocyte of the green sea urchin, *Strongylocentrotus droebachiensis*, similar to defensins with potent activities against Gram + ve and Gram − ve bacteria [81]. AMPs from the *Cucumaria frondosa* coelomic fluid (mw <6 kDa) have also shown broad spectrum antibacterial activity [82]. A detailed report about AMPs from echinoderms can be read from Scheibling and Hatcher (2013) [83].

## 11.3 Antimicrobial peptides from marine microorganisms

### 11.3.1 Antimicrobial peptides from marine bacteria

The AMPs produced by bacteria can be categorized based on their mode of synthesis—whether by the ribosomal (bacteriocin) or the nonribosomal pathway [84].

### 11.3.1.1 Ribosomal antimicrobial peptides (bacteriocins) from marine bacteria

Bacteriocins are ribosomally synthesized proteinaceous compounds which are toxic to bacteria closely related to the producer strain [85] and to which the producer has a specific immunity mechanism. While they are best characterized in the lactic acid bacteria and *Bacillus* species, it has been projected that the bacteriocins are produced by most of the bacteria, and hence a vast diversity of potential compounds could be isolated to be exploited for therapeutic purposes [86,87]. The potential of bacteriocins as probiotics, preservatives, and as antibiotic substitutes has been widely studied [88–90]. Harveyicin, the first marine bacteriocin was isolated from *Vibrio harveyi* [91] in the study in which 795 strains of *Vibrio* spp. from Galveston Island, Texas were screened [92]. The studies on the marine animal-associated microorganisms have shown the bacteriocins producers belong to the genera *Aeromonas, Vibrio, Pseudoalteromonas, Alteromonas,* and *Flavobacterium* [93,94]. Ahmad and Hamid [95] isolated *Pseudomonas putida* from shark skin that produced a potent bacteriocin effective against a wide range of Gram + ve and Gram − ve bacteria. The bacteriocin was partially purified and had an apparent molecular weight of ~32 kDa. Sarika et al. [96] reported the broad spectrum activity of a marine bacteriocin PSY2 of *Lactococcus lactis* PSY2 isolated from grouper fish. The *Bacillus subtilis* strains from marine sponges possessed subtilomycin biosynthetic cluster; the subtilomycin bacteriocin inhibited several pathogens [97].

### 11.3.1.2 Nonribosomal antimicrobial peptides from marine bacteria

Marine microbes contribute major part of discovery of non-ribosomal antimicrobial peptides with promising bioactivities [98]. These peptides are the secondary metabolites synthesized by nonribosomal peptide synthetases (NRPSs) found only in bacteria, cyanobacteria, and fungi [99,100]. An extensive review on the nonribosomal peptides with antimicrobial potential was made by Agrawal et al. [101]. Table 11.2 depicts the antimicrobial NRPs isolated from marine bacteria. Bogorol A, an AMP active against methicillin-resistant *Staphylococcus aureus* (MRSA) and vancomycin-resistant Enterococcal strains (VRE), was isolated from marine *Bacillus laterosporus* collected from Papua New Guinea [115]. The broad spectrum thiazolyl peptide antibiotics, Nocathiacins I–III from *Nocardia* sp. which shared structural similarities to glycothiohexide-alpha [116,117]. Two cyclic thiopeptide antibiotics, YM-266183 and YM-266184 from marine sponge isolate *Bacillus cereus*, were active against *Staphylococci* and *Enterococci* including multiple drug-resistant strains, with no activity against Gram − ve bacteria [102].

### 11.3.2 Antimicrobial peptides from marine actinomycetes

Actinomycetes are Gram + ve bacteria with high G + C ratio and they are one of the most studied and exploited class of bacteria for their potential to produce a

**Table 11.2** Antimicrobial NRPs from marine bacteria.

| S. no. | NRPs | Chemical structure | Source | Biological target | References |
|---|---|---|---|---|---|
| 1. | YM-266183 and YM-266184 | Cyclic peptide | *B. cereus* | *Staphylococci, Enterococci* | [102,103] |
| 2. | Cyclo-(glycyl-l-seryl-l-prolyl-l-glutamyl) and cyclo-(glycyl-l-prolyl-l-glutamyl) | Cyclic peptide | *Ruegeria* sp. | *B. subtilis* | [104] |
| 3. | Tauramamide | Lipopeptide | *B. laterosporus* | *Enterococcus* sp. | [105] |
| 4. | Tetrapeptide cyclo-isoleucyl-prolyl-leucyl alanyl and cyclophenyl alanyl-prolyl-leucyl prolyl | Cyclic tetrapeptide | *Pseudomonas* sp. | Marine bacteria | [106] |
| 5. | Unnarmicin A and C | Depsipeptide | *Photobacterium*sp. | Pseudovibrio | [107] |
| 6. | Thiopeptide TP-1161 | Cyclic peptide | *Nocardiopsis* sp. | Gram- + ve bacteria | [108] |
| 7. | Solonamide A and B | Cyclodepsipeptide | *Photobacterium halotolerans* | *S. aureus* | [109] |
| 8. | Peptidolipins B–F | Lipopeptide | *Nocardia* sp. | MRSA, MSSA | [110] |
| 9. | Kocurin | Cyclic peptide | *Kocuria palustris* | MRSA | [111] |
| 10. | Ngercheumicin F–I | Cyclodepsipeptide | *P. halotolerans* | *S. aureus* | [112] |
| 11. | Gageostatins A–C | Linear lipopeptides | *B. subtilis* | Antibacterial and antifungal | [113] |
| 12. | Gageopeptides A–D | Lipopeptides | *Bacillus* strain 109GGC020 | Antibacterial *Rhizoctonia solani, Botrytis cinerea,* and *Colletotrichum acutatum* | [114] |

*Source:* Adapted from S. Agrawal, D. Acharya, A. Adholeya, C.J. Barrow, S.K. Deshmukh, Nonribosomal peptides from marine microbes and their antimicrobial and anticancer potential, Frontiers in Pharmacology 8 (2017) 828, https://doi.org/10.3389/fphar.2017.00828.

wide array of biological molecules [118,119]. Lobophorin G, Lobophorin A, and Lobophorin B with molecular weights 1.23, 1.16, and 1.19 kDa, respectively, were isolated from a marine *Actinomycete* strain, with moderate to high activity against *Mycobacterium tuberculosis* and *Mycobacterium bovis* BCG; more antimycobacterial potency shown by Lobophorin G [120].

Mathermycin, lantibiotic from a marine-derived *Marinactinospora thermotolerans* strain from deep-sea sediment of the northern South China Sea was reported by Chen et al. [121] which exhibited antimicrobial activity toward a *Bacillus* indicator strain. Mohangamide A from marine *Streptomyces* sp. exhibited strong activity against *Candida albicans* [122]. Metagenomic methods are currently being used to isolate the genes of microbes that are unable to be cultivated [123].

## 11.3.3 Antimicrobial peptides from marine fungi

Numerous AMPs have so far been isolated from marine fungi. Youssef et al. [113] have detailed about 131 marine peptides isolated from 17 marine fungi having multitude of bioactivities. Unguisins A and B are two cyclic heptapeptides from the fungus *Emericella unguis* isolated from a *Stomolopus meliagris* (medusa) collected in Venezuelan water [124] showed activity against *S. aureus* and *Vibrio parahaemolyticus*. Lajollamide A (a pentapeptide) with known compounds was isolated from filamentous marine fungus *Asteromyces cruciatus* 763 [125]. Guangomides A and B (cyclic depsipeptides) with a new destruxin derivative from sponge-derived fungus [126] showed weak antibacterial activity against *Staphylococcus epidermidis* and *Enterococcus durans*. A novel peptide of mixed polyketide synthase-nonribosomal peptide synthetase (PKS/NRPS) origin and a linear peptide 11-O-methylpseurotin A, which was active against *Saccharomyces cerevisiae*, was isolated from a marine *Aspergillus fumigatus* [127]. Emericellamides A and B (cyclic depsipeptides) isolated from certain *Emericella* sp. cocultured with marine actinomycete *Salinispora arenicola* showed modest activity against MRSA [128].

The marine fungal strain *Aspergillus* sp. Af119 produced a new cyclic heptapeptide containing aminobutyric acid in the ring, unguisin E [129]. One of the isolated peptide derivatives from marine *Penicillium citrinum* showed potent antibacterial activity against *S. aureus* [130]. Scopularides A and B (cyclodepsipeptides) from the marine sponge-derived fungus *Scopulariopsis brevicaulis* showed weak activity against Gram + ve bacteria [131]. Alternaramide, a cyclic pentadepsipeptide from the marine fungus *Alternaria* sp. SF-5016, showed weak antibacterial activity against *B. subtilis* and *S. aureus* [132]. Trichoderins A, A1, and B aminolipopeptides from marine sponge-derived *Trichoderma* sp. possessed antimycobacterial activity against *Mycobacterium smegmatis*, *M. bovis* BCG, and *M. tuberculosis* H37Rv [133]. Sclerotides A and B (cyclic hexapeptides) from marine halotolerant *Aspergillus sclerotiorum* PT06−1 showed moderate antifungal activity; compound 2 also showed weak antibacterial activity [134].

## 11.4 Antimicrobial peptides from marine vertebrates

### 11.4.1 Antimicrobial peptides from marine fishes

The fishes have acquired defense mechanisms and secrete biologically powerful bioactive molecules to protect themselves from diverse microbes in the aquatic environment. Various natural endogenous AMPs have been from fishes. Several studies examined the antimicrobial properties of skin mucus of fishes [135,136]. Epinecidin-1, AMP from the orange-spotted grouper (*Epinephelus coioides*), possessed several bioactivities including antimicrobial activity [137,138]. Collagencin, a broad spectrum AMP isolated from collagen hydrolysate of Atlantic mackerel by-products inhibited most of the tested bacteria with potent activity against *S. aureus* [139]. Oncorhyncin II, isolated from rainbow trout skin secretions [140], showed activity against Gram + ve as well as Gram − ve bacteria. Acosta et al. [141] identified three AMPs—Oreoch-1, (25.24 kDa); Oreoch-2 (29.81 kDa), and Oreoch-3 (36.54 kDa), from the Teloest fish *Oreochromis niloticus* which showed relatively different antimicrobial and cytotoxicity activity.

## 11.5 Antimicrobial peptides from marine algae

Microalgae include the unicellular eukaryotic photosynthetic organisms which are ubiquitous. They are sources of diverse compounds with different biological properties and potent activities including the primary and secondary metabolites such as phytopigments (xanthophylls and carotenoids), carbohydrates, polyunsaturated fatty acids, phenols, docosahexaenoic acid, vitamins, tannins, terpenoids, and peptides. The peptides from microalgae are the products of proteolytic enzyme digestion [142,143] and are shown to possess different activities like antimicrobial, antioxidant, anticancer, and antihypertensive with immense potential for neutraceutical and therapeutic applications. Guzman et al. [144] reported AMPs from the marine green microalgae *Tetraselmis suecica*, with activity against Gram + ve and Gram − ve human pathogenic bacteria. SP-1 was the first AMP derived from *Spirulina platensis* protein hydrolysates with mw of 18.79 kDa and it inhibited *Escherichia coli* and *S. aureus* [145].

## 11.6 Conclusions

The numerous works described in this chapter provides a profound understanding of the presence of diverse potent AMPs which were isolated from marine flora and fauna. This point to the fact that the prospect of yielding more novel products from the sea is enormous. Several lead AMPs purified from the marine sources are in different stages of preclinical and clinical testing and a few could reach the market. Further, more intense research on to marine AMPS would be required toward producing cost-effective, safe drugs out of these peptides. The research and technology should be targeted toward developing these peptides and their derivatives as potential therapeutics which could cater to providing novel antimicrobials to combat the menace of antibiotic resistance in the near future.

# References

[1] K. Kolanjinathan, P. Ganesh, P. Saranraj, Pharmacological importance of seaweeds: a review, World Journal of Fish and Marine Sciences 6 (1) (2014) 01−15.
[2] J.C. Noro, J.A. Kalaitzis, B.A. Neilan, Bioactive natural products from Papua New Guinea marine sponges, Chemistry & Biodiversity 9 (10) (2012).
[3] M. Zanetti, Cathelicidins, multifunctional peptides of the innate immunity, Journal of Leukocyte Biology 75 (2004) 39−48. Available from: https://doi.org/10.1189/jlb.0403147.
[4] R. Ghanbari, Review on the bioactive peptides from marine sources: indication for health effects, International Journal of Peptide Research and Therapeutics 25 (2019) 1187−1199.
[5] R.I. Lehrer, M.E. Selsted, D. Szklarek, J. Fleischmann, Antibacterial activity of microbicidal cationic proteins 1 and 2, natural peptide antibiotics of rabbit lung macrophages, Infection and Immunity 42 (1) (1983) 10−14.
[6] T. Ganz, M.E. Selsted, D. Szklarek, S.S. Harwig, K. Daher, D.F. Bainton, et al., Defensins. Natural peptide antibiotics of human neutrophils, Journal of Clinical Investigation 76 (4) (1985) 1427−1435.
[7] M. Zasloff, Magainins, a class of antimicrobial peptides from Xenopus skin: isolation, characterization of two active forms, and partial cDNA sequence of a precursor, Proceedings of the National Academy of Sciences (PNAS) USA 84 (15) (1987) 5449−5453.
[8] R.E.W. Hancock, A. Patrzykat, Clinical development of cationic antimicrobial peptides: from natural to novel antibiotics, Current Drug Targets - Infectious Disorders 2 (2002) 79−83.
[9] M. Pushpanathan, P. Gunasekaran, J. Rajendhran, Antimicrobial peptides: versatile biological properties, International Journal of Peptides (2013) 1−15. Available from: https://doi.org/10.1155/2013/675391.
[10] R.E. Hancock, Cationic peptides: effectors in innate immunity and novel antimicrobials, Lancet Infectious Diseases 1 (3) (2001) 156−164.
[11] K. Matsuzaki, Why and how are peptide-lipid interactions utilized for self-defense? Magainins and tachyplesins as archetypes, Biochimica et Biophysica Acta 1462 (1−2) (1999) 1−10.
[12] R.E.W. Hancock, H.G. Sahl, Antimicrobial and host-defense peptides as new anti-infective therapeutic strategies, Nature Biotechnology 24 (2006) 1551−1557.
[13] U.R. Abdelmohsen, K. Bayer, U. Hentschel, Diversity, abundance and natural products of marine sponge-associated actinomycetes, Natural Product Reports 31 (2014) 381−399. Available from: https://doi.org/10.1039/C3NP70111E.
[14] S.V. Sperstad, T. Haug, V. Paulsen, T.M. Rode, G. Strandskog, et al., Characterization of crustins from the haemocytes of the spider crab, *Hyas araneus*, and the red king crab, *Paralithodes camtschaticus*, Developmental & Comparative Immunology 33 (2009) 583−591.

[15] W.B. Lee, C.Y. Fu, W.H. Chang, H.L. You, C.H. Wang, M.S. Lee, et al., A microfluidic device for antimicrobial susceptibility testing based on a broth dilution method, Biosensors and Bioelectronics 87 (2017) 669−678. Available from: https://doi.org/10.1016/j.bios.2016.09.008.

[16] A. Vitali, Antimicrobial peptides derived from marine sponges, American Journal of Clinical Microbiology and Antimicrobials 1 (1) (2018) 1006.

[17] N. Fusetani, Antifungal substances from marine invertebrates, Annals of the New York Academy of Sciences 544 (1988) 113−127.

[18] H.Y. Li, S. Matsunaga, N. Fusetani, Antifungal metabolites from marine sponges, Current Organic Chemistry 2 (1998) 649−682.

[19] N. Fusetani, S. Matsunaga, Bioactive sponge peptides, Chemical Reviews 93 (1993) 1793−1806.

[20] C.A. Bewley, H. He, D.H. Williams, D.J. Faulkner, Aciculitins A-C: cytotoxic and antifungal cyclic peptides from the lithistid sponge *Aciculites orientalis*, Journal of the American Chemical Society 118 (18) (1996) 4314−4321.

[21] A. Plaza, M. Bifulco, J.R. Lloyd, J.L. Keffer, P.L. Colin, J.N.A. Hooper, et al., Peptide inhibitors of HIV-1 entry from different *Theonella* species, The Journal of Organic Chemistry 75 (13) (2010) 4344−4355.

[22] W.R. Chan, W.F. Tinto, P.S. Manchand, L.J. Todaro, Stereostructures of geodiamolides A and B, novel cyclodepsipeptides from the marine sponge *Geodia* sp, The Journal of Organic Chemistry 52 (14) (1987) 3091−3093.

[23] C.A. Bewley, D.J. Faulkner, Theonegramide, an antifungal glycopeptide from the philippine lithistid sponge *Theonella swinhoei*, The Journal of Organic Chemistry 59 (17) (1994) 4849−4852.

[24] S. Matsunaga, N. Fusetani, S. Konosu, Bioactive marine metabolites, VI isolation and the amino acid composition of discodermin A, an antimicrobial peptide, from the marine sponge Discodermia kiiensis, Journal of Natural Products 48 (2) (1985) 236−241.

[25] I. Takayuki, H. Yasumasa, S. Takayuki, Efficient syntheses of geodiamolide A and jaspamide, cytotoxic and antifungal cyclic depsipeptides of marine sponge origin, Tetrahedron Letters 35 (4) (1994) 591−594.

[26] M.J. Martín, R. Rodríguez-Acebes, Y. García-Ramos, V. Martínez, C. Murcia, I. Digón, et al., Stellatolides, a new cyclodepsipeptide family from the sponge *Ecionemia acervus*: isolation, solid-phase total synthesis, and full structural assignment of stellatolide A, Journal of the American Chemical Society 136 (18) (2014) 6754.

[27] D.T. Youssef, L.A. Shaala, G.A. Mohamed, J.M. Badr, F.H. Bamanie, S.R. Ibrahim, et al., Theonellamide G a potent antifungal and cytotoxic bicyclic glycopeptide from the red sea marine sponge *Theonella swinhoei*, Marine Drugs 12 (4) (2014) 1911−1923.

[28] D.P. Clark, J. Carrol, S. Naylor, P. Crews, An antifungal cyclodepsipeptide, cyclolithistide A, from the sponge *Theonella swinhoei*, The Journal of Organic Chemistry 63 (24) (1998) 8757−8763.

[29] Y. Okada, S. Matsunaga, R.W. Van Soest, N. Fusetani, Nagahamide A, an antibacterial depsipeptide from the marine sponge *Theonella swinhoei*, Organic Letters 4 (18) (2002) 3039−3042.

[30] H.Y. Li, S. Matsunaga, N. Fusetani, Halicylindramides A-C, antifungal and cytotoxic depsipeptides from the marine sponge *Halichondria cylindrata*, Journal of Medicinal Chemistry 38 (2) (1995) 338−343.

[31] H.Y. Li, S. Matsunaga, N. Fusetani, Halicylindramides D and E, antifungal peptides from the marine sponge *Halichondria cylindrata*, Journal of Natural Products 59 (2) (1996) 163–166.
[32] A. Plaza, Mirabamides A-D, depsipeptides from the sponge *Siliquariaspongia mirabilis* that inhibit HIV-1 fusion, Journal of Natural Products 70 (11) (2007) 1753–1760.
[33] Z. Lu, R.M. Van Wagoner, M.K. Harper, H.L. Baker, J.N.A. Hooper, C.A. Bewley, et al., Mirabamides E-H, HIV-inhibitory depsipeptides from the sponge *Stelletta clavosa*, Journal of Natural Products 74 (2) (2011) 185–193.
[34] A. Zampella, V. Sepe, P. Luciano, F. Bellotta, M.C. Monti, M.V. D'Auria, et al., Homophymine A, an anti-HIV cyclodepsipeptide from the sponge *Homophymia* sp, The Journal of Organic Chemistry 73 (14) (2008) 5319–5327.
[35] V.A. D'Auria, O.U. Alomal, A. Uigim Inale, N. Ampella, Neosiphoniamolide A, A novel cyclodepsipeptide, with antifungal activity from the sponge *Neosiphonia superstes*, Journal of Natural Product 58 (1995) 121–123.
[36] A. Plaza, G. Bifulco, J.L. Keffer, J.R. Lloyd, H.L. Baker, C.A. Bewley, Celebesides A-C and theopapuamides B-D, depsipeptides from an Indonesian sponge that inhibit HIV-1 entry, The Journal of Organic Chemistry 74 (2) (2009) 504–512.
[37] M.A. Rashid, K.R. Gustafson, L.K. Cartner, N. Shigematsu, L.K. Pannell, M.R. Boyd, Microspinosamide, a new HIV-inhibitory cyclic depsipeptide from the marine sponge *Sidonops microspinosa*, Journal of Natural Product 64 (1) (2001) 117–121.
[38] N. Oku, K.R. Gustafson, L.K. Cartner, J.A. Wilson, N. Shigematsu, S. Hess, et al., A new HIV-inhibitory depsipeptide from the Papua New Guinea marine sponge *Neamphius huxleyi*, Journal of Natural Product 67 (8) (2004) 1407–1411.
[39] S.P. Gunasekera, S.A. Pomponi, P.J. McCarthy, Discobahamins A and B, new peptides from the Bahamian deep water marine sponge *Discodermia* sp, Journal of Natural Product 57 (1) (1994) 79–83.
[40] H.J. Shin, M.A. Rashid, H.R. Bokesch, Stellettapeptins A and B, HIV-inhibitory cyclic depsipeptides from the marine sponge *Stelletta* sp, Tetrahedron Letters 56 (28) (2015).
[41] G. Daletos, R. Kalscheuer, H. Koliwer-Brandl, R. Hartmann, N.J. Voogd, V. Wray, et al., Callyaerins from the marine sponge *Callyspongia aerizusa*: cyclic peptides with antitubercular activity, Journal of Natural Product 78 (8) (2015) 1910–1925.
[42] M.V. D'Auria, V. Sepe, R. D'Orsi, F. Bellotta, C. Debitus, A. Zampella, Isolation and structural elucidation of callipeltins J–M: antifungal peptides from the marine sponge *Latrunculia* sp, Tetrahedron 63 (2007) 131–140. Available from: https://doi.org/10.1016/j.tet.2006.10.032.
[43] C. Cychon, M. Kock, Stylissamides E and F, cyclic heptapeptides from the Caribbean sponge *Stylissa caribica*, Journal of Natural Product 73 (4) (2010) 738–742.
[44] X. Zhang, M. Jacob, R.R. Rvau, Y. Wamg, Antifungal cyclic peptides from the marine sponge *Microscleroderma herdmani*, Research and Reports in Medicinal Chemistry 2 (2012) 7–14.
[45] R. Dahiya, S. Singh, A. Sharma, S.V. Chennupati, S. Maharaj, First total synthesis and biological screening of a proline-rich cyclopeptide from Caribbean marine sponge, Marine Drugs 18 (8) (2016) 396.
[46] G. Rosenberg, A new critical estimate of named species-level diversity of the recent Mollusca, American Malacological Bulletin 32 (2) (2014) 308–322. Available from: https://doi.org/10.4003/006.032.0204.
[47] G. Mitta, F. Vandenbulcke, P. Roch, Original involvement of antimicrobial peptides in mussel innate immunity, FEBS Letters 486 (2000) 185–190.

[48] L. Cheng-Hua, Z. Jian-Min, S. Lin-Sheng, Review of advances in research on marine molluscan antimicrobial peptides and their potential application in aquaculture, Molluscan Research 29 (2009) 17−26.

[49] Z. Liao, X.-C. Wang, H.-H. Liu, M.-H. Fan, J.-J. Sun, W. Shen, Molecular characterization of a novel antimicrobial peptide from *Mytilus coruscus*, Fish and Shellfish Immunology 34 (2) (2013) 610−616.

[50] R. Oh, M.J. Lee, Y.O. Kim, B.H. Nam, H.J. Kong, J.W. Kim, et al., Myticusin-beta, antimicrobial peptide from the marine bivalve, *Mytilus coruscus*, Fish Shellfish Immunology 99 (2020) 342−352. Available from: https://doi.org/10.1016/j.fsi.2020.02.020. Epub 2020 Feb 14. PMID: 32061872.

[51] T. Yao, J. Lu, L. Ye, J. Wang, Molecular characterization and immune analysis of a defensin from small abalone, *Haliotis diversicolor*, Comparative Biochemistry and Physiology Part B 235 (2019) 1−7.

[52] I.K.Ø. Hansen, J. Isaksson, A.G. Poth, K.Ø. Hansen, A.J. CAndersen, C.S.M. Richard, et al., Isolation and characterization of antimicrobial peptides with unusual disulphide connectivity from the colonial ascidian *Synoicum turgens*, Marine Drugs 18 (2020) 51.

[53] I.K.Ø. Hansel, T. Lövdahl, D. Simonovic, K.Ø. Hansen, A.J.C. Andersen, H. Devold, et al., Antimicrobial activity of small synthetic peptides based on the marine peptide turgencin A: prediction of antimicrobial peptide sequences in a natural peptide and strategy for optimization of potency, International Journal of Molecular Sciences 21 (2020) 5460.

[54] R. Galinier, E. Roger, P.E. Sautiere, A. Aumelas, B. Banaigs, G. Mitta, Halocyntin and papillosin, two new antimicrobial peptides isolated from haemocytes of the solitary tunicate, *Halocynthia papillosa*, Journal of Peptide Science 15 (1) (2009) 48−55. Available from: https://doi.org/10.1002/psc.1101. PMID: 19085906.

[55] I.H. Lee, Y.S. Lee, C.H. Kim, C.R. Kim, T. Hong, L. Menzel, et al., Dicynthaurin: an antimicrobial peptide from haemocytes of the solitary tunicate, *Halocynthia aurantium*, Biochimica et Biophysica Acta (BBA) - General Subjects1527 (3)(2001) 141−148.

[56] R.D. Rosa, M.A. Barracco, Antimicrobial peptides in crustaceans, Invertebrate Survival Journal. 7 (2010) 262−284.

[57] T. Becking, C. Delaunay, R. Cordaux, J.M. Berjeaud, C. Braquart-Varnier, J. Verdon, Light on the antimicrobial peptide arsenal of terrestrial isopods: focus on Armadillidins, a new crustacean AMP, family, Genes (Basel). 11 (1) (2020) 93. Available from: https://doi.org/10.3390/genes11010093. PMID: 31947541; PMCID: PMC7017220.

[58] D. Schnapp, G.D. Kemp, V.J. Smith, Purification and characterization of a proline-rich antibacterial peptide, with sequence similarity to bactenecin-7, from the haemocytes of the shore crab, *Carcinus maenas*, European Journal of Biochemistry 240 (3) (1996) 532−539. Available from: https://doi.org/10.1111/j.1432-1033.1996.0532h.x. PMID: 8856051.

[59] J. Herbiniere, C. Braquart-Varnier, P. Greve, J.M. Strub, J. Frere, A. Van Dorsselaer, et al., Armadillidin: a novel glycine-rich antibacterial peptide directed against gram-positive bacteria in the woodlouse *Armadillidium vulgare* (terrestrial isopod, crustacean), Developmental & Comparative Immunology 29 (2005) 489−499. Available from: https://doi.org/10.1016/j.dci.2004.11.001.

[60] K. Stensvåg, T. Haug, S.V. Sperstad, O. Rekdal, B. Indrevoll, O.B. Styrvold, Arasin 1, a proline-arginine-rich antimicrobial peptide isolated from the spider crab, *Hyas araneus*, Developmental & Comparative Immunology 32 (3) (2008) 275−285. Available from: https://doi.org/10.1016/j.dci.2007.06.002.

[61] V.S. Paulsen, H.M. Blencke, M. Benincasa, T. Haug, J.J. Eksteen, O.B. Styrvold, et al., Structure-activity relationships of the antimicrobial peptide arasin 1 — and mode of action studies of the N-terminal, proline-rich region, PLoS One 8 (2013) e53326.

[62] E.J. Noga, K.L. Stone, A. Wood, W.L. Gordon, D. Robinette, Primary structure and cellular localization of callinectin, an antimicrobial peptide from the blue crab, Developmental & Comparative Immunology 35 (4) (2011) 409−415. Available from: https://doi.org/10.1016/j.dci.2010.11.015.

[63] S. Sperstad, T. Haug, T. Vasskog, K. Stensvag, Hyastatin, a glycine-rich multi-domain antimicrobial peptide isolated from the spider crab (*Hyas araneus*) haemocytes, Molecular Immunology 46 (13) (2009) 2604−2612.

[64] Y. Xie, H. Wan, X. Zeng, Z. Zhang, Y. Wang, Characterization and antimicrobial evaluation of a new Spgly-AMP, glycine-rich antimicrobial peptide from the mud crab *Scylla paramamosain*, Fish and Shellfish Immunology 106 (2020) 384−392.

[65] A. Falanga, L. Lombardi, G. Franci, M. Vitiello, M. Rosaria Iovene, G. Morelli, et al., Marine antimicrobial peptides: nature provides templates for the design of novel compounds against pathogenic bacteria, International Journal of Molecular Sciences 17 (5) (2016) 785. Available from: https://doi.org/10.3390/ijms17050785.

[66] T. Nakamura, H. Furunaka, T. Miyata, F. Tokunaga, T. Muta, S. Iwanagall, Tachyplesin, a class of antimicrobial peptide from the haemocytes of the horseshoe crab (*Tachypleus tridentatus*), Journal of Biological Chemistry 263 (32) (1988) 16709−16713.

[67] W.S. Huang, K.J. Wang, M. Yang, J.J. Cai, S.J. Li, G.Z. Wang, Purification and part characterization of a novel antibacterial protein Scygonadin, isolated from the seminal plasma of mud crab, *Scylla serrata* (Forskål, 1775), Journal of Experimental Marine Biology and Ecology 339 (1) (2006) 37−42.

[68] K. Qiao, W. Xu, H. Chen, H. Peng, Y. Zhang, W. Huang, et al., A new antimicrobial peptide SCY2 identified in *Scylla paramamosain* exerting a potential role of reproductive immunity, Fish and Shellfish Immunology 51 (2016) 251−262. Available from: https://doi.org/10.1016/j.fsi.2016.02.022.

[69] Y. Yang, F. Chen, H.-Y. Chen, H. Peng, H. Hao, K.-J. Wang, A novel antimicrobial peptide scyreprocin from mud crab *Scylla paramamosain* showing potent antifungal and anti-biofilm activity, Frontiers in Microbiology (2020). Available from: https://doi.org/10.3389/fmicb.2020.01589.

[70] D. Destoumieux, P. Bulet, D. Loew, A. Van Dorsselaer, J. Rodriguez, E. Bachère, Penaeidins, a new family of antimicrobial peptides isolated from the shrimp *Penaeus vannamei* (Decapoda), Journal of Biological Chemistry 272 (1997) 28398−28406.

[71] Jenn, et al., Novel antimicrobial peptide isolated from *Penaeus monodon*, Academia Sinica, 2004 (Patent). https://patentimages.storage.googleapis.com/3f/ac/cc/af2589938ce31f/US20040235738A1.pdf.

[72] T. Saito, S.-I. Kawabata, T. Shigenaga, Y. Takayenoki, J. Cho, H. Nakajima, et al., Novel big defensin identified in horseshoe crab haemocytes: isolation, amino acid sequence, and antibacterial activity, The Journal of Biochemistry 117 (5) (1995) 1131−1137.

[73] W. Pan, X. Liu, F. Ge, J. Han, T. Zheng, Perinerin, a novel antimicrobial peptide purified from the clamworm Perinereis aibuhitensis grube and its partial characterization, The Journal of Biochemistry 135 (3) (2004) 297−304. Available from: https://doi.org/10.1093/jb/mvh036. PMID: 15113828.

[74] J.K. Seo, B.H. Nam, H.J. Go, M. Jeong, K.Y. Lee, S.M. Cho, et al., Hemerythrin-related antimicrobial peptide, msHemerycin, from the body of the lugworm, *Marphysa sanguinea*, Fish and Shellfish Immunology 57 (2016) 49−59.
[75] Y. López, V. Cepas, S.M. Soto, The marine ecosystem as a source of antibiotics, in: P. H. Rampelotto, A. Trincone (Eds.), Grand Challenges in Marine Biotechnology, Biology and Biotechnology, Springer Nature, 2018, pp. 3−48.
[76] B. Mason, I. Cooke, A. Moya, R. Augustin, M.F. Lin, N. Satoh, et al., AmAMP1 from *Acropora millepora* and damicornin define a family of coral-specific antimicrobial peptides related to the Shk toxins of sea anemones, Developmental & Comparative Immunology, 114, 2021.
[77] J. Vidal-Dupiol, O. Ladrière, D. Destoumieux-Garzón, P.E. Sautière, A.L. Meistertzheim, E. Tambutté, et al., Innate immune responses of a scleractinian coral to vibriosis, Journal of Biological Chemistry 286 (25) (2011) 22688−22698. Available from: https://doi.org/10.1074/jbc.m110.216358. PMID: 21536670; PMCID: PMC3121412.
[78] Z.O. Shenkarev, P.V. Panteleev, S.V. Balandin, A.K. Gizatullina, D.A. Altukhov, E.I. Finkina, et al., Recombinant expression and solution structure of antimicrobial peptide aurelin from jellyfish *Aurelia aurita*, Biochemical and Biophysical Research Communications 429 (2012) 63−69.
[79] C. Canicatti, G. D'Ancona, Cellular aspects of *Holothuria polii* immune response, Journal of Invertebrate Pathology 53 (2) (1989) 152−158.
[80] G.P. Kaaya, Inducible humoral antibacterial immunity in insects, in: J.P.N. Pathak (Ed.), Insect Immunity, Oxford & IBH Publishing Co. Pvt. Ltd., New Delhi, 1993, pp. 69−89.
[81] C. Li, T. Haug, O.B. Styrvold, T.Ø. Jørgensen, K. Stensvåg, Strongylocins, novel antimicrobial peptides from the green sea urchin, *Strongylocentrotus droebachiensis*, Developmental & Comparative Immunology 32 (12) (2008) 1430−1440. Available from: https://doi.org/10.1016/j.dci.2008.06.013. Epub 2008 Jul 24. PMID: 18656496.
[82] K.A. Beauregard, N.T. Truong, H. Zhang, W. Lin, G. Beck, The detection and isolation of a novel antimicrobial peptide from the echinoderm, *Cucumaria frondosa*, Advances in Experimental Medicine and Biology 484 (2001) 55−62. Available from: https://doi.org/10.1007/978-1-4615-1291-2_5. PMID: 11419006.
[83] R.E. Scheibling, B.G. Hatcher, Sea urchins: biology and ecology. *Strongylocentrotus droebachiensis*, Developments in Aquaculture and Fisheries Science 38 (2013) 381−412.
[84] F. Desriac, C. Jégou, E. Balnois, B. Brillet, P. Le Chevalier, Y. Fleury, Antimicrobial peptides from marine proteobacteria, Marine Drugs 11 (10) (2013) 3632−3660. Available from: https://doi.org/10.3390/md11103632.
[85] R.D. Joerger, Alternatives to antibiotics: bacteriocins, antimicrobial peptides and bacteriophages, Poultry Science 82 (4) (2003) 640−647. Available from: https://doi.org/10.1093/ps/82.4.640. PMID: 12710486.
[86] A.B. Snyder, R.W. Worobo, Chemical and genetic characterization of bacteriocins: antimicrobial peptides for food safety, Journal of the Science of Food and Agriculture 94 (1) (2014) 28−44. Available from: https://doi.org/10.1002/jsfa.6293. Epub 2013 Aug 9. PMID: 23818338.
[87] H. Mathur, D. Field, M.C. Rea, P.D. Cotter, C. Hill, R.P. Ross, Bacteriocin-antimicrobial synergy: a medical and food perspective, Frontiers in Microbiology 8 (2017) 1205. Available from: https://doi.org/10.3389/fmicb.2017.01205.
[88] T. Abee, Pore-forming bacteriocins of Gram-positive bacteria and self-protection mechanisms of producer organisms, FEMS Microbiology Letters 129 (1995) 1−10.

[89] H. Einarsson, H.L. Lauzon, Biopreservation of brined shrimp (*Pandalus borealis*) by bacteriocins from lactic acid bacteria, Applied and Environmental Microbiology 61 (1995) 669−676.

[90] O. Gillor, L. Ghazaryan, Recent advances in bacteriocin application as antimicrobials, Recent Patents on Anti-Infective Drug Discovery 2 (2) (2007) 115−122. Available from: https://doi.org/10.2174/157489107780832613. PMID: 18221167.

[91] J.O. McCall, R.K. Sizemore, Description of a bacteriocinogenic plasmid in *Beneckea harveyi*, Applied and Environmental Microbiology 38 (5) (1979) 974−999. Available from: https://doi.org/10.1128/AEM.38.5.974-979.1979. PMID: 317423; PMCID: PMC243617.

[92] E.S. Bindiya, S.G. Bhat, Marine bacteriocins: a review, Journal of Bacteriology and Mycology, Open Access 2 (5) (2016) 140−147. Available from: https://doi.org/10.15406/jbmoa.2016.02.00040.

[93] L.A. Romanenko, M. Uchino, N.I. Kalinovskaya, V.V. Mikhailov, Isolation, phylogenetic analysis and screening of marine mollusc-associated bacteria for antimicrobial, haemolytic and surface activities, Microbiological Research 163 (2008) 633−644.

[94] G.S. Wilson, D.A. Raftos, S.L. Corrigan, S.V. Nair, Diversity and antimicrobial activities of surface-attached marine bacteria from Sydney Harbour, Australia, Microbiology Research (2009).

[95] A. Ahmad, R. Hamid, A.C. Christopher Dada, G. Usup, *Pseudomonas putida* strain FStm2 isolated from shark skin: a potential source of bacteriocin, Probiotics and Antimicrobial Proteins. 5 (2013) 165−175.

[96] A.R. Sarika, A.P. Lipton, M.S. Aishwarya, R.S. Dhivya, Isolation of a bacteriocin-producing *Lactococcus lactis* and application of its bacteriocin to manage spoilage bacteria in high-value marine fish under different storage temperatures, Applied Biochemistry and Biotechnology 167 (2012) 1280−1289. Available from: https://doi.org/10.1007/s12010-012-9701-0.

[97] R.W. Phelan, M. Barret, P. Cotter, P.M. O'Connor, Subtilomycin: a new lantibiotic from *Bacillus subtilis* strain MMA7 isolated from the marine sponge *Haliclona simulans*, Marine Drugs 11 (6) (2013) 1878−1898.

[98] S. Vinothkumar, P.S. Parameswaran, Recent advances in marine drug research, Biotechnology Advances 31 (2013) 1826−1845.

[99] S. Matsunaga, N. Fusetani, Nonribosomal peptides from marine sponges, Current Organic Chemistry 7 (10) (2003) 945−966.

[100] K. Nikolouli, D. Mossialos, Bioactive compounds synthesized by non-ribosomal peptide synthetases and type-I polyketide synthases discovered through genome-mining and metagenomics, Biotechnology Letters 8 (2012) 1393−1403.

[101] S. Agrawal, D. Acharya, A. Adholeya, C.J. Barrow, S.K. Deshmukh, Nonribosomal peptides from marine microbes and their antimicrobial and anticancer potential, Frontiers in Pharmacology 8 (2017) 828. Available from: https://doi.org/10.3389/fphar.2017.00828.

[102] K. Nagai, K. Kamigiri, N. Arao, K.-I. Suzumura, Y. Kawano, M. Yamaoka, et al., YM-266183 and YM-266184, novel thiopeptide antibiotics produced by *Bacillus cereus* isolated from a marine sponge. I. Taxonomy, fermentation, isolation, physicochemical properties and biological properties, The Journal of Antibiotics 56 (2003) 123−128. Available from: https://doi.org/10.7164/antibiotics.56.123.

[103] K. Suzumura, T. Yokoi, M. Funatsu, K. Nagai, K. Tanaka, H. Zhang, et al., YM-266183 and YM-266184, novel thiopeptide antibiotics produced by *Bacillus cereus*

isolated from a marine sponge II. Structure elucidation, The Journal of Antibiotics 56 (2) (2003) 129−134.
[104] M. Mitova, S. Popov, S. De Rosa, Cyclic peptides from a Ruegeria strain of bacteria associated with the sponge *Suberites domuncula*, Journal of Natural Products 67 (7) (2004) 1178−1181.
[105] K. Desjardine, A. Pereira, A. Wright, T. Matainaho, M. Kelly, R.J. Andersen, Tauramamide, a lipopeptide antibiotic produced in culture by *Brevibacillus laterosporus* isolated from a marine habitat: structure elucidation and synthesis, Journal of Natural Products 70 (12) (2007) 1850−1853.
[106] W. Rungprom, R.O. Eric Siwu, L.K. Lambert, C. Dechsakulwatana, Cyclic tetrapeptides from marine bacteria associated with the seaweed *Diginea* sp. and the sponge *Halisarca ectofibrosa*, Tetrahedron. 64 (14) (2008) 3147−3152.
[107] N. Oku, K. Kawabata, K. Adachi, A. Katsuta, Y. Shizuri, Unnarmicins A and C, new antibacterial depsipeptides produced by marine bacterium *Photobacterium* sp. MBIC06485, The Journal of Antibiotics 61 (1) (2008) 11−17.
[108] K. Engelhardt, K.F.M. Degnes, H. Bredholt, E. Fjaervik, G. Klinkenberg, H. Sletta, et al., Production of a new thiopeptide antibiotic, TP-1161, by a marine *Nocardiopsis* species, Applied and Environmental Microbiology 76 (15) (2010) 4969−4976.
[109] M. Mansson, A. Nielsen, L. Kjærulff, C.H. Gotfredsen, M. Wietz, H. Ingmer, et al., Inhibition of virulence gene expression in *Staphylococcus aureus* by novel depsipeptides from marine *Photobacterium*, Marine Drugs 9 (12) (2011) 2537−2552.
[110] T.P. Wyche, Y. Hou, E. Vazquez-Rivera, D. Braun, T.S. Bugni, Peptidolipins B-F, antibacterial lipopeptides from an ascidian-derived *Nocardia* sp, Journal of Natural Products 75 (4) (2012) 735−740.
[111] J. Martín, T.Da. S. Sousa, G. Crespo, S. Palomo, I. González, J.R. Tormo, et al., Kocurin, the true structure of PM181104, an anti-methicillin-resistant *Staphylococcus aureus* (MRSA) thiazolyl peptide from the marine-derived bacterium *Kocuria palustris*, Marine Drugs 11 (2013) 387−398. Available from: https://doi.org/10.3390/md11020387.
[112] L. Kjaerulff, A. Nielsen, M. Manson, L. Gram, T.O. Larsen, H. Ingmer, et al., Identification of four new agr quorum sensing-interfering cyclodepsipeptides from a marine *Photobacterium*, Marine Drugs 11 (12) (2013) 5051−5062.
[113] F.S. Tareq, M.A. Lee, H.S. Lee, J.S. Lee, Y.J. Lee, H.J. Shin, et al., Antimicrobial linear lipopeptides from a marine *Bacillus subtilis*, Marine Drugs 12 (2) (2014) 871−885.
[114] F.S. Tareq, M.A. Lee, H.S. Lee, Y.J. Lee, J.S. Lee, C.M. Hasan, et al., Non-cytotoxic antifungal agents: isolation and structures of gageopeptides A-D from a *Bacillus* strain 109GGC020, Journal of Agricultural and Food Chemistry 62 (24) (2014) 5565−5572.
[115] T. Barsby, M.T. Kelly, S.M. Gagné, R.J. Andersen, M. Stéphane, J. Raymond, Bogorol A produced in culture by a marine *Bacillus* sp. reveals a novel template for cationic peptide antibiotics, Organic Letters 3 (3) (2001) 437−440.
[116] W. Li, J.E. Leet, H.A. Ax, D.R. Gustavson, D.M. Brown, L. Turner, et al., Nocathiacins, new thiazolyl peptide antibiotics from *Nocardia* sp. I. Taxonomy, fermentation and biological activities, The Journal of Antibiotics 56 (2003) 226−231. Available from: https://doi.org/10.7164/antibiotics.56.226.
[117] J.E. Leet, W. Li, H.A. Ax, J.A. Matson, S. Huang, R. Huang, et al., Nocathiacins, new thiazolyl peptide antibiotics from *Nocardia* sp. II. Isolation, characterization, and structure determination, The Journal of Antibiotics 56 (2003) 232−242. Available from: https://doi.org/10.7164/antibiotics.56.232.

[118] H. Ikeda, J. Ishikawa, A. Hanamoto, et al., Complete genome sequence and comparative analysis of the industrial microorganism *Streptomyces avermitilis*, Nature Biotechnology 21 (2003) 526–531.

[119] N. Tamilselvan, E. David, D. Dhanasekaran, K. Saurav, Marine actinobacteria as potential drug storehouses: a future perspective on antituberculosis compounds, in: D. Dhanasekaran, N. Thajuddin, A. Panneerselvam (Eds.), Antimicrobials Synthetic and Natural Compounds, CRC Press, 2015.

[120] C. Chen, J. Wang, H. Guo, W. Hou, N. Yang, B. Ren, et al., Three antimycobacterial metabolites identified from a marine-derived *Streptomyces* sp. MS100061, Applied Microbiology and Biotechnology 97 (9) (2013) 3885–3892.

[121] E. Chen, Q. Chen, S. Chen, B. Xu, J. Ju, H. Wang, Mathermycin, a lantibiotic from the marine actinomycete *Marinactinospora thermotolerans* SCSIO 00652, Applied Microbiology and Biotechnology 83 (2017) e00926-17.

[122] M. Bae, H. Kim, K. Moon, S.J. Nam, J. Shin, K.B. Oh, et al., Mohangamides A and B, new dilactone-tethered pseudo-dimeric peptides inhibiting *Candida albicans* isocitrate lyase, Organic Letters 17 (2015) 712–715. Available from: https://doi.org/10.1021/ol5037248.

[123] S.G. Tringe, C. von Mering, A. Kobayashi, A.A. Salamov, K. Chen, H.W. Chang, et al., Comparative metagenomics of microbial communities, Science 308 (5721) (2005) 554–557. Available from: https://doi.org/10.1126/science.1107851. PMID: 15845853.

[124] J. Malmstrom, Unguisins A and B: new cyclic peptides from the marine-derived fungus *Emericella unguis*, Journal of Natural Products 62 (5) (1999) 787–789.

[125] T.A.M. Gulder, H. Hong, J. Correa, E. Egereva, Isolation, structure elucidation and total synthesis of lajollamide A from the marine fungus *Asteromyces cruciatus*, Marine Drugs 10 (12) (2012) 2912–2935.

[126] T.B.I. Amagata, A. Amagata, K. Tenney, F.A. Valeriote, E. Lobkovsky, J. Clardy, et al., A chemical study of cyclic depsipeptides produced by a sponge-derived fungus, Journal of Natural Products 69 (11) (2006) 1560–1565.

[127] C.M. Boot, N.C. Gassner, J.J.E. Compton, Pinpointing pseurotins from a marine-derived Aspergillus as tools for chemical genetics using a synthetic lethality yeast screen, Journal of Natural Products 70 (10) (2007) 1672–1675.

[128] D.C. Oh, C.A. Kauffman, P.R. Jensen, W. Fenical, Induced production of emericellamides A and B from the marine-derived fungus *Emericella* sp. in competing coculture, Journal of Natural Products 70 (2007) 515–520.

[129] S. Liu, Y. Shen, A new cyclic peptide from the marine fungal strain *Aspergillus* sp. AF119, Chemistry of Natural Compounds 47 (2011) 786–788.

[130] W. Gu, M. Cueto, P.R. Jensen, W. Fenical, R.B. Silverman, A and B: new histone deacetylase inhibitors from the marine-derived fungus *Microsporum* cf. gypseum and the solid-phase synthesis of microsporin A, Tetrahedron 63 (2007) 6535–6541. Available from: https://doi.org/10.1016/j.tet.2007.04.025.

[131] Z. Yu, G. Lang, I. Kajahn, R. Schmaljohann, J.F. Imhoff, Scopularides A and B, cyclodepsipeptides from a marine sponge-derived fungus, *Scopulariopsis brevicaulis*, Journal of Natural Products. 71 (6) (2008) 1052–1054. Available from: https://doi.org/10.1021/np070580e. Epub 2008 Apr 16. PMID: 18412398.

[132] M.Y. Kim, J.H. Sohn, J.S. Ahn, H. Oh, Alternaramide, a cyclic depsipeptide from the marine-derived fungus *Alternaria* sp. SF-5016, Journal of Natural Products 72 (11) (2009) 2065–2068.

[133] P. Pruksakorn, M. Arai, N. Kotoku, C. Vilcheze, A.D. Baughn, P. Moodley, et al., Trichoderins, novel aminolipopeptides from a marine sponge-derived *Trichoderma* sp., are active against dormant mycobacteria, Bioorganic and Chemistry Letter (2010).

[134] J. Zheng, H. Zhu, K. Hong, Y. Wang, P. Liu, X. Wang, et al., Novel cyclic hexapeptides from marine-derived fungus, *Aspergillus sclerotiorum* PT06-1, Organic Letters 11 (22) (2009) 5262–5265.

[135] F. Tiralongo, G. Messina, B.M. Lombardo, L. Longhitano, G. Li Volti, D. Tibullo, Skin mucus of marine fish as a source for the development of antimicrobial agents, Frontiers in Marine Science 7 (2020) 541853. Available from: https://doi.org/10.3389/fmars.2020.541853.

[136] S. Dash, S.K. Das, J. Samal, H.N. Thatoi, Epidermal mucus, a major determinant in fish health: a review, Iranian Journal of Veterinary Research. 19 (2) (2018) 72–81. PMID: 30046316; PMCID: PMC6056142.

[137] P.Y. Chee, M. Mang, E.S. Lau, L.T. Tan, Y.W. He, W.L. Lee, et al., Epinecidin-1, an antimicrobial peptide derived from grouper (*Epinephelus coioides*): pharmacological activities and applications, Frontiers in Microbiology 10 (2019) 2631. Available from: https://doi.org/10.3389/fmicb.2019.02631.

[138] W.J. Lin, Y.L. Chien, C.Y. Pan, T.L. Lin, J.Y. Chen, S.J. Chiu, et al., Epinecidin-1, an antimicrobial peptide from fish (*Epinephelus coioides*) which has an antitumor effect like lytic peptides in human fibrosarcoma cells, Peptides 30 (2) (2009) 283–290. Available from: https://doi.org/10.1016/j.peptides.2008.10.007. Epub 2008 Oct 21. PMID: 19007829.

[139] N. Ennaas, R. Hammami, A. Gomaa, F. Bédard, E. Biron, M. Subirade, et al., Collagencin, an antibacterial peptide from fish collagen: activity, structure and interaction dynamics with membrane, Biochemical and Biophysical Research Communications 473 (2) (2016) 642–647.

[140] J.M.O. Fernandes, G. Molle, G.D. Kemp, V.J. Smith, Isolation and characterisation of oncorhyncin II, a histone H1-derived antimicrobial peptide from skin secretions of rainbow trout, *Oncorhynchus mykiss*, Developmental & Comparative Immunology 28 (2) (2004) 127–138.

[141] J. Acosta, V. Montero, Y. Carpio, J. Velázquez, H. Garay, E. Reyes, et al., Cloning and functional characterization of three novel antimicrobial peptides from tilapia (*Oreochromis niloticus*), Aquaculture. 9–18 (2013) 372–375.

[142] A.C. Guedes, H.M. Amaro, F.X. Malcata, Microalgae as sources of high added-value compounds—a brief review of recent work, Biotechnology Progress 27 (3) (2011) 597–613. Available from: https://doi.org/10.1002/btpr.575. Epub 2011 Mar 30. PMID: 21452192.

[143] V. Mimouni, L. Ulmann, V. Pasquet, M. Mathieu, L. Picot, G. Bougaran, et al., The potential of microalgae for the production of bioactive molecules of pharmaceutical interest, Current Pharmaceutical Biotechnology 13 (2012) 2733–2750.

[144] F. Guzmán, G. Wong, T. Román, C. Cárdenas, C. Alvárez, P. Schmitt, et al., Identification of antimicrobial peptides from the microalgae *Tetraselmis suecica* (Kylin) butcher and bactericidal activity improvement, Marine Drugs 17 (2019) 453.

[145] Y. Sun, R. Chang, Q. Li, B. Li, Isolation and characterization of an antibacterial peptide from protein hydrolysates of *Spirulina platensis*, European Food Research and Technology 242 (2016) 685–692.

# Peptides with antiviral activities

Anjali Jayasree Balakrishnan, Aswathi Kodenchery Somasundaran, Prajit Janardhanan and Rajendra Pilankatta
Department of Biochemistry and Molecular Biology, Central University of Kerala, Periye, Kasaragod, Kerala, India

## 12.1 Introduction

Viral epidemics and pandemics create global health threats resulting in significant mortality and morbidity. Emerging and re-emerging viral epidemics and pandemics have had profound impacts on the cultural as well as economic aspects of the human population for centuries. The last century has witnessed a series of epidemics of emerging infections due to Human Immunodeficiency Virus (HIV), Ebola Virus (EV), Dengue Virus (DENV), Chikungunya, Nipah, Avian Influenza Virus (AIV), Yellow Fever Virus (YFV) and Severe Acute Respiratory Syndrome Corona Virus (SARS-CoV) in various parts of the world. Notably, the first two decades of the 21$^{st}$ century witnessed a series of outbreaks of respiratory viral infections starting from SARS-CoV in 2002 to SARS-CoV-2, resulting in the Corona Virus Disease-19 (COVID-19) pandemic, which led to millions of deaths across the globe. Such viral episodes lead to substantial economic loss and a decline in the growth and development of the affected countries, thereby creating chaos in the society, especially in the regions where they are endemic. The lack of specific drugs or universal vaccine strategies and risks of resistance to already available antiviral medications for the emerging and re-emerging viral diseases call for a novel antiviral approach to combat them [1,2]. The advent of scientific development, especially in modern science, helped to surveillance the virus transmission quickly and to develop the measures to curb the spreading along with the development of therapeutics.

About six decades ago, the first antiviral drug was elaborated; still, no effective antiviral drugs are available for most viral infections. The effectiveness of chemotherapeutics against viruses has setbacks as the susceptibility of viruses to drugs changes, which may be due to the use of host cellular machinery for virus replication and mutations in the viral genome. Moreover, cross-species transmission of viruses has also been a cause of concern since previously unknown zoonotic infections such as COVID-19 have emerged in the decade. These features make it challenging to develop selectively toxic drugs for viruses [3,4].

The mode of action of antiviral drugs can be of two types, either targeting the viruses or the host machinery required for the replication of viruses. Viral targets

Antimicrobial Peptides. DOI: https://doi.org/10.1016/B978-0-323-85682-9.00002-7
© 2023 Elsevier Inc. All rights reserved.

can be broadly classified as (1) viral entry inhibitors such as hydroxychloroquine, chloroquine [4], lacidipine, and phenothrin [5], (2) viral uncoating inhibitors such as rimantadine and amantadine, (3) nucleic acid polymerases inhibitors such as acyclovir, valacyclovir, ganciclovir, penciclovir, ribavirin, lamivudine, remdesivir, foscarnet [4,6], (4) protease inhibitors such as atazanavir, darunavir, lopinavir, and ritonavir [4] and (5) integrase inhibitors such as dolutegravir, elvitegravir, and raltegravir [7]. There are many host-targeted antiviral drugs approved by the US Food and Drug Administration (FDA) such as cabozantinib, R42853, nanchangmycin, mycophenolic acid, ribavirin, difluoro methyl ornithine, suramin, bortezomib, ezetimibe, alisporivir, sanglifehrin, captopril, lisinopril, camostat, nafamostat, chloramphenicol, tigecycline, linezolid, silmitasertib, merimepodib, loratadine, pimozide, tall oil fatty acid, ivermectin, diethylnorspermine, celgosivir of which some are in preclinical and some in clinical trials [8]. The primary pitfall of these drugs is that since it affects the host machinery, the chance of side effects from these drugs cannot be ruled out.

Taking that in mind, it is high time we used novel molecules with antiviral potential [1,2]. One such molecule is antiviral peptides. Peptides are biologically active molecules containing a minimum of two amino acids linked through a peptide bond. They are smaller and readily exist in humans, comprising less than a hundred amino acid residues. Small peptides are highly selective and relatively safe. Due to their diverse biological roles in various physiological processes, they are considered potential antiviral candidates [9].

## 12.2 Viral life cycle

Seven stages of replication are being performed by the viruses inside the target cell to create new infectious viral particles that can infect neighboring cells or hosts. They are attachment, penetration, uncoating, replication, assembly, maturation, and release. Depending on the virus type, some of these stages may coincide with others, whereas some may occur randomly [10,11]. Understanding each stage of viral replication is crucial for identifying potential antiviral agents [12].

Attachment is the stage in which an attachment protein of the virus binds explicitly to the receptor on the cell surface of the host cell. Some viruses, such as HIV, also need a coreceptor for infection. In the case of enveloped viruses, the attachment protein extends from the envelope whereas, in non-enveloped viruses, capsid proteins interact with the receptors. Some of the cell surface receptors are intercellular adhesion molecule-1 for 90% of Rhinoviruses and low-density lipoprotein (LDL) receptor, VLDL for 10% of Rhinoviruses; poliovirus receptor, PVR clusters of differentiation 155 (CD155) for Polioviruses; CD4 as receptor and chemokine receptors CC-chemokine receptor 5 (CCR5) or C-X-C chemokine receptor type 4 (CXCR4) as coreceptors for HIV; sialic acid for Influenza A virus; CD46 and CD150 for Measles virus; heparan sulfate, herpes virus entry mediator and nectin-1 for Herpes Simplex Virus-1 (HSV-1); dendritic cell-specific intercellular adhesion molecule 3-grabbing nonintegrin for DENV; sodium taurocholate−cotransporting polypeptide for Hepatitis B virus (HBV); heparan sulfate and integrins for Human Papillomavirus (HPV) [10,11], obligate receptor angiotensin-converting enzyme 2 for SARS-CoV-2 [13−15] and the transmembrane dipeptidylpeptidase 4/CD26 for the Middle East Respiratory Syndrome Coronavirus (MERS-CoV) [15−17].

Penetration is a stage that involves crossing the plasma membrane to gain entry into the cytoplasm by a virus that requires energy contributed by the host cell. Penetration methods include receptor-mediated endocytosis, bulk-phase endocytosis, phagocytosis and fusion. Both enveloped and non-enveloped viruses commonly use receptor-mediated endocytosis for penetration. In receptor-mediated endocytosis, ligands bind to the receptors on the cell surface and are engulfed in clathrin-coated pits, which then form endocytic vesicles. They, later on, lose their outer coat and fuse with vesicles with a pH of 6.0−6.5, called early endosomes. As their pH shifts to a range of 5.0−6.0, early endosomes become late endosomes. Late endosomes then pass on their content to larger vesicles called lysosomes. Viral protein configuration is altered when the pH of the endosome reduces. This takes place in different vesicles depending on the virus type. A few examples of human viruses exhibiting clathrin-mediated endocytosis are DENV, West Nile virus, Adenovirus, and Hepatitis C Virus (HCV). Polyomavirus Simian Virus, HPV, and HBV exhibit caveolin-mediated endocytosis. Some viruses even make use of receptor-mediated endocytosis, which is clathrin-independent. The cells form a vesicle engulfing viruses and other molecules in the extracellular fluid in bulk-phase endocytosis. Some specialized cells carry out phagocytosis to internalize entire cells. HSV-1 and Mimivirus exhibit a phagocytosis-like mode of penetration. Fusion is solely shown by viruses such as HIV, Influenza, Respiratory Syncytial Virus (RSV), HSV, DENV, and EV, during which viral envelope attaches to the plasma membrane. It is mediated either by virus attachment protein or a different viral protein based on the virus. It can occur within endocytosed vesicles or cell membranes [11].

Uncoating is the stage where the capsid is removed, which in turn leads to the release of the viral genome into the host cell. It can be seen either as a separate process in the case of Rhinoviruses or linked with penetration, as seen in Influenza Viruses. Rhinoviruses expand their size by about 4% within the acidic endosome, and the capsid protein, viral protein 1, helps the viral genome to enter the host cell. In respiratory epithelial cells, penetration of the Influenza Virus occurs through receptor-mediated endocytosis. Hemagglutinin (HA) on the envelope of Influenza Virus fuses with the endosomal membrane and helps in viral attachment as well as uncoating. The viral capsid of Polioviruses induces a conformational change on binding to the receptor, thereby allowing the transport of the viral genome. Some viruses, such as the Reoviruses, remain primarily intact even after penetration. Complete uncoating does not take place in this case. This partially intact capsid becomes a home base for replication in many viruses [10,11].

The next stage in the viral life cycle is replication. Viral replication is classified based on the structure and composition of the genome. They are as follows:

1. Double-stranded DNA

   In this case, replication is of two types. One type of replication depends on the cellular factors and takes place within the nuclear or nucleoid. In contrast, the other occurs in the cytoplasm and does not depend on cellular factors as they acquire necessary factors from the hosts.

2. Single-stranded DNA

   Replication involves a double-stranded intermediate. It acts as a template for new strand synthesis, and this process takes place within the nucleus.

3. Double-stranded RNA

   Replication takes place within the cytoplasm and is primarily independent of cellular factors. Each segmented genome produces monocistronic messenger RNAs (mRNAs).

4. Single-stranded (+)-sense RNA

   After infection, viruses with polycistronic mRNA are translated into a polyprotein, eventually cleaved into mature proteins. Viruses such as Togaviruses and Coronaviruses undergo two sets of translations to produce sub-genomic and full-length RNA. RNA template is required for replication, and the replicase enzyme is produced after infection.

5. Single-stranded (−)-sense RNA

   In viruses with a segmented genome, monocistronic mRNAs are produced by the RNA-dependent RNA polymerase enzyme (RdRp) by transcribing a (−)-sense RNA in the virus particles. In viruses with a segmented genome, transcriptase produces monocistronic mRNAs as a full-length genome. The (+)-sense genome then made, acts as the template in synthesizing (−)-sense progeny genomes.

6. Single-stranded RNA with DNA intermediate

   The (+)-sense RNA of the retrovirus genome acts as a template for reverse transcription by the reverse transcriptase enzyme. The provirus, thus formed, integrates into the host chromatin.

7. Double-stranded DNA with RNA intermediate

   This takes place within the virus particle at the time of maturation. During infection, the gapped genome is repaired, followed by transcription. Reverse transcriptase converts viral RNA into the DNA inside the virus particle [10,11,18].

Assembly can occur along with maturation and release in some viruses. All the newly synthesized components required for virus formation are collected and assembled at a particular site, depending on the type of virus.

Maturation is the stage in which an infectious virion becomes an infectious viral particle. This often requires structural changes in the capsid that are mediated by viral or host enzymes.

In the final stage of virus replication, the virion is released into the extracellular environment. This happens by exocytosis or cell lysis, depending on the virus [10,11,18].

## 12.3 Peptides as viral inhibitors

The first antiviral peptide that received approval for therapy by the FDA in 2003 was T20/enfuvirtide [19]. T20 peptides are derived from the membrane-proximal external region of the transmembrane glycoprotein of HIV-1, gp41 [20,21]. It belongs to the class of direct-acting antiviral drugs and has anti-HIV activity [19].

In the current chapter, antiviral peptides are classified based on the stages at which antiviral peptides block the viral infection.

## 12.4 Mechanism of inhibition

Antiviral peptides impede viral infection at one or many stages of viral infection, ranging from attachment to egress of the virus after completion of its life cycle inside the host cell [22].

### 12.4.1 Viral attachment inhibitors

Many antiviral drugs in use and clinical trials target viral attachment proteins, as viral infections can be effectively prevented by inhibiting attachment [11]. One way to suppress viral infections is by disrupting the interaction between glycoprotein and receptors. Antiviral peptides can be used effectively to achieve this goal [23]. In DENV infection, viral entry can be inhibited by targeting envelope proteins, precursor-membrane/membrane proteins, and capsid proteins or host cells [9]. A similar approach is effective in the case of other viruses also. HIV is one such virus against which most peptide inhibitors have been developed. Maraviroc is a small molecule CCR5 inhibitor that can bind to the CCR5 coreceptor and has been reported to have successfully inhibited HIV-1 infection [9,24,25]. Further, analogs of Maraviroc, Met-R and UK484900 have been reported to prevent CCR5 activation and have reduced DENV load [26].

Inhibitors of hemagglutination activity block virus attachment onto host cells by binding to surface protein HA. P1 peptide with a heptapeptide sequence, NDFRSKT, has been reported to show in vitro and in ovo inhibition of hemagglutination activity of the H9N2 subtype of AIV, which in turn leads to inhibition of viral replication [27,28]. This peptide has also been proven inhibitory in other virus infections. P1 binds competitively to the E protein of the Japanese Encephalitis Virus (JEV), thereby blocking its interaction with the cellular receptor. This, in turn, inhibits the entry of the virus into the host cell [29]. Another peptide against JEV is NSK tripeptide. Different strains of JEV have been found to have a conserved NSK region in the envelope glycoprotein domain III sequence. NSK blocks the entry of JEV into the cells by inhibiting the attachment of the virus and host cell [30,31].

The Pep-RTYM peptide prevents DENV infection by directly attaching to DENV particles and blocking the association of the virus with host cell receptors. Further nucleic acid release into the host cell is effectively prevented by Pep-RTYM [32,33].

Toxins produced by plants, animals, insects, and marine organisms have been proven to contain peptides that show remarkable antiviral activity [34]. Mellitin derived from honey bee venom disrupts viral envelop and prevents the entry of HIV into host cells [35].

Venom from the snake species *Bungarus fasciatus* hosts a peptide named BF-30, which is known to exhibit both antibacterial and antitumor activity [36]. In vitro experiments proved the antiviral nature of the peptide through effective inhibition of Influenza Viruses H1N1 and H3N2. BF-30 causes virus particles to fuse, changing the structural conformation of the virus, and inhibiting infection. The peptide has also been proven effective against resistant strains to the antiviral drug oseltamivir [37].

## 12.4.2 Plasma membrane and viral fusion inhibitors

Membrane fusion is a critical step in infecting enveloped viruses effectively [38]. Designing peptides against the fusion of viral membrane with the host is one of the most widely used strategies [39]. Four viral fusion proteins have been identified [40]. Peptides are derived from these four types of fusion proteins called fusion inhibitors. They block the entry of the virus into cells by competing in attachment to the fusion proteins [41]. Fusion inhibitors target fusion proteins of the virus and disrupt the conformational changes occurring to fusion proteins during membrane fusion. Among fusion inhibitors, class I viral fusion inhibitors impede the six-helix bundle (6-HB) structure formation during the virus—host membrane fusion. Fusion inhibitors minimize the damage caused by the virus to the host by blocking virus entry into the cytoplasm at the initial stage of infection [42].

In flaviviruses, membrane fusion is facilitated by envelope fusion protein (E). Peptides obtained from the stem region of E bind to a recombinant E in its low pH conformation [43]. In the case of DENV infection, currently, there is no vaccine or antiviral drug available that can neutralize all the four serotypes of DENV [44]. In DENV2 infection, a fusion intermediate in the late stage is inhibited by the stem peptides obtained from the DENV2 E. It was reported that inhibition of all four serotypes by these DENV2 peptides was more potent for DENV2 infection followed by DENV1, DENV3, and DENV4, respectively [21,43].

A combination of cyclic peptides targeting the antibodies against surface glycoproteins, Gn and Gc of Andes Virus (ANDV) have been reported to show a significant increase in the inhibition of ANDV infection in vitro [45].

The fusion inhibitor, OC34-HR2P, is a peptide consisting of 36 amino acid residues that target the HR1 region of the S2 subunit of S protein. It is derived from the S2 subunit of OC34-type human coronavirus spike (S) glycoprotein. During virus-cell fusion, the S2 subunit forms a 6-HB structure. Since the sequence is conserved in different strains of human coronaviruses, OC34-HR2P can elicit broad-spectrum antiviral activity. The salt bridge-forming residues Glu, Lys, and mutating amino acids of OC34-HR2P were further modified by replacing amino acids at the i, i + 3 or i + 4 positions with Gln4, Tyr14, Asp32, and Leu36 to obtain EK1. EK1 peptide effectively represses corona virus-cell adhesion and exhibits no toxicity at higher levels [46]. The effectiveness of EK1 leads to further modification of the peptide with cholesterol. The newly amended peptide, EK1C4, showed heightened activity even at nanomolar levels against live virus infections such as MERS, OC43, HCoV-229E, and HCoV-NL63 which requires S protein-mediated membrane fusion. EKC41 has also been found to be non-toxic to target cells at active concentrations [47].

In the case of HIV infection, an ester peptide LP-19 showed effective inhibition against HIV-1 and HIV-2 infection both in vitro and in vivo. LP-19 is engineered from the C-terminal heptad repeat (CHR) region of gp41. The peptide functions by binding to the N-terminal heptad repeat region of gp41, blocking the generation of the 6-HB structure. To ensure activity against HIV-2, vital amino acids required for their attachment to the virus were also inserted. Subsequently, to ensure stability and be protected from protease degradation, various other modifications were added to LP-19 [48]. The PIE12 peptide was obtained through a mirror-image phage display library using the D peptide (D-IZN17). D peptide acts like the hydrophobic pocket region of gp41. In vivo, PIE12 has higher stability and avoids protease degradation [49]. Individual units of PIE12 trimerize to form PIE12 trimer and can bind to gp41 at its hydrophobic pocket [50]. Various modifications of the trimer were performed, and the modification with cholesterol-generated peptide CPT31 exhibited notably enhanced activity, half-life time, and solubility in water. CPT31 effectively prevented SHIVAD8-EO, an infectious molecular clone obtained from simian-HIV chimeric virus AD8. The peptide protected rhesus monkeys from being infected in animal experiments and also inhibited the virus in vivo [51]. As the peptide exhibits good pharmacokinetic stability, it need not be administered frequently like other formulations. As the functional part of CPT31 with antiviral activity is PIE12, CPT31 and PIE12 have similar modes of action, making them potent drug candidates for treating HIV-1 infection. Considering the effectiveness, the FDA planned to conduct phase I clinical trials for CPT31 [52].

A conformation-locked peptide modified with cholesterol, EBOV-7, was designed in 2019 against the EV and Marburg virus. EBOV-7 sequence is derived from the transmembrane glycoprotein subunit 2 (GP2) of EV, specifically, the CHR region, which is a class I fusion protein. Like all class I fusion proteins, it requires generating a 6-HB structure to mediate virus and cell membrane fusion. EBOV-7 binds to the C-terminus of the HR2 domain of GP2 and impedes the generation of the 6-HB structure. This in turn inhibits the fusion of the EV membrane and the endosomal membrane [53].

SAH-RSVFBD is a fusion inhibitor which prevents the infection of RSV and was derived from an HR2 region sequence of the RSV F fusion protein. The presence of stable alpha ($\alpha$)-helical structures imparts the peptide with a clear advantage over linear peptides, in terms of increased protease stability and target binding ability. SAH-RSVFBD competitively binds to the RSV F protein at the HR1 trimer and blocks the membrane fusion of the virus with the cell by inhibiting the generation of a 6-HB structure [54]. Another fusion inhibitor for RSV infection, peptide 4ca, was developed using a hydrocarbon stapling strategy. As it is derived from the HR2 region of the RSV F protein, the antiviral mechanisms of 4ca and SAH-RSVFBD are exactly the same since the SAH-RSVFBD sequence area covers 4ca [55]. Human cathelicidin (LL-37), an antimicrobial peptide, was also used to combat RSV infection. LL-37 disrupts virus particles by causing damage to the viral envelope. Once the envelope is damaged, the virus cannot infect epithelial cells [56].

An antiviral peptide was generated from the Newcastle Disease Virus and Infectious Bronchitis Virus, using the HR region of the fusion glycoproteins, namely NOVEL-2 [57]. The N-terminal end of NOVEL-2 was modified to form a cholesterol-tagged peptide CAU, exhibiting a higher antiviral activity. CAU also exhibited increased half-life and also had the ability to be transported and enter the central nervous system of birds. In addition, the formation of the 6-HB structure is blocked by the competitive binding of CAU to the HR2 region of fusion glycoprotein [58].

Toxic skin secretions of the Indian frog, *Hydrophylax bahuvistara*, host 32 identified host defense peptides. Among them, urumin has been reported to inhibit the Human Influenza Virus H1N1. The antiviral property is against the H1 influenza virus, and the target is found to be the conserved stalk region of H1 type HA. Upon administration, oseltamivir-resistant H1N1 strain titers were lowered significantly. Urumin binds to the region of HA that is conserved, disrupting viral particles, causing a conformational change, and interacts with the membrane aided by positive charges on the peptide [59].

## 12.4.3 Endosomal acidification inhibitors

Viruses require endosomal or lysosomal acidification to infect host cells [60–62]. Before releasing nucleic acids into the cytoplasm, some viruses must undergo endosomal acidification to initiate replication [58]. Endosome acidification aids in the fusion of the viral membrane with the host membrane inside the cell, like in the case of the influenza virus. One of the challenges in developing antivirals against the influenza virus is membrane penetration by the developed drug [63].

Tat-HA2Ec3 is a fusion inhibitor that effectively resists the influenza virus in vivo. The peptide consists of three parts: (1) a cell-penetrating peptide, derived from the transactivator of transcription (Tat) protein of HIV, conjugated to the N-terminus of the inhibitor. This ensures that the inhibitor enters the host cell, (2) a middle peptide from the C-terminus of influenza virus which inhibits membrane fusion, and (3) a cell membrane targeting lipid which is added at the C-terminal. During pH-induced membrane fusion, Tat-HA2Ec3 traps transient intermediates of HAs, preventing viral fusion [64].

A peptide of murine beta-defensin-4, P9 has broad-spectrum inhibitory activity against respiratory viruses. The P9 enters the host cell along with the virus by binding to the surface glycoprotein of the virus. Through the net positive charge rested on it, the peptide prevents endosome acidification, thereby inhibiting membrane fusion and the release of RNA into the host cell. P9 exhibits commendable antiviral activity against subtypes of influenza viruses such as H3N2, H5N1, H7N7, H7N9, and coronaviruses such as SARS-CoV and MERS-CoV, both in vitro and in vivo [65].

P9 was later re-engineered by replacing three amino acid residues—His21, Lys23, and Lys28, with arginine to obtain P9R. With its increased net charge, P9R effectively inhibits enveloped viruses (HIN1, N7N9, SARS-CoV/CoV-2, and MERS-CoV) and non-enveloped Rhinovirus, proving to be a broad-spectrum antiviral peptide with higher viral inhibitory activity. Being a cationic peptide, P9R binds directly to the virus as a cationic peptide, preventing endosomal acidification and ribonucleoprotein release [66].

Subsequent modifications on P9R resulted in an 8-branch P9R (8P9R), which could suppress SARS-CoV-2 infection more efficiently than P9R. 8P9R prevents entry of SARS-CoV-2 through the endocytic pathway by inhibiting endosome acidification. Also, the peptide causes SARS-CoV-2 to clump together and the entry into the cells is prevented via the TMPRSS2-mediated surface pathway [67].

Viruses use the motor protein dynein for its intracellular transport and internalization. P54, a significant protein on the virion membranes of the African Swine Fever Virus (ASFV), interacts with the dynein light chain, DLC8, which leads to the intracellular transport of the virus prior to replication and protein synthesis. Hernáez et al. [68] designed a peptide compound, interactionstop1 to contain the DLC8. Dynblock1 with an added arginine tail at the N-terminal end has been reported to help develop antivirals against viruses that share the exact transport mechanism as ASFV [68].

Meliacine is a glycopeptide isolated from *Melia azedarach* [69] that can prevent the hulling process of the Foot-and-Mouth Disease Virus by inhibiting vacuolar acidification [70].

### 12.4.4 Replication and translation inhibitors

The drugs targeting the assembled and mature viral proteins are less studied. Antiviral drug research on the viral replication cycle mainly focuses on the viral entry into host cells, viral and host proteases, and RdRp [71]. In DENV infection, viral replication and translation are inhibited by targeting non-structural (NS) proteins [9]. A cationic antimicrobial protein, CAP37-derived peptide, that has established antibacterial activity has been reported to inhibit HSV-1 and serotypes Ad3 and Ad5 of respiratory adenovirus in vitro. Inhibition disrupts the envelope and capsid, thereby targeting viral assembly [72]. LL-37 has also shown similar antiviral inhibitory activity against HSV-1 [73]. Other antimicrobial peptides, human defensins ($\alpha$1), have demonstrated antiviral activity against recombinant Ad5 vectors, HSV-1, HSV-2, and HIV. Defensin-like chemokines (IP-10, I-TAC), along with $\alpha$-1, have been reported to inhibit the common respiratory adenovirus serotypes Ad3 and Ad5 [74].

Another cationic antimicrobial peptide, derived from neutrophils of bovine origin, APB-13, has been reported to inhibit Transmissible Gastroenteritis Virus (TGEV) infection in vitro and in vivo. Even though the target of APB-13 has not been determined, the peptide decreased the mRNA levels and protein expression levels of the TGEV nucleocapsid gene [75]. Some antiviral peptides show multiple target activities such as hesperidin, which inhibits NS3 protease of HCV genotype 3a. Hesperidin acts by interacting with active site residues and also with the catalytic triad of NS3-NS4A protease. This dual activity made hesperidin a potent inhibitor of HCV genotype 3a [76].

Plants produce small cationic polypeptides roughly 10 kDa in size as a nonspecific innate immune system response. These peptides are termed botany-derived antimicrobial peptides and are found to exhibit anticancerous and antiviral activity [77]. One such peptide is nonspecific lipid transfer protein (NTP) obtained from *Narcissus tazetta*. NTP inhibits H1N1 by blocking the neuraminidase on the virus envelope, thereby disrupting the cytopathic effect of the virus. NTP also binds to viral glycoproteins of RSV and blocks viral entry into the cells. The peptide can also prevent the proliferation of viruses by inhibiting viral replication and assembly [78]. Other plant-derived peptides such as ginkbilobin, ascalin, lunatusin, and vulgarinin suppress reverse transcriptase of HIV-1 [79–82].

## 12.5 Peptides as therapeutics

Besides being highly target-specific and high in activity, peptides also have a better safety margin than small molecules since they are made of amino acid residues [83,84]. With the advancements in the biopharmaceutical sector, peptides and proteins have found new avenues of application in therapeutics [85]. The FDA has approved 239 therapeutic proteins and peptides since the 1980s. A total of 60 peptide-based drugs are available after clinical trials, and various others are clinically evaluated [86]. The first comprehensive database of antiviral peptides, AVPdb [87], and the database for HIV-inhibiting peptides, HIPdb [88], are manually curated databases containing information on peptide sequences, sources, viruses targeted, cell lines, experimental assays used, qualitative or quantitative efficacies, PubMed references, target proteins, properties, and structure of peptides. The former provides information on 2683 experimentally verified antiviral peptides, including 624 modified peptides, targeting more than 60 viruses. The latter includes information on experimentally validated anti-HIV peptides with data on 981 peptides, 87 modified peptides, and 179 low/nil active peptides [89,90]. FDA maintains a manually curated repository, THPdb [91], for approved peptides and peptides as therapeutics. The repository provides information on the 239 US FDA-approved therapeutic peptides and proteins, and their 380 drug variants [92]. Nearly 150 antiviral peptides are in the advanced stages of clinical trials [93]. Some of the antiviral peptides in clinical research are sifuvirtide for HIV [94], myrcludex B [95] and hepatize (L47) [96] for HBV, SCV-07 for HCV [97].

## 12.6 Challenges and future scope

Emerging and re-emerging viral pandemics warrant an effective antiviral strategy. The antiviral peptide-based therapeutic approach was highly promising apart from the small molecule-based antiviral drugs and vaccines. We need a multipronged strategy to combat future viral epidemics and pandemics. In this context, more research must be conducted to solve the challenges involved in the clinical usage of antiviral peptides, which are found to be effective in laboratory-based in vitro studies.

The significant challenges that arise while using antiviral peptides as therapeutic drugs are due to the bioavailability, cytotoxicity, immunogenicity, low and relative selectivity, high cost of synthesis, allergic and inflammatory responses, and low serum stability [2,9,29]. Peptides getting cleaved by trypsin and their relatively short half-lives are another primary concern [9,29]. They could easily be degraded by at least 569 proteases in the human body.

In order to overcome the challenges, we need to (1) develop cost-effective peptide synthesis methods by alternative strategies like natural sources or by recombinant engineering. The development of a low-cost peptide synthesis method will also augment the basic research in the area, (2) usage of D-enantiomers and modified peptides so that the problems such as bioavailability, reduced half-life, and other related pharmacodynamic properties of antiviral peptides can be improved, and (3) application of nanoformulated antiviral peptides for the delivery so that the bioavailability, as well as toxicity of the peptides, can be minimized. Alternatively, peptides may be conjugated with carbohydrates and fatty acids to deliver antiviral peptides effectively.

Modified peptides are one of the technological breakthroughs and advancements that allow alterations in the physicochemical properties of peptides and can be an alternative to these limitations. These man-made variants have improved stability and are capable of overcoming pharmacodynamic weaknesses [9]. Moreover, targets for synthetic peptides are very diverse, such as inhibiting virus entry into the cell, inhibition of enzymes relevant to the virus life cycle, immune system modulation, and mimicking and competing receptors for the virus [2]. In addition, peptide-oriented virtual screening strategies such as molecular docking can provide a reliable and rapid platform for antiviral drug development [98]. Virus-targeted therapy with nanotechnology is a novel research platform. Nanoparticle—peptide conjugates have been reported to be active against infections of viruses such as HIV and influenza A virus [99,100]. Nanoparticles can also be applied as targeted drug carriers [71].

In conclusion, the development of novel, cost-effective peptide synthesis and improved antiviral peptide delivery mechanisms, along with the advancement in the discovery of novel antiviral peptides, will help in the combat against future viral epidemics and pandemics.

# Acknowledgments

This work was supported by the Council of Scientific and Industrial Research, India (CSIR Ref. No: 09/1108(0005)/2015-EMR-I), Kerala State Council for Science Technology and Environment, India (KSCSTE Ref. No: KSCSTE/2078/ 2019-FSHP-LS), Indian Council of Medical Research (ICMR No: 67/6/2020-DDI/BMS) and Central University of Kerala, India.

# References

[1] Y. Fu, A.H. Jaarsma, O.P. Kuipers, Antiviral activities and applications of ribosomally synthesized and post-translationally modified peptides (RiPPs), Cellular and Molecular Life Sciences 78 (2021) 3921–3940.

[2] M.S. Mousavi Maleki, M. Rostamian, H. Madanchi, Antimicrobial peptides and other peptide-like therapeutics as promising candidates to combat SARS-CoV-2, Expert Review of Anti-infective Therapy 19 (2021) 1205–1217.

[3] A.S. Skwarecki, M.G. Nowak, M.J. Milewska, Amino acid and peptide-based antivirala, ChemMedChem 16 (2021) 3106–3135.

[4] S. Kausar, F. Said Khan, M. Ishaq Mujeeb Ur Rehman, M. Akram, M. Riaz, G. Rasool, et al., A review: mechanism of action of antiviral drugs, International Journal of Immunopathology and Pharmacology 35 (2021). 20587384211002621.

[5] P. Wang, Y. Liu, G. Zhang, S. Wang, J. Guo, J. Cao, et al., Screening and identification of Lassa virus entry inhibitors from an FDA-approved drug library, Journal of Virology 92 (2018) e00954–18.

[6] P. Intharathep, C. Laohpongspaisan, T. Rungrotmongkol, A. Loisruangsin, M. Malaisree, P. Decha, et al., How amantadine and rimantadine inhibit proton transport in the M2 protein channel, Journal of Molecular Graphics and Modelling 27 (2008) 342–348.

[7] J.L. Blanco, G. Whitlock, A. Milinkovic, G. Moyle, HIV integrase inhibitors: a new era in the treatment of HIV, Expert Opinion on Pharmacotherapy 16 (2015) 1313–1324.

[8] S. Mahajan, S. Choudhary, P. Kumar, S. Tomar, Antiviral strategies targeting host factors and mechanisms obliging + ssRNA viral pathogens, Bioorganic & Medicinal Chemistry 15 (2021) 116356.

[9] M.F. Chew, K.S. Poh, C.L. Poh, Peptides as therapeutic agents for dengue virus, International Journal of Medical Sciences 14 (2017) 1342.

[10] A.J. Cann, Replication of viruses, Encyclopedia of Virology (2008) 406.

[11] J. Louten, Virus replication, Essential Human Virology (2016) 49.

[12] J. Magden, L. Kääriäinen, T. Ahola, Inhibitors of virus replication: recent developments and prospects, Applied Microbiology and Biotechnology 66 (2005) 612–621.

[13] C.B. Jackson, M. Farzan, B. Chen, H. Choe, Mechanisms of SARS-CoV-2 entry into cells, Nature Reviews Molecular Cell Biology 23 (2022) 3–20.

[14] E. Hartenian, D. Nandakumar, A. Lari, M. Ly, J.M. Tucker, B.A. Glaunsinger, The molecular virology of coronaviruses, Journal of Biological Chemistry 295 (2020) 12910–12934.

[15] P.K. Raghav, K. Kalyanaraman, D. Kumar, Human cell receptors: potential drug targets to combat COVID-19, Amino Acids 53 (2021) 813–842.

[16] Y. Li, Z. Zhang, L. Yang, X. Lian, Y. Xie, S. Li, et al., The MERS-CoV receptor DPP4 as a candidate binding target of the SARS-CoV-2 spike, iScience 23 (2020) 101160.

[17] V.S. Raj, H. Mou, S.L. Smits, D.H. Dekkers, M.A. Müller, R. Dijkman, D. Muth, J.A. Demmers, A. Zaki, R.A. Fouchier, V. Thiel, Dipeptidyl peptidase 4 is a functional receptor for the emerging human coronavirus-EMC, Nature 495 (2013) 251–254.

[18] D. Baltimore, Viral RNA-dependent DNA polymerase: RNA-dependent DNA polymerase in virions of RNA tumour viruses, Nature 226 (1970) 1209–1211.

[19] B. Berkhout, D. Eggink, R.W. Sanders, Is there a future for antiviral fusion inhibitors? Current Opinion in Virology 2 (2012) 50–59.

[20] D.M. Lambert, S. Barney, A.L. Lambert, K. Guthrie, R. Medinas, D.E. Davis, et al., Peptides from conserved regions of paramyxovirus fusion (F) proteins are potent inhibitors of viral fusion, Proceedings of the National Academy of Sciences 93 (1996) 2186−2191.
[21] A.G. Schmidt, P.L. Yang, S.C. Harrison, Peptide inhibitors of dengue-virus entry target a late-stage fusion intermediate, PLoS Pathogens 6 (2010) e1000851.
[22] S. Skalickova, Z. Heger, L. Krejcova, V. Pekarik, K. Bastl, J. Janda, et al., Perspective of use of antiviral peptides against influenza virus, Viruses 7 (2015) 5428−5452.
[23] M. Krepstakies, J. Lucifora, C.H. Nagel, M.B. Zeisel, B. Holstermann, H. Hohenberg, et al., A new class of synthetic peptide inhibitors blocks attachment and entry of human pathogenic viruses, The Journal of Infectious Diseases 205 (2012) 1654−1664.
[24] P. Pugach, T.J. Ketas, E. Michael, J.P. Moore, Neutralizing antibody and anti-retroviral drug sensitivities of HIV-1 isolates resistant to small molecule CCR5 inhibitors, Virology 377 (2008) 401−407.
[25] S.S. Lieberman-Blum, H.B. Fung, J.C. Bandres, Maraviroc: a CCR5-receptor antagonist for the treatment of HIV-1 infection, Clinical Therapeutics 30 (2008) 1228−1250.
[26] R.E. Marques, R. Guabiraba, J.L. Del Sarto, R.F. Rocha, A.L. Queiroz, D. Cisalpino, et al., Dengue virus requires the CC-chemokine receptor CCR5 for replication and infection development, Immunology 145 (2015) 583−596.
[27] M. Rajik, F. Jahanshiri, A.R. Omar, A. Ideris, S.S. Hassan, K. Yusoff, Identification and characterisation of a novel anti-viral peptide against avian influenza virus H9N2, Virology Journal 6 (2009) 1−2.
[28] M. Rajik, A.R. Omar, A. Ideris, S.S. Hassan, K. Yusoff, A novel peptide inhibits the influenza virus replication by preventing the viral attachment to the host cells, International Journal of Biological Sciences 5 (2009) 543.
[29] J. Wei, M. Hameed, X. Wang, J. Zhang, S. Guo, M.N. Anwar, et al., Antiviral activity of phage display-selected peptides against Japanese encephalitis virus infection in vitro and in vivo, Antiviral Research 174 (2020) 104673.
[30] C. Li, L.Y. Zhang, M.X. Sun, P.P. Li, L. Huang, J.C. Wei, et al., Inhibition of Japanese encephalitis virus entry into the cells by the envelope glycoprotein domain III (EDIII) and the loop3 peptide derived from EDIII, Antiviral Research 94 (2012) 179−183.
[31] C. Li, L.L. Ge, Y.L. Yu, L. Huang, Y. Wang, M.X. Sun, et al., A tripeptide (NSK) inhibits Japanese encephalitis virus infection in vitro and in vivo, Archives of Virology 159 (2014) 1045−1055.
[32] A. Panya, P. Yongpitakwattana, P. Budchart, N. Sawasdee, S. Krobthong, A. Paemanee, et al., Novel bioactive peptides demonstrating anti-dengue virus activity isolated from the Asian medicinal plant *Acacia catechu*, Chemical Biology & Drug Design 93 (2019) 100−109.
[33] A. Panya, N. Sawasdee, P. Songprakhon, Y. Tragoolpua, S. Rotarayanont, K. Choowongkomon, et al., A synthetic bioactive peptide derived from the asian medicinal plant *Acacia Catechu* binds to dengue virus and inhibits cell entry, Viruses 12 (2020) 1267.
[34] L.C. Vilas Boas, M.L. Campos, R.L. Berlanda, N. de Carvalho Neves, O.L. Franco, Antiviral peptides as promising therapeutic drugs, Cellular and Molecular Life Sciences 76 (2019) 3525−3542.
[35] H. Memariani, M. Memariani, H. Moravvej, M. Shahidi-Dadras, Melittin: a venom-derived peptide with promising anti-viral properties, European Journal of Clinical Microbiology & Infectious Diseases 39 (2020) 5−17.
[36] Y. Bao, S. Wang, H. Li, Y. Wang, H. Chen, M. Yuan, Characterization, stability and biological activity in vitro of cathelicidin-BF-30 loaded 4-arm star-shaped PEG-PLGA microspheres, Molecules 23 (2018) 497.

[37] J. Xu, S. Chen, J. Jin, L. Ma, M. Guo, C. Zhou, et al., Inhibition of peptide BF-30 on influenza A virus infection in vitro/vivo by causing virion membrane fusion, Peptides 112 (2019) 14–22.
[38] S.C. Harrison, Viral membrane fusion, Virology 479 (2015) 498–507.
[39] D.M. Eckert, P.S. Kim, Mechanisms of viral membrane fusion and its inhibition, Annual Review of Biochemistry 70 (2001) 777–810.
[40] B. Podbilewicz, Virus and cell fusion mechanisms, Annual Review of Cell and Developmental Biology 30 (2014) 111–139.
[41] C. Li, Q. Ba, A. Wu, H. Zhang, T. Deng, T. Jiang, A peptide derived from the C-terminus of PB 1 inhibits influenza virus replication by interfering with viral polymerase assembly, The FEBS Journal 280 (2013) 1139–1149.
[42] J. Pu, J.T. Zhou, P. Liu, F. Yu, X. He, L. Lu, et al., Viral entry inhibitors targeting six-helical bundle core against highly pathogenic enveloped viruses with class I fusion proteins, Current Medicinal Chemistry 29 (2022) 700–718.
[43] A.G. Schmidt, P.L. Yang, S.C. Harrison, Peptide inhibitors of flavivirus entry derived from the E protein stem, Journal of Virology 84 (2010) 12549–12554.
[44] M.N. Reddy, R. Dungdung, L. Valliyott, R. Pilankatta, Occurrence of concurrent infections with multiple serotypes of dengue viruses during 2013–2015 in northern Kerala, India, PeerJ 5 (2017) e2970.
[45] R. Pamela, C. David, A. Kathleen, Phage display selection of cyclic peptides that inhibit Andes virus infection, Journal of Virology 83 (2009) 8965–8969.
[46] S. Xia, L. Yan, W. Xu, A.S. Agrawal, A. Algaissi, C.T. Tseng, et al., A pan-coronavirus fusion inhibitor targeting the HR1 domain of human coronavirus spike, Science Advances 5 (2019) eaav4580.
[47] S. Xia, M. Liu, C. Wang, W. Xu, Q. Lan, S. Feng, et al., Inhibition of SARS-CoV-2 (previously 2019-nCoV) infection by a highly potent pan-coronavirus fusion inhibitor targeting its spike protein that harbors a high capacity to mediate membrane fusion, Cell Research 30 (2020) 343–355.
[48] H. Chong, J. Xue, S. Xiong, Z. Cong, X. Ding, Y. Zhu, et al., A lipopeptide HIV-1/2 fusion inhibitor with highly potent in vitro, ex vivo, and in vivo antiviral activity, Journal of Virology 91 (2017) e00288–17.
[49] B.D. Welch, J.N. Francis, J.S. Redman, S. Paul, M.T. Weinstock, J.D. Reeves, et al., Design of a potent D-peptide HIV-1 entry inhibitor with a strong barrier to resistance, Journal of Virology 84 (2010) 11235–11244.
[50] J.N. Francis, J.S. Redman, D.M. Eckert, M.S. Kay, Design of a modular tetrameric scaffold for the synthesis of membrane-localized D-peptide inhibitors of HIV-1 entry, Bioconjugate Chemistry 23 (2012) 1252–1258.
[51] Y. Nishimura, J.N. Francis, O.K. Donau, E. Jesteadt, R. Sadjadpour, A.R. Smith, et al., Prevention and treatment of SHIVAD8 infection in rhesus macaques by a potent d-peptide HIV entry inhibitor, Proceedings of the National Academy of Sciences 117 (2020) 22436–22442.
[52] B. Gao, D. Zhao, L. Li, Z. Cheng, Y. Guo, Antiviral peptides with in vivo activity: development and modes of action, ChemPlusChem 86 (2021) 1547–1558.
[53] A. Pessi, S.L. Bixler, V. Soloveva, S. Radoshitzky, C. Retterer, T. Kenny, et al., Cholesterol-conjugated stapled peptides inhibit Ebola and Marburg viruses in vitro and in vivo, Antiviral Research 171 (2019) 104592.
[54] G.H. Bird, S. Boyapalle, T. Wong, K. Opoku-Nsiah, R. Bedi, W.C. Crannell, et al., Mucosal delivery of a double-stapled RSV peptide prevents nasopulmonary infection, The Journal of Clinical Investigation 124 (2014).

[55] V. Gaillard, M. Galloux, D. Garcin, J.F. Eléouët, R. Le Goffic, T. Larcher, et al., A short double-stapled peptide inhibits respiratory syncytial virus entry and spreading, Antimicrobial Agents and Chemotherapy 61 (2017). 02241-02216.
[56] S.M. Currie, E. Gwyer Findlay, A.J. McFarlane, P.M. Fitch, B. Böttcher, N. Colegrave, et al., Cathelicidins have direct antiviral activity against respiratory syncytial virus in vitro and protective function in vivo in mice and humans, The Journal of Immunology 196 (2016) 2699−2710.
[57] X.J. Wang, C.G. Li, X.J. Chi, M. Wang, Characterisation and evaluation of antiviral recombinant peptides based on the heptad repeat regions of NDV and IBV fusion glycoproteins, Virology 416 (2011) 65−74.
[58] C.G. Li, W. Tang, X.J. Chi, Z.M. Dong, X.X. Wang, X.J. Wang, A cholesterol tag at the N terminus of the relatively broad-spectrum fusion inhibitory peptide targets an earlier stage of fusion glycoprotein activation and increases the peptide's antiviral potency in vivo, Journal of Virology 87 (2013) 9223−9232.
[59] D.J. Holthausen, S.H. Lee, V.T. Kumar, N.M. Bouvier, F. Krammer, A.H. Ellebedy, et al., An amphibian host defense peptide is virucidal for human H1 hemagglutinin-bearing influenza viruses, Immunity 46 (2017) 587−595.
[60] A. Marzi, R. Yoshida, H. Miyamoto, M. Ishijima, Y. Suzuki, M. Higuchi, et al., Protective efficacy of neutralizing monoclonal antibodies in a nonhuman primate model of Ebola hemorrhagic fever, PLoS One 7 (2012) e36192.
[61] M.A. Al-Bari, Targeting endosomal acidification by chloroquine analogs as a promising strategy for the treatment of emerging viral diseases, Pharmacology Research & Perspectives 5 (2017) e00293.
[62] K. Chandran, N.J. Sullivan, U. Felbor, S.P. Whelan, J.M. Cunningham, Endosomal proteolysis of the Ebola virus glycoprotein is necessary for infection, Science 308 (2005) 1643−1645.
[63] C.J. Stevens, N.B. Dise, J.O. Mountford, D.J. Gowing, Impact of nitrogen deposition on the species richness of grasslands, Science 303 (2004) 1876−1879.
[64] T.N. Figueira, M.T. Augusto, K. Rybkina, D. Stelitano, M.G. Noval, O.E. Harder, et al., Effective in vivo targeting of influenza virus through a cell-penetrating/fusion inhibitor tandem peptide anchored to the plasma membrane, Bioconjugate Chemistry 29 (2018) 3362−33676.
[65] H. Zhao, J. Zhou, K. Zhang, H. Chu, D. Liu, V.K. Poon, et al., A novel peptide with potent and broad-spectrum antiviral activities against multiple respiratory viruses, Scientific Reports 6 (2016) 1−3.
[66] H. Zhao, K.K. To, K.H. Sze, T.T. Yung, M. Bian, H. Lam, et al., A broad-spectrum virus-and host-targeting peptide against respiratory viruses including influenza virus and SARS-CoV-2, Nature Communications 11 (2020) 4252.
[67] H. Zhao, K.K. To, H. Lam, X. Zhou, J.F. Chan, Z. Peng, et al., Cross-linking peptide and repurposed drugs inhibit both entry pathways of SARS-CoV-2, Nature Communications 12 (2021) 1−9.
[68] B. Hernáez, T. Tarragó, E. Giralt, J.M. Escribano, C. Alonso, Small peptide inhibitors disrupt a high-affinity interaction between cytoplasmic dynein and a viral cargo protein, Journal of Virology 84 (2010) 10792−10801.
[69] G.M. Andrei, F.C. Coulombie, M.C. Courreges, R.A. de Torres, C.E. Coto, Meliacine, an antiviral compound from *Melia azedarach* L., inhibits interferon production, Journal of Interferon Research 10 (1990) 469−475.
[70] M.B. Wachsman, V. Castilla, C.E. Coto, Inhibition of foot and mouth disease virus (FMDV) uncoating by a plant-derived peptide isolated from *Melia azedarach* L leaves, Archives of Virology 143 (1998) 581−590.

[71] Y. Zhang, L.V. Tang, Overview of targets and potential drugs of SARS-CoV-2 according to the viral replication, Journal of Proteome Research 20 (2020) 49−59.

[72] Y.J. Gordon, E.G. Romanowski, R.M. Shanks, K.A. Yates, H. Hinsley, H.A. Pereira, CAP37-derived antimicrobial peptides have in vitro antiviral activity against adenovirus and herpes simplex virus type 1, Current Eye Research 34 (2009) 241−249.

[73] Y.J. Gordon, L.C. Huang, E.G. Romanowski, K.A. Yates, R.J. Proske, A.M. McDermott, Human cathelicidin (LL-37), a multifunctional peptide, is expressed by ocular surface epithelia and has potent antibacterial and antiviral activity, Current Eye Research 30 (2005) 385−394.

[74] S.A. Harvey, E.G. Romanowski, K.A. Yates, Y.J. Gordon, Adenovirus-directed ocular innate immunity: the role of conjunctival defensin-like chemokines (IP-10, I-TAC) and phagocytic human defensin-$\alpha$, Investigative Ophthalmology & Visual Science 46 (2005) 3657−3665.

[75] X. Liang, X. Zhang, K. Lian, X. Tian, M. Zhang, S. Wang, et al., Antiviral effects of Bovine antimicrobial peptide against TGEV in vivo and in vitro, Journal of Veterinary Science 21 (2020).

[76] M. Khan, W. Rauf, F.E. Habib, M. Rahman, S. Iqbal, A. Shehzad, et al., Hesperidin identified from citrus extracts potently inhibits HCV genotype 3a NS3 protease, BMC Complementary Medicine and Therapies 22 (2022) 1−8.

[77] N.L. Van der Weerden, M.R. Bleackley, M.A. Anderson, Properties and mechanisms of action of naturally occurring antifungal peptides, Cellular and Molecular Life Sciences 70 (2013) 3545−3570.

[78] L.S. Ooi, L. Tian, M. Su, W.S. Ho, S.S. Sun, H.Y. Chung, et al., Isolation, characterization, molecular cloning and modeling of a new lipid transfer protein with antiviral and antiproliferative activities from Narcissus tazetta, Peptides 29 (2008) 2101−2109.

[79] H. Wang, T.B. Ng, Ginkbilobin, a novel antifungal protein from Ginkgo biloba seeds with sequence similarity to embryo-abundant protein, Biochemical and Biophysical Research Communications 279 (2000) 407−411.

[80] H.X. Wang, T.B. Ng, Ascalin, a new anti-fungal peptide with human immunodeficiency virus type 1 reverse transcriptase-inhibiting activity from shallot bulbs, Peptides 23 (2002) 1025−1029.

[81] J.H. Wong, T.B. Ng, Lunatusin, a trypsin-stable antimicrobial peptide from lima beans (Phaseolus lunatus L.), Peptides 26 (2005) 2086−2092.

[82] J.H. Wong, T.B. Ng, Vulgarinin, a broad-spectrum antifungal peptide from haricot beans (*Phaseolus vulgaris*), The International Journal of Biochemistry & Cell Biology 37 (2005) 1626−1632.

[83] E. Gwyer Findlay, S.M. Currie, D.J. Davidson, Cationic host defence peptides: potential as antiviral therapeutics, BioDrugs 27 (2013) 479−493.

[84] A. Henninot, J.C. Collins, J.M. Nuss, The current state of peptide drug discovery: back to the future? Journal of Medicinal Chemistry 61 (2018) 1382−1414.

[85] Z. Antosova, M. Mackova, V. Kral, T. Macek, Therapeutic application of peptides and proteins: parenteral forever? Trends in Biotechnology 27 (2009) 628−635.

[86] K. Fosgerau, T. Hoffmann, Peptide therapeutics: current status and future directions, Drug Discovery Today 20 (2015) 122−128.

[87] AVPdb. <http://crdd.osdd.net/servers/avpdb> (2014).

[88] HIPdb. <http://crdd.osdd.net/servers/hipdb> (2013).

[89] A. Qureshi, N. Thakur, H. Tandon, M. Kumar, AVPdb: a database of experimentally validated antiviral peptides targeting medically important viruses, Nucleic Acids Research 42 (2014) D1147−D1153.

[90] A. Qureshi, N. Thakur, M. Kumar, HIPdb: a database of experimentally validated HIV inhibiting peptides, PLoS One 8 (2013) e54908.
[91] THPdb. <http://crdd.osdd.net/raghava/thpdb/> (2017).
[92] S.S. Usmani, G. Bedi, J.S. Samuel, S. Singh, S. Kalra, P. Kumar, et al., THPdb: database of FDA-approved peptide and protein therapeutics, PLoS One 12 (2017) e0181748.
[93] N.A. Murugan, K. Raja, N.T. Saraswathi, Peptide-based antiviral drugs, Advances in Experimental Medicine and Biology (2021) 261–284.
[94] X. Yao, H. Chong, C. Zhang, S. Waltersperger, M. Wang, S. Cui, et al., Broad antiviral activity and crystal structure of HIV-1 fusion inhibitor sifuvirtide, Journal of Biological Chemistry 287 (2012) 6788–6796.
[95] T. Volz, L. Allweiss, M.B. MBarek, M. Warlich, A.W. Lohse, J.M. Pollok, et al., The entry inhibitor Myrcludex-B efficiently blocks intrahepatic virus spreading in humanized mice previously infected with hepatitis B virus, Journal of Hepatology 58 (2013) 861–867.
[96] S. Chaudhuri, J.A. Symons, J. Deval, Innovation and trends in the development and approval of antiviral medicines: 1987–2017 and beyond, Antiviral Research 155 (2018) 76–88.
[97] I. Gentile, A.R. Buonomo, E. Zappulo, G. Borgia, Discontinued drugs in 2012–2013: hepatitis C virus infection, Expert Opinion on Investigational Drugs 24 (2015) 239–251.
[98] S. Mahmud, S. Biswas, G.K. Paul, M.A. Mita, S. Afrose, M.R. Hasan, et al., Antiviral peptides against the main protease of SARS-CoV-2: a molecular docking and dynamics study, Arabian Journal of Chemistry 14 (2021) 103315.
[99] M. Chakravarty, A. Vora, Nanotechnology-based antiviral therapeutics, Drug Delivery and Translational Research 11 (2021) 748–787.
[100] Z.K. Alghrair, D.G. Fernig, B. Ebrahimi, Enhanced inhibition of influenza virus infection by peptide–noble-metal nanoparticle conjugates, Beilstein Journal of Nanotechnology 10 (2019) 1038–1047.

# Antimicrobial peptide antibiotics against multidrug-resistant ESKAPE pathogens

*Guangshun Wang, Atul Verma and Scott Reiling*
Department of Pathology and Microbiology, College of Medicine, University of Nebraska Medical Center, Omaha, NE, United States

## 13.1 Introduction

Bacterial antibiotic resistance is an increasingly difficult challenge for healthcare providers worldwide, due to inherent mechanisms of pathogenic bacteria that allow them to outpace novel antibiotic development. According to a 2019 CDC threat report, greater than 2.8 million cases of antibiotic-resistant infections arise each year, resulting in more than 35,000 deaths [1]. A UK report projects mortality of 10 million by 2050 [2]. The ESKAPE pathogens, including *Enterococcus faecium*, *Staphylococcus aureus*, *Klebsiella pneumoniae*, *Acinetobacter baumannii*, *Pseudomonas aeruginosa*, and *Enterobacter* species, are of special concern due to their wide-range and synergistic mechanisms of resistance, and wide representation in nosocomial infections [3]. They account for high annual morbidity rates, increased healthcare costs, and the highest mortality risk within the hospital setting, which cause treatment dilemmas stemming from ever evolving multidrug-resistant (MDR) strains [4,5]. The global priority pathogen list was collated by the World Health Organization (WHO) as a means to stratify the danger-level of pathogens as "critical," "high," or "medium" [6]. *A. baumannii*, *P. aeruginosa*, and *Enterobacteriaceae* species are within the "critical" priority tier, while methicillin-resistant *S. aureus* (MRSA) and vancomycin resistant *E. faecium* (VRE) are in the "high" tier. Since 2015, the WHO has initiated the world antimicrobial awareness week to better conserve existing antibiotics. Based on the One Health approach, there are efforts to establish a global antimicrobial resistance surveillance system, global antimicrobial research and development partnership, and interagency coordination group on antimicrobial resistance [6]. These research and education efforts will lead to more useful methods to manage antibiotic resistance.

As a complementary strategy, it is also important to understand the resistance mechanisms of ESKAPE pathogens (Section 13.2). Equally important is to identify various antimicrobial strategies that are potentially useful to combat the ESKAPE pathogens (Section 13.3). Our subsequent discussion focuses on antimicrobial peptides (AMPs). We discuss their advantages and disadvantages (Section 13.4). We then highlight AMP discovery methods, their mimics, conjugates, combinations, formulated nanoparticles (NPs), and surface immobilized forms that have the potential to combat the ESKAPE pathogens (Section 13.5). We also mention the major mechanisms of bacterial killing by AMPs, ranging from the classic membranes to nonmembrane targets, including ribosomes (Section 13.6). Finally, we comment on the efficacy testing of AMPs for systemic use although most of the peptides are limited to topical applications. It is important to note that some AMPs are already in use, inspiring the development of more novel antibiotics. In particular, we need consider personalized medicine so that novel antimicrobials selectively eliminate the targeted pathogens without harming the functions of human microbiota.

## 13.2 Antibiotic resistance of ESKAPE pathogens

Bacteria develop antibiotic resistance through two general mechanisms involving mutation of virulence factor genes, and horizontal transfer of such genes between bacteria within a community via plasmids and transposons, which may then be chromosomally integrated. The functions of bacterial virulence factor genes are extensive, as many genes do not directly interact with antibiotics, but rather serve to increase the fitness of the pathogen for survival within the host. Herein, we discuss ESKAPE gene products that closely interact with antibiotics by direct binding or inactivation, pumping drug out of cells, the alteration of bacterial membrane permeability, and the development of an external physical barrier in the form of biofilm.

### 13.2.1 Direct drug interaction

The overall outcome of bacterial resistance is to mitigate the killing effects of antibiotics. This can be achieved physically and chemically. Physically, bacteria can pump drugs out of cells so that intracellular concentrations of drugs are below the threshold of killing. ESKAPE bacteria also have genotypes exhibiting a variety of membrane efflux pumps, each of which displays polyspecificity. The most common families are the ATP-binding cassette (ABC), the small multidrug resistance, and the resistance-nodulation division (RND) families [7]. The polyselective efflux pump, which is common in gram-negative bacteria, is a member of the RND family. *P. aeruginosa* chromosomally encodes four potent RND MDR "Mex" pumps conferring protection against aminoglycosides, fluoroquinolones, and carbapenems [8]. Similarly, *Enterobacter aerogenes* and *K. pneumoniae* isolates from patients contain other RNDs that also eject antibiotics normally unspecific to the Mex pumps, including tetracycline and chloramphenicol. The AdeABC RNDs are hallmark broad-specificity pumps found in *A. baumannii*. There are other molecular mechanisms (e.g., production of porin proteins) that affect the uptake of antibiotics. The outer membranes of gram-negative bacteria have porins, which allow transport of hydrophilic molecules from the extracellular environment to the cytoplasm. Downregulation or loss of outer membrane proteins such as the Opr cluster in *P. aeruginosa*, or the Omp proteins in *A. baumannii* and *K. pneumoniae*, are found in clinical isolates, contributing to MDR from many -lactam drugs [9]. Alteration or downregulation of porin proteins work synergistically with efflux pumps to keep intracellular concentrations of -lactams at concentrations too low to inhibit or kill their target pathogens.

There are also multiple chemical mechanisms of bacterial resistance. Some bacteria may alter their molecules being targeted to reduce the affinity of drug. Penicillin-binding proteins (PBPs) are best characterized. The PBPs are cell-surface anchored proteins that serve as conduits for intracellular transport of many -lactamases; mutation of these genes creates some of the most virulent serotypes within ESKAPE. For example, the unique PBP2a is found in MRSA variants, which has a low affinity for -lactam drugs [10]. Gram-positive bacteria are susceptible to another drug-binding site: the peptidoglycan layer of cell walls. However, the *E. faecalis* and *E. faecium* bacteria have evolved to render many glycopeptides antibiotics ineffective, by altering the cross-linking residues within peptidoglycan from D-Ala-D-Ala to D-Ala-D-Lac or D-Ala-D-Ser [11,12]. It is known that the anionic membranes of MRSA can be modified with a basic lysine to repel the effect of cationic antibiotics [13]. Likewise, the lipid A moiety of gram-negative bacteria outer membrane lipopolysaccharides (LPS) is a target for the cationic peptide polymyxins, of which colistin is a last-line treatment option for MDR strains. The *mcr-1* cassette is a plasmid-mediated resistance gene against the polymyxins, and works similarly by altering the lipid A structure such that overall negative charge is diminished, through removal of phosphate groups. The *mcr-1* product grants colistin resistance to *K. pneumoniae* and *A. baumannii* [14]. A related gene, *mcr-2*, is also plasmid-borne and able to integrate into other genetic elements via transposase domains [15]. Both are of concern due to demonstrated ease of lateral transfer into the gram-negative ESKAPE members, and the consequent increase in colistin MIC.

Finally, the drug may be destroyed or inactivated. The primary bacterial antibiotic defense mechanism to produce enzymes that bind and cleave antibiotics into inactive products. The -lactamases are the hallmark class of proteolytic enzymes that bind to the -lactam rings of respective antibiotics such as the penicillins, cephalosporins, and carbapenems [16]. Multiple classes of enzymes have been found in gram-negative pathogens such as *Escherichia coli*, *P. aeruginosa*, *K. pneumoniae*, and *Enterobacter* spp. [11].

## 13.2.2 Indirect drug resistance

Some bacterial defenses employ proteins that do not directly bind antibiotics, but lessen their potency by building a "house." Biofilms are formed by multispecies bacterial colonies living within a complex house coated with extracellular DNA, polysaccharides, lipids, proteins, and inorganic molecules. Biofilm structure benefits bacteria by providing channels for molecular communication via two-component systems, consisting of both sensor histidine kinases and response regulators [17]. Within the context of antibiotic resistance, biofilms confer a passive physical defense against antibiotic penetration to its respective target, and a microenvironment that reduces drug efficacy due to the lowered pH, water and oxygen content, and the increased $CO_2$ concentrations [11]. These environmental variables may attenuate drug interaction, but their values within a biofilm also shift the encapsulated bacteria away from a planktonic state, and toward a nutrient-scarce phenotype, which increases antibiotic tolerance [18]. The biofilms within the healthcare setting most commonly consist of ESKAPE constituents *S. aureus*, *P. aeruginosa*, *A. baumannii*, and *K. pneumonia* [19].

## 13.3 Strategies to combat the ESKAPE pathogens

To meet the challenge of antibiotic resistance, scientists around the world are constantly searching new therapies. Some alternatives are vaccines, phages, antibiotic derivatives, and AMPs.

### 13.3.1 Vaccines

Vaccines are one of the best strategies to combat invading pathogens such as viruses and bacteria. Traditionally, they were generated by utilizing inactivated bacteria or their essential pathogenic components such as proteins. However, vaccines are generally successful against viral infections but more challenging to develop against bacterial diseases. Tefibazumab was developed against *S. aureus*. It showed a positive response in animal studies but failed in clinical trials [20]. To date, all clinical trials of vaccines against *S. aureus* have failed. It appears that different strategies or bacterial virulence factors such as toxins may be required to generate more useful antibacterial vaccines [21,22].

### 13.3.2 Phage therapy

With the development of antibiotic resistance, phages are now considered as another alternative [23,24]. Phage therapy is not a new concept. It was discovered before the use of antibiotics. Bacteriophages and their products, such as lysins, are specific bacteria-infecting viruses leading to the lysis and disruption of the bacterial integrity. Bacteriophages have several advantages over antibiotics, including specificity, safety, efficacy against bacteria with multiple antibiotic resistances, and the capability of biofilm degradation. Phages can even penetrate through the protective shield of biofilms in bacteria [25]. The decoding of their genomes was exploited for production against antibiotic-resistant bacteria [26], as seen with N1M2 and phage K. While N1M2, isolated from maize silage, lyses the *K. pneumoniae*, phage K has also been observed to remove the biofilm of *S. aureus* [27,28]. Various phages, such as AP22, AB1, Abp1 and 9, are identified in hospital sewages against the *A. baumannii*. PD6A3, and its endolysin Ply6A3 [29] have been found to have lytic activity against *A. baumannii*. Lysins are bacteriophage-encoded peptidoglycan hydrolases and may serve as another alternative to antibiotics. Endolysins, LysAm24, LysECD7, and LysSi3, isolated from Myxoviridae family, are found to be against *P. aeruginosa* [24].

However, bacteria can also develop resistance to phages. A combination of phages with antibiotics may lead to more effective treatments [23,30].

### 13.3.3 Antibiotic derivatives

It is also possible to improve antibiotic treatment by combining them in clinical applications. Two such combinations were approved during 2018–19 [31]. There are also efforts in making derivatives of antibiotics or expanding the treatment scope of existing antibiotics [31]. This approach temporarily extends the clinical life of antibiotics, but may lead to wide spread of resistance as well. Formulation can also improve activity of antibiotics. Vancomycin was incorporated into NP to increase peptide activity [32]. In addition, activity of antibiotics may also be boosted in the presence of adjuvants. Cefepime, colistin, ofloxacin, rifampicin, tetracycline, and vancomycin become more effective in the presence of an AMP, which targets bacterial membranes [33].

### 13.3.4 Antimicrobial peptides

AMPs are small host defense proteins that are regarded as an important alternative. The interest in these peptides, in part, results from their antimicrobial potency against bacteria, fungi, viruses, and parasites, including drug-resistant species. Since these peptides usually target membranes, it is difficult for pathogens to develop resistance [34]. As of October 2021, the antimicrobial peptide database (APD) houses 3283 AMPs discovered from six life kingdoms, including bacteria, archaea, protists, fungi, plants, and animals [35–37]. The APD (website: https://aps.unmc.edu) has recently been reprogrammed to enhance cybersecurity [38]. Naturally occurring AMPs are frequently amphipathic and cationic (most net charge +1 to +5) with less than 60 amino acids [39]. They are diverse in amino acid sequence and structural scaffold. In the current APD, the majority of AMPs are synthesized in ribosomes. However, some are made via nonribosomal mechanisms. Because of the continued interest in these host defense peptides, we focus on AMPs in the rest of this chapter.

## 13.4 Advantages and disadvantages of cationic antimicrobial peptides

Since an increased interest in AMPs in the 1980s, several aspects of such peptides are accepted as advantages [40,41]. First, AMPs are potent against antibiotic-resistant bacteria, including the ESKAPE pathogens and their persisters in biofilms. Second, AMPs can eliminate bacteria rapidly. They are the key components of the innate immune systems. Macrophages deficient in mouse CRAMP is unable to eliminate engulfed *E. coli* [42]. The expression of AMPs starts from pathogen recognition via Toll-like receptors (TLRs). In the case of bacteria, TLR4 can recognize

LPS in the outer membranes of gram-negative bacteria [43], whereas TLR1/2 can recognize cell wall components of gram-positive bacteria [44]. TLR5 can recognize flagellin in bacterial tail [45], whereas TLR9 is located in the endosome to recognize bacterial DNA [46]. It is important to note that the Toll pathway was initially discovered in insects by Hoffmann and colleagues [47]. Such molecular recognition can initiate the signal transduction and expression of AMPs to control invading bacteria. The innate immune system is clearly significant. One on hand, bacteria are able to replicate rapidly in minutes. On the other hand, it takes days or even weeks for the adaptive immune system to respond. Thus innate and adaptive immune systems work together in vertebrates to combat pathogen infections.

Third, bacteria usually do not develop resistance to AMPs. The multiple hits may be part of the reason. In addition to direct killing, AMPs may regulate immune system to initiate action against invading pathogens, including recruiting immune cells to clear infection [48]. However, some bacterial resistance mechanisms described above may apply to AMPs [49]. *S. aureus* secretes proteases to cleave and inactivate human cathelicidin LL-37 [50,51]. By screening the Nebraska Transposon Mutant Library of *S. aureus* USA300 LAC [52], we identified multiple bacterial response genes that make MRSA more resistant to the killing of human cathelicidin LL-37 [53]. These include the *mprF* gene that places a lysine on anionic phosphatidylglycerol (PG) [13]. In addition, pathogens are able to modulate the host response to favor infections. Using a CRISPR knockout library of human macrophage THP-1, 183 host factors were found to be involved in Salmonella–macrophage interactions. These host molecules may become therapeutic targets to block pathogen invasion [54]. For instance, an NHLRC2 mutant shows >85% resistance to Salmonella infection. In summary, rapid killing, antimicrobial potency, and immune regulation of AMPs are important reasons for continued research interest in developing them into a new generation of antimicrobials.

There are other aspects regarded as disadvantages of AMPs. Some AMPs such as melittin are toxic to both bacteria and host cells, disqualifying their direct use due to a lack of cell selectivity [55]. However, many AMPs are selective (i.e., preferentially killing pathogens rather than human cells). Toxic peptides may be made useful via peptide engineering. For example, some selective hybrid peptides involving melittin have been designed [56,57]. Reducing peptide hydrophobicity is proposed to be a useful strategy to increase therapeutic index. This can be achieved by amino acid substitution, peptide truncation, or partial incorporation of D-amino acids [58].

Another concern is peptide stability. While many linear peptides can be cleared in less than 1-hour, cyclic AMPs are more durable. Linear peptides can be made more stable by partially or fully incorporating D-amino acids. In collaboration with Merrifield, Boman and his team reported the first peptide hybrids made entirely of D-amino acids [56]. Shai and colleagues partially incorporated D-amino acids into AMPs to improve both peptide selectivity and stability [59]. Other common approaches include C-terminal amidation, N-terminal acetylation, the incorporation of other types of nonstandard amino acids, backbone changes, and peptide cyclization [60,61]. Star-shaped antimicrobials, consisting of unnatural amino acid ornithines, are more resistant to proteases [62]. However, we found similar systemic efficacy for susceptible and stable peptides in a neutropenic mouse model [63,64]. Therefore further studies are required to establish the relationship between peptide stability and therapeutic efficacy in vivo. It is likely that peptide stability requirements depend on treatment situations. In treating Ebola virus, peptide stability was found to be critical since the natural forms were cleaved in endosomes [65].

AMPs can lose activity under various conditions (e.g., pH, salts, and serum). Human LL-37 loses its antibacterial activity in the presence of human serum by binding to apolipoprotein A-I (apoA-I) [66]. Thus the interaction with host components may be one of the major reasons why some AMPs are not as effective as antibiotics [67]. One useful method to identify more robust AMPs is to compare antimicrobial robustness of a library of peptides under different conditions. By expanding our in silico filtering to in vitro filtering, we obtained such peptides that retain activity in rich media, at a reduced pH, or in the presence of salts or serum [63].

Peptide cost is frequently listed as a disadvantage. There are two major methods for peptide production. The first is to express it in biological systems, ranging from bacteria to plants. Nisin, which is used as a food preservative, is best produced in bacteria due to numerous posttranslational modifications [68]. Similarly, there is an advantage to produce cyclotides or their analogs in plants or suspension cells [69]. A second method is chemical synthesis, which has been greatly facilitated due to the discovery of automated solid-phase synthesis [70–72]. To reduce the production cost, it is useful to simplify the peptides by removing nonessential amino acids [73].

## 13.5 Antimicrobial peptides to stop ESKAPE pathogens

There are two methods for AMP discovery: rational design and library approach. Rational design, also known as structure-based design, plays an important role in designing antibiotics against a defined target once the three-dimensional structure is available [74–76]. Library approaches are conceptually broad, ranging from a collection of peptide-producing bacteria in a defined niche to a collection of peptides generated in silico for virtual screening or chemically synthesized in laboratories for antimicrobial assays [77–79]. There are other strategies (e.g., induction, delivery, prodrug, conjugates, and combination) to improve the therapeutic potential of AMPs [80].

### 13.5.1 Structure-based design

AMPs with different sequences can be candidates to combat the ESKAPE pathogens [81–85]. However, cationic and hydrophobic amino acids in numerous peptides are frequently arranged in such a manner that the structure is amphipathic. Amino acid rearrangement (e.g., sequence reversal, permutation, and shuffling), substitution, deletion, or addition of any residue in the sequence is known to alter peptide properties owing to changes in composition and structure. Here we utilize human cathelicidin LL-37 (37 residues) as an example to illustrate structure-based design. First, we identified the major antimicrobial region corresponding to residues

17–32 (GF-17) based on structure–activity relationship studies of peptide fragments [58]. GF-17 is more potent against MRSA than LL-37 [74]. Second, we improved peptide stability by incorporating D-amino acids at positions 20, 24, and 28 (as numbered in LL-37). Structural determination revealed a novel amphipathic structure for GF-17d3, drastically different from the canonical amphipathic helix of GF-17 [58]. Third, we enhanced the activity of GF-17d3 by substituting two phenylalanines (F17 and F27) with two biphenylalanines. The resulting peptide 17BIPHE2 (a 17-residue peptide with two biphenylalanines) is able to kill the ESKAPE pathogens in vitro and to prevent biofilm formation of MRSA in catheters embedded in mice [74]. When formulated into nanofiber, it also removed biofilms of MRSA in chronic wounds [86]. It is possible to further reduce hemolysis of 17BIPHE2 by altering the sidechain moieties [87]. The peptide 17tF-W was obtained by replacing the first biphenyl of 17BIPHE2 with 4-t-butylphenylalanine (tF), and the second biphenyl with a Trp side chain. The new peptide analog has an MIC value in the range of 3.1–6.2 M against the ESKAPE pathogens, while its 50% hemolytic concentration ($HC_{50}$) is greater than 440 M. The cell selectivity can be calculated as $HC_{50}/MIC = 71–142$. In other words, no hemolysis is anticipated at the bacterial killing concentrations of 3.1–6.2 M [87]. Therefore it is feasible to engineer LL-37 into peptides that are stable, selective, and potent against the ESKAPE pathogens.

### 13.5.2 Library-based search and peptide mimetics

Using the library approach, we screened representative AMPs collected in the APD [88]. This led to the identification of multiple peptides (e.g., DASamP1 and DASamP2) with minimal inhibitory concentrations (MIC) in the range of 3.1–6.1 M against MRSA. As an alternative approach, we also developed the database-filtering technology [89]. This method is referred to as *ab initio* since it derives each peptide design parameter based on the APD database from the beginning. This *ab initio* designed peptide DFTamP1, with a high hydrophobic ratio and low cationicity, is primarily active against gram-positive MRSA. Such peptide features turned out to be important for in vivo efficacy [63]. It is feasible to synthesize a small molecule mimic for DFTamP1. The small molecule mimic shows a similar activity spectrum [90]. Our study illustrates a practical antimicrobial discovery path from database construction, method development, peptide design, to peptide mimics synthesis.

Other labs have made various peptide mimetics, usually by following the amphipathic pattern of natural peptides. Since these mimicries are antimicrobial, they substantiate the concept that the peptide backbone is not an absolute requirement. This discovery further expands the molecular space for antimicrobials. Success has been made in synthesizing nonclassic peptide backbone such as antimicrobial peptoids and star-shaped dendrimeric AMPs [91–94].

### 13.5.3 Peptide conjugates

There are cases where an AMP is covalently connected with a different moiety that may not be antimicrobial by itself. Costa et al. [95] conjugated ferrocene with

RP-1, derived from platelet factor-4 protein, which increased peptide activity against both bacteria and parasites. Proportionally, the conjugate becomes more toxic to murine macrophages than the peptide alone. As a consequence, a similar therapeutic index is observed. Straus and colleagues conjugated poly(ethyleneglycol) (PEG) with aurein 2.2. It appears that both conjugate size and peptide density affect peptide biocompatibility [96]. There are also efforts to conjugate peptides with antibiotics to gain the desired potency [97,98].

Both daptomycin and colistin contain a lipid tail. These in-clinical-use AMPs inspired the search of other lipopeptides as potential novel antimicrobials [99]. It is evident that lipidation (addition of a hydrophobic tail) confers activity to cationic amino acids such as lysine, since the combination generates a different amphipathic construction [100]. Also, the longer the fatty acid chain, the more hydrophobic the lipopeptide, leading to increased antibacterial activity as well as toxicity. As a consequence, many ultrashort peptides (usually four amino acids or less) conjugated with long fatty acids have a low cell selectivity. Kamysz et al. combined KR-12 (the minimal antibacterial peptide of LL-37) with different fatty acids C4-C14 [101]. C8-KR-12 is effective against the ESKAPE pathogens (MIC 1–4 g/mL). Prior to the above study, we have systematically compared the antimicrobial and hemolytic activities of a series of KR-12 truncated peptides conjugated with fatty acids with different chain lengths, allowing us to identify highly selective lipopeptides. Since the identified lipoLL-37 peptides can retain antibacterial activity under various conditions, they are superior to the parent peptides (KR-12 or LL-37). Indeed, the optimized peptide demonstrated efficacy in different mice models [102]. Some pathogens can hide in host cells to escape elimination. Coupling of KR-12 with a cell-penetrating peptide TAT is found to kill intracellular *S. aureus* [103]. This conjugated peptide also inhibited bacterial growth in a subcutaneous infection mouse model.

### 13.5.4 Combined treatment

Human cathelicidin LL-37 is a broad-spectrum AMP by itself [73,104–106]. Other AMPs, such as defensins and amyloid peptides, may also be expressed in humans [107]. It is possible that some AMPs, when present at the same niche, work together to more effectively combat invading pathogens. In airways, human AMPs, including LL-37, defensin, lysozyme, lactoferrin, secretory leukocyte protease inhibitor (SLPI), can work synergistically or additively [108]. Recently, LL-37 is shown to enhance the activity of AMPs expressed by skin commensal bacteria [109,110]. Also, LL-37 and human -defensin 3 (hBD-3) can synergistically regulate cytokines in a 3D cell culture model [111]. Cationic AMPs may be used together with antibiotics for improved efficacy [112–114]. We found antibiotics were ineffective against the biofilms of *P. aeruginosa*. However, both database and structure-designed peptides are effective to different extents. When combined with antibiotics, the LL-37 engineered peptide, 17BIPHE2, is more effective in eliminating the biofilms of *P. aeruginosa* [115]. In combination with DNase and antibiotics kanamycin and ciprofloxacin, AMP Gad-1 can even clear *P. aeruginosa* biofilms under cystic fibrosis conditions [116].

It is proposed that the combination of AMPs with antibiotics offers a practical avenue to combating antibiotic-resistant pathogens [80].

### 13.5.5 Formulated antimicrobial peptides

Cationic AMPs, especially linear ones, are susceptible to degradation and can readily lose activity due to association with host factors. The stability of such peptides may be enhanced by engineering (see above). Another approach is formulation where the peptide is included within an NP to avoid being degraded or lost before pathogen killing. Stem cells with LL-37-expression ability can promote wound healing when incorporated into hydrogel [117]. The NPs retain antibiofilm and immune modulation activities [118]. It is also possible to include engineered LL-37 peptides. Su et al. incorporated 17BIPHE2 into nanofiber that can deliver the peptide in a sustainable manner for a month. The initial rapid release of AMPs removes antibiotic-resistant pathogens, while the subsequent antimicrobial release prevents infection [86]. It is feasible to further improve the delivery efficiency of the matrix by coupling with microneedles [119]. Microneedles enable deeper penetration of peptides into biofilms, leading to pathogen elimination without need of debridement, a common surgery practice to control biofilms in wounds in the United States.

### 13.5.6 Surface immobilized antimicrobial peptides

Medical implants are multibillion dollars business that plays an important role in supporting the life quality of patients. In certain cases, these implants can get infected. Although contaminated implants may be treated [120], such pathogens are usually difficult to remove probably due to the formation of biofilms on the surfaces [121]. The matrix of biofilms reduces the entrance of antibiotics as well as the infiltration of immune cells. This frequently necessitates replacement of the implants, an undesired process. Alternatively, researchers have thought about coating the implants with antibiotics [122]. Such surfaces may be useful to avoid the infection of antibiotic-susceptible pathogens, but they cannot stop the colonization of ESKAPE pathogens. Under such a circumstance, AMPs may be useful [123]. However, there are few studies that demonstrate antibacterial activity against all the ESKAPE pathogens. FK-16, the major AMP from human cathelicidin LL-37, when immobilized to the titanium surface, is effective against these pathogens [124]. Compared to GF-17 (17 residues), FK-16 (16 residues) does not have the N-terminal glycine. In the free state, GF-17 and FK-16 show similar antibacterial activity [73]. There are still challenging problems before implementation in practice. How long will the coated surface remain useful? The peptides may be cleaved by proteases or become ineffective due to biofouling. One solution is to incorporate peptides into a multilayer structure of nanofibers coated on titanium to control the antimicrobial release, which prevents early and delayed infections for 6 weeks [125].

## 13.6 Mechanisms of bacterial killing by antimicrobial peptides

The mechanisms of action of AMPs are complex and not limited to bacterial membranes. In the following sections, we briefly describe three main molecular targets. Additional killing mechanisms have been depicted elsewhere [126].

### 13.6.1 Bacterial membranes

It is well known that cationic AMPs kill bacterial pathogens by targeting and damaging anionic membranes. This results from the long-range recognition via electrostatic attractions between positively charged amino acids of AMPs and negatively charged bacterial surface. After this initial docking, the peptide transforms its conformation to an amphipathic structure perfect for membrane binding. The peptide is not static and is able to reorient itself to reach bacterial membranes to achieve optimized binding with anionic phophatidylglycerols (PGs). The existence of such an electrostatic interaction is supported by several lines of evidence: (1) reduced antimicrobial activity when the basic amino acids of the peptide is substituted by alanine [127,128]; (2) reduced physical interactions when neutral zwitterionic lipids such as phosphocholines (PCs) are utilized to replace PGs [129]; (3) direct observation of dipolar–dipolar interactions between interfacial arginine head group and anionic PGs by NMR spectroscopy [130]; and (4) increased antimicrobial susceptibility of a bacterial mutant where the key enzyme MprF that places basic lysine on PGs has been genetically knocked out or disabled via transposon insertion [13,53]. It appears that not all the basic amino acids play an equal role. Among the five basic amino acids (K18, R19, R23, K25, and R29) in the major AMP (residues 17–32) of human cathelicidin LL-37, R23 is most important in lipid clustering, membrane permeation, and bacterial killing [127,128]. Subsequently, the hydrophobic moieties of peptides, in particular aromatic rings of phenylalanines (F5, F6, F17, and F27) of LL-37, find their way to the acyl chain of bacterial lipids to achieve van der Waals hydrophobic interactions [131]. The motion of anionic lipids toward cationic amino acids causes membrane reorganization, a process termed lipid clustering [132,133]. This process usually does not lead to membrane damage to the extent it generates a pore. It differs from the carpet model [134] due to a lack of membrane lysis. Pore formation [135,136], however, can occur for certain peptides. We have recently proposed a phase model [38] by considering the dynamic processes of peptide–membrane interactions. Each state may correspond to one or more current membrane damaging models. At a low peptide concentration, one only sees peptide binding to the membrane surface and the initiation of lipid clustering and even domain formation [132]. With an increase in concentration, peptides may also cause phase changes, leading to pore formation [135,136]. With a further increase in peptide concentration, the bacterial membranes could entirely be fragmented and disintegrated into small micelle-like particles. The entire disruption of bacterial membranes is evidenced by a clear solution (bacteria lysis) [127]. This model can incorporate other factors such as peptide oligomerization [137] and synergy with other peptides or components [138]. This dynamic phase model for AMPs unifies the existing models (peptide aggregation, lipid clustering, carpet model, pore model, and membrane disruption), each emphasizing part of the bacterial membrane remodeling and damaging processes.

### 13.6.2 Cell wall

Not all AMPs act on bacterial membranes. The first indication is that the MIC values are rather different for the L- and D-forms of the same peptide. An alternative mode of action is for the peptide to act on cell wall. The APD has annotated 28 AMPs with a known action on cell walls [39,139]. These include nine defensins, human HNP-1, hBD-3, fungal plectasin, copsin, eurocin, insect lucifensin, oyster defensins Cg-Def, Cg-Defh1, and Cg-Defh2. Also, 15 bacterial lantibiotics inhibit cell wall synthesis (see these entries in the APD3 database for original references). Nisin appears to remove lipid II from the gram-positive bacterial division site (or septum) to block cell wall synthesis [140].

### 13.6.3 Bacterial ribosomes

Proline-rich AMPs (PrAMPs) are active mainly against gram-negative pathogens. They do not disrupt membranes. Earlier, it was observed that PrAMPs could associate with heat shock proteins [141]. This was proposed as a possible killing mechanism. However, the bacteria were shown to be vulnerable to killing after knocking out the gene for that protein. This implies a different target for inhibition. More recently, ribosomes are established as a target for PrAMPs [142,143]. In the current APD [40], 71 AMPs are annotated as PrAMPs based on the definition for the word "rich" at 30% [41], including astacidin 2, Cg-Prp, lebocin-1/2, P9, PP30, and PR-bombesin. The binding of PrAMPs to ribosomes may not be the same as some traditional antibiotics, offering an alternative mechanism to fight such pathogens. However, this picture may be incomplete. Some newly discovered PrAMPs are able to kill most of the ESKAPE pathogens, including gram-positive *S. aureus* (MIC 16 M) and *E. faecium* (MIC 4 M), by disrupting membranes [144]. Also, the multimeric peptide, Chex1-Arg20, designed based on a proline-rich template displays effects on membranes as well [145]. These observations expand both activity spectrum and mechanism of action of PrAMPs. It appears that the key is an increase in peptide hydrophobicity that could give the peptide more time to act on bacterial membranes.

Compared to planktonic cells, the killing mechanisms may be more complex in the case of biofilms. In particular, the outer matrix of biofilms poses a challenge for antibiotics to penetrate physically, rendering them ineffective. AMPs, however, can disrupt the outer layer to some extent, allowing the entry of antibiotics to kill persisters. This may explain the improved efficacy of combined treatment [115]. In line with this reasoning, Su et al. observed improved antibiofilm efficacy by combining AMP-delivering nanofibers with microneedles that penetrate the outer layers of biofilms [119].

## 13.7 Efficacies in animal models and clinical use of antimicrobial peptides

*Efficacies in animal models.* At present, most AMPs are developed for topical use. This is important to treat patients for whom traditional antibiotics have failed. Infections involving biofilms are such problems. There are animal models for both catheters and chronic wounds. We observed clear in vivo efficacy of an LL-37 engineered peptides and DASamP1 from database screening in preventing biofilm formation in a catheter-associated murine model [74,87,88]. In addition, 17BIPHE2 displayed outstanding efficacy in a chronic wound-healing model [86]. Likewise, verine is able to eliminate biofilms in the same mouse model when formulated with nanofiber attached to microneedles [119].

There are, however, also other situations where resistant pathogens found their way into blood streams. Therefore it is useful to test systemic efficacy of new candidates as well. While LL-37 engineered peptides are effective in eliminating pathogens in topical catheter or wound models [74,86,87], they are unable to reduce bacterial burdens in a systemic mouse model [63]. We reasoned that the peptide might become active if they could pass a series of in vitro filters (acidic pH, salts, and serum) where AMPs lose activity. In this manner, we have recently obtained database designed peptides (e.g., DFT503, horine, and verine). Horine, a short helical peptide, displays systemic efficacy against mainly gram-positive pathogens such as MRSA, while verine, with a novel spiral structure, can inhibit all the ESKAPE pathogens with demonstrated efficacy against both MRSA and Klebsiella in mice [63,64]. These results imply that some AMPs may find systemic use in medical settings.

*Clinical use of antimicrobial peptides.* There are already AMP examples that have utilized in clinical settings [80]. Gramicidin A is a linear peptide and was first used clinically. Daptomycin and colistin are both sidechain-backbone linked peptides. Gramicidin S is a circular peptide used to treat wounds. Other FDA-approved peptide antibiotics include bacitracin from *Bacillus subtilis* and bleomycin from a fungus *Streptomyces verticillus*. These successfully examples with different structural scaffolds [146] may inspire the design of additional peptides as a new generation of antibiotics.

## 13.8 Concluding remarks

The ESKAPE pathogens pose a major threat to the current medical system as a consequence of antibiotic resistance. They have developed a variety of resistance mechanisms, including biofilm formation to escape the attack of conventional antibiotics and immune systems. To meet this challenge, numerous strategies, including vaccines and phages, are under development or screening. Antibiotics may also be enhanced by modification or conjugation. Antibiotics have also been used in combination to improve efficacy. One component can be AMPs. AMPs are important candidates because of their lasting antimicrobial potency. In addition, the membrane targeting of AMPs may facilitate the entry of antibiotics and renders it difficult for pathogens to develop resistance. Different methods have been utilized to improve their therapeutic potential of AMPs. These include peptide end capping, D-amino acids, backbone alteration, sidechain modification, cyclization in different ways (sidechain-linked sidechain-backbone linked, and head-to-tail cyclization). With a future emphasis on personalized medicine, it may not be necessary to produce all the peptides in large quantity. It appears that the application scope (topical vs systemic) is peptide dependent. It is encouraging that some AMPs, such as nisin, gramicidin, daptomycin, and colistin, are already in medical use or as food preservatives [80]. While many more are under active development, new ideas and concepts are required for future success. The classic view of the peptide stability is to ensure the peptide to have sufficient time for bacterial killing. However, this may not be a general requirement. Under certain situations, peptide stability may not be the limiting factor, enabling direct use of AMPs in natural forms to minimize bacterial resistance and potential toxic effects to infected subjects [64]. With the evolving idea of precise medicine, it is necessary for future antibiotics to achieve high selectivity so that only the targeted pathogen will be eliminated but not commensal microbiota. A more complete measurement and annotation of antimicrobial activity spectrum in the APD database would lay a foundation for this application. Future studies will further our understanding of AMPs in innate immunity in general and explore their application scopes in particular. To expand the horizon, one should also consider nonantimicrobial functions and applications of innate immune peptides [41]. To conclude, we anticipate a bright future for AMPs because they have a variety of structural scaffolds created for different biological functions and applications.

## Acknowledgment

This study was supported by the NIH grant R01GM138552.

## References

[1] Centers for Disease Control and Prevention. Antibiotic-resistant germs: new threats., <https://www.cdc.gov/drugresistance/biggest-threats.html> 2021 (accessed 03.01.21).
[2] J. O'Neil, The Review on Antimicrobial Resistance: Tracking Drug resistant Infections Globally, UK, 2014.
[3] L.B. Rice, Progress and challenges in implementing the research on ESKAPE pathogens, Infection Control and Hospital Epidemiology 31 (2010) S7–S10.
[4] H.W. Boucher, G.H. Talbot, D.K. Benjamin Jr, J. Bradley, R.J. Guidos, R.N. Jones, et al., Infectious Diseases Society of America. 10 '20 Progress–development of new drugs active against gram-negative bacilli: an update from the Infectious Diseases Society of America, Clinical Infectious Diseases 56 (2013) 1685–1694.

[5] R.C. Founou, L.L. Founou, S.Y. Essack, Clinical and economic impact of antibiotic resistance in developing countries: a systematic review and *meta*-analysis, PLoS One 12 (2017) e0189621.
[6] World Health Organization. Global priority list of antibiotic resistant bacteria to guide research, discover, and development of new antibiotics, <https://www.who.int/medicines/publications/WHO-PPL-Short_Summary_25Feb-ET_NM_WHO.pdf> 2021 (accessed January 2021).
[7] J. Sun, Z. Deng, A. Yan, Bacterial multidrug efflux pumps: mechanisms, physiology and pharmacological exploitations, Biochemistry and Biophysical Research Communications 453 (2014) 254–267.
[8] H. Vaez, J. Faghri, B.N. Isfahani, S. Moghim, S. Yadegari, H. Fazeli, et al., Efflux pump regulatory genes mutations in multidrug resistance *Pseudomonas aeruginosa* isolated from wound infections in Isfahan hospitals, Advanced Biomedical Research 3 (2014) 117.
[9] J.M. Thomson, R.A. Bonomo, The threat of antibiotic resistance in gram-negative pathogenic bacteria: beta-lactams in peril!, Current Opinion in Microbiology 8 (2005) 518–524.
[10] M.J. Pucci, T.J. Dougherty, Direct quantitation of the numbers of individual penicillin-binding proteins per cell in *Staphylococcus aureus*, Journal of Bacteriology 184 (2002) 588–591.
[11] S. Santajit, N. Indrawattana, Mechanisms of antimicrobial resistance in ESKAPE pathogens, Biomed Research International 2016 (2016) 2475067.
[12] A. Giedraitienė, A. Vitkauskienė, R. Naginienė, A. Pavilonis, Antibiotic resistance mechanisms of clinically important bacteria, Medicina 47 (2011) 19.
[13] C.M. Ernst, P. Staubitz, N.N. Mishra, S.J. Yang, G. Hornig, H. Kalbacher, et al., The bacterial defensin resistance protein MprF consists of separable domains for lipid lysinylation and antimicrobial peptide repulsion, PLoS Pathogens 5 (2009) e1000660.
[14] Y.Y. Liu, C.E. Chandler, L.M. Leung, C.L. McElheny, R.T. Mettus, R.M.Q. Shanks, et al., Structural modification of lipopolysaccharide conferred by *mcr-1* in gram-negative ESKAPE pathogens, Antimicrobial Agents and Chemotherapy 61 (2017) e00580-17.
[15] J.H. Moffatt, M. Harper, J.D. Boyce, Mechanisms of polymyxin resistance, Advances in Experimental Medicine and Biology 1145 (2009) 55–71.
[16] G.A. Jacoby, L.S. Munoz-Price, The new beta-lactamases, New England Journal of Medicine 352 (2005) 380–391.
[17] S. Tiwari, S.B. Jamal, S.S. Hassan, P.V.S.D. Carvalho, S. Almeida, D. Barh, et al., Two-component signal transduction systems of pathogenic bacteria as targets for antimicrobial therapy: an overview, Frontiers in Microbiology 8 (2017) 1878.
[18] A. Hussain, A. Ansari, R. Ahmad, Chapter 4—Microbial biofilms: human mucosa and intestinal microbiota, in: M.K. Yadav, B.P. Singh (Eds.), New and Future Developments in Micorbial Biotechnology and Bioengineering. Microbial Biofilms: Current Research and Future Trends, Elsevier, 2020, pp 47–60.
[19] N. Høiby, T. Bjarnsholt, M. Givskov, S. Molin, O. Ciofu, Antibiotic resistance of bacterial biofilms, International Journal of Antimicrobial Agents 35 (2010) 322–332.
[20] V.K. Ganesh, X. Liang, J.A. Geoghegan, A.L.V. Cohen, N. Venugopalan, T.J. Foster, et al., Lessons from the crystal structure of the *S. aureus* surface protein clumping factor A in complex with tefibazumab, an inhibiting monoclonal antibody, EBioMedicine 13 (2016) 328–338.
[21] N.D. Cohen, C. Cywes-Bentley, S.M. Kahn, A.I. Bordin, J.M. Bray, S.G. Wehmeyer, et al., Vaccination of yearling horses against poly-N-acetyl glucosamine fails to protect against infection with *Streptococcus equi* subspecies equi, PLoS One 15 (2020) e0240479.

[22] L.S. Miller, V.G. Fowler, S.K. Shukla, W.E. Rose, R.A. Proctor, Development of a vaccine against *Staphylococcus aureus* invasive infections: evidence based on human immunity, genetics and bacterial evasion mechanisms, FEMS Microbiology Reviews 44 (2020) 123−153.

[23] K.E. Kortright, B.K. Chan, J.L. Koff, P.E. Turner, Phage therapy: a renewed approach to combat antibiotic-resistant bacteria, Cell Host and Microbe 25 (2019) 219−232.

[24] D. Romero-Calle, R. GuimarãesBenevides, A. Góes-Neto, C. Billington, Bacteriophages as alternatives to antibiotics in clinical care, Antibiotics 8 (2019) 138.

[25] S. González, L. Fernández, D. Gutiérrez, A.B. Campelo, A. Rodríguez, P. García, Analysis of different parameters affecting diffusion, propagation and survival of staphylophages in bacterial biofilms, Frontiers in Microbiology 9 (2018) 2348.

[26] T.L. Tagliaferri, M. Jansen, H.P. Horz, Fighting pathogenic bacteria on two fronts: phages and antibiotics as combined strategy, Frontiers in Cellular and Infection Microbiology 9 (2019) 22.

[27] S.T. Abedon, Bacteriophages and Biofilms, Nova Science Publishers, Inc, 2011.

[28] M. Wu, K. Hu, Y. Shi, Y. Liu, D. Mu, H. Guo, et al., A novel phage PD-6A3, and its endolysin Ply6A3, with extended lytic activity against *Acinetobacter baumannii*, Frontiers in Microbiology 9 (2019) 3302.

[29] N.P. Antonova, D.V. Vasina, A.M. Lendel, E.V. Usachev, V.V. Makarova, A.L. Gintsburg, et al., Broad bactericidal activity of the *Myoviridae* bacteriophage lysins LysAm24, LysECD7, and LysSi3 against gram-negative ESKAPE pathogens, Viruses. 11 (2019) 284.

[30] R. Lewis, A.G. Clooney, S.R. Stockdale, C. Buttimer, L.A. Draper, R.P. Ross, et al., Isolation of a novel jumbo bacteriophage effective against *Klebsiella* aerogenes, Frontiers in Medicine 7 (2020) 67.

[31] S. Andrei, G. Droc, G. Stefan, FDA approved antibacterial drugs: 2018–2019, Discoveries (Craiova) 7 (2019) e102.

[32] W. Zhang, R. Taheri-Ledari, Z. Hajizadeh, E. Zolfaghari, M.R. Ahghari, A. Maleki, et al., Enhanced activity of vancomycin by encapsulation in hybrid magnetic nanoparticles conjugated to a cell-penetrating peptide, Nanoscale 12 (2020) 3855−3870.

[33] Y. Liu, X. Huang, S. Ding, Y. Wang, J. Shen, K. Zhu, A broad-spectrum antibiotic adjuvant reverses multidrug-resistant gram-negative pathogens, Nature Microbiology 5 (2020) 1040−1050.

[34] H.G. Boman, Antibacterial peptides: basic facts and emerging concepts, Journal of Internal Medicine 254 (2003) 197−215.

[35] Z. Wang, G. Wang, APD: the antimicrobial peptide database, Nucleic Acids Research 32 (2004) D590−D592.

[36] G. Wang, X. Li, Z. Wang, APD2: the updated antimicrobial peptide database and its application in peptide design, Nucleic Acids Research 37 (2009) D933−D937.

[37] G. Wang, X. Li, Z. Wang, APD3: the antimicrobial peptide database as a tool for research and education, Nucleic Acids Research 44 (2016) D1087−D1093.

[38] G. Wang, C.M. Zietz, A. Mudgapalli, S. Wang, Z. Wang, The evolution of the antimicrobial peptide database over 18 years: milestones and new features, Protein Science 31 (2022) 92−106. Available from: https://doi.org/10.1002/pro.4185.

[39] G. Wang, Database-guided discovery of potent peptides to combat HIV-1 or superbugs, Pharmaceuticals (Basel), 6, 2013, pp. 728−758.

[40] G. Wang, Boinformatic analysis of 1000 amphibian antimicrobial peptides uncovers multiple length-dependent correlations for peptide design and prediction, Antibiotics (Basel), 9, 2020, p. 491.

[41] X. Dang, G. Wang, Spotlight on the selected new antimicrobial innate immune peptides discovered during 2015–2019, Current Topical Medicinal Chemistry 20 (2020) 2984–2998.

[42] T. Yoshimura, W. Gong, C. Tian, J. Huang, G. Trinchieri, J.M. Wang, Requirement of CRAMP for mouse macrophages to eliminate phagocytosed *E. coli* through an autophagy pathway, Journal of Cell Sciences 134 (2021) jcs.252148.

[43] A. Poltorak, X. He, I. Smirnova, M.-Y. Liu, C.V. Huffel, X. Du, et al., Defective LPS signaling in C3H/HeJ and C57BL/10ScCr mice: mutations in Tlr4 gene, Science (New York, N.Y.) 282 (1998) 2085–2088.

[44] C. Trichot, S. Korniotis, L. Pattarini, V. Soumelis, TLR1/2 orchestrate human plasmacytoid predendritic cell response to gram + bacteria, PLoS Biology 17 (2019) e3000209.

[45] K.D. Smith, A. Ozinsky, T.R. Hawn, E.C. Yi, D.R. Goodlett, J.K. Eng, et al., The innate immune response to bacterial flagellin is mediated by toll-like receptor 5, Nature 410 (2001) 1099–1103.

[46] S. Bauer, C.J. Kirschning, H. Häcker, V. Redecke, S. Hausmann, S. Akira, et al., Human TLR9 confers responsiveness to bacterial DNA via species-specific CpG motif recognition, Proceedings of the National Academy of Sciences of the United States of America 98 (2001) 9237–9242.

[47] J.L. Imler, J.A. Hoffmann, Toll and toll-like proteins: an ancient family of receptors signaling infection, Reviews in Immunogenetics 2 (2000) 294–304.

[48] M. Magana, M. Pushpanathan, A.L. Santos, L. Leanse, M. Fernandez, A. Ioannidis, et al., The value of antimicrobial peptides in the age of resistance, Lancet Infectious Diseases 20 (2020) e216–e230.

[49] D.I. Andersson, D. Hughes, J.Z. Kubicek-Sutherland, Mechanisms and consequences of bacterial resistance to antimicrobial peptides, Drug Resistance Updates 26 (2016) 43–57.

[50] P.N. Oliveira, D.S. Courrol, R.M. Chura-Chambi, L. Morganti, G.O. Souza, M.R. Franzolin, et al., Inactivation of the antimicrobial peptide LL-37 by pathogenic Leptospira, Microbial Pathogenesis 150 (2020) 104704.

[51] M. Sieprawska-Lupa, P. Mydel, K. Krawczyk, K. Wójcik, M. Puklo, B. Lupa, et al., Degradation of human antimicrobial peptide LL-37 by *Staphylococcus aureus*-derived proteinases, Antimicrobial Agents and Chemotherapy 48 (2004) 4673–4679.

[52] J.L. Endres, V.K. Yajjala, P.D. Fey, K.W. Bayles, Construction of a sequence-defined transposon mutant library in *Staphylococcus aureus*, Methods in Molecular Biology 2019 (2016) 29–37.

[53] R.M. Golla, B. Mishra, X. Dang, J.L. Narayana, A. Li, L. Xu, et al., Resistome of *Staphylococcus aureus* in response to human cathelicidin LL-37 and its engineered antimicrobial peptides, ACS Infectious Diseases. 6 (2020) 1866–1881.

[54] A.T.Y. Yeung, Y.H. Choi, A.H.Y. Lee, C. Hale, H. Ponstingl, D. Pickard, et al., A genome-wide knockout screen in human macrophages identified host factors modulating *Salmonella* infection, mBio 10 (2019) e02169-19.

[55] G. Kreil, H. Bachmayer, Biosynthesis of melittin, a toxic peptide from bee venom. Detection of a possible precursor, European Journal of Biochemistry 20 (1971) 344–350.

[56] E.L. Merrifield, S.A. Mitchell, J. Ubach, H.G. Boman, D. Andreu, R.B. Merrifield, D-enantiomers of 15-residue cecropin A-melittin hybrids, International Journal of Peptides and Protein Research 46 (1995) 214–220.

[57] A.R. Ferreira, C. Teixeira, C.F. Sousa, L.J. Bessa, P. Gomes, P. Gameiro, How insertion of a single tryptophan in the N-terminus of a cecropin A-melittin hybrid peptide changes its antimicrobial and biophysical profile, Membranes (Basel), 11, 2021, p. E48.

[58] X. Li, H. Han, D.W. Miller, G. Wang, Solution structures of human LL-37 fragments and NMR-based identification of a minimal membrane-targeting antimicrobial and anticancer region, Journal of American Chmical Society 128 (2006) 5776–5785.
[59] N. Papo, Z. Oren, U. Pag, H.-G. Sahl, Y. Shai, The consequence of sequence alteration of an amphipathic alpha-helical antimicrobial peptide and its diastereomers, Journal of Biological Chemistry 277 (2002) 33913–33921.
[60] G. Wang, Post-translational modifications of natural antimicrobial peptides and strategies for peptide engineering, Current Biotechnology 1 (2012) 72–79.
[61] S. Gunasekera, T. Muhammad, A.A. Strömstedt, K.J. Rosengren, U. Göransson, Backbone cyclization and dimerization of LL-37-derived peptides enhance antimicrobial activity and proteolytic stability, Frontiers in Microbiology 11 (2020) 168.
[62] M. Pan, C. Lu, M. Zheng, W. Zhou, F. Song, W. Chen, et al., Unnatural amino-acid-based star-shaped poly(L-ornithine)s as emerging long-term and biofilm-disrupting antimicrobial peptides to treat *Pseudomonas aeruginosa*-infected burn wounds, Advanced Healthcare Materials 9 (2020) e2000647.
[63] B. Mishra, J.L. Narayana, T. Lushnikova, X. Wang, G. Wang, Low cationicity is important for systemic in vivo efficacy of database-derived peptides against drug-resistant gram-positive pathogens, Proceedings of the National Academy of Sciences of the United States of America 116 (2019) 13517–13522.
[64] J.L. Narayana, B. Mishra, T. Lushnikova, Q. Wu, Y.S. Chhonker, Y. Zhang, et al., Two distinct amphipathic peptide antibiotics with systemic efficacy, Proceedings of the National Academy of Sciences of the United States of America 117 (2020) 19446–19454.
[65] Y. Yu, C.L. Cooper, G. Wang, M.J. Morwitzer, K. Kota, J.P. Tran, et al., Engineered human cathelicidin antimicrobial peptides inhibit Ebola virus infection, iScience 23 (2020) 100999.
[66] Y. Wang, J. Johansson, B. Agerberth, H. Jörnvall, W.J. Griffiths, The antimicrobial peptide LL-37 binds to the human plasma protein apolipoprotein A-I, Rapid Communications in Mass Spectrometry 18 (2004) 588–589.
[67] C.G. Starr, J. He, W.C. Wimley, Host cell interactions are a significant barrier to the clinical utility of peptide antibiotics, ACS Chemical Biology 11 (2016) 3391–3399.
[68] I. Mierau, M. Kleerebezem, 10 years of the nisin-controlled gene expression system (NICE) in *Lactococcus lactis*, Applied Microbiology and Biotechnology 68 (2005) 705–717.
[69] H. Qu, M.A. Jackson, K. Yap, P.J. Harvey, E.K. Gilding, D.J. Craik, Production of a structurally validated cyclotide in rice suspension cells is enabled by a supporting biosynthetic enzyme, Planta 252 (2020) 97.
[70] R.B. Merrifield, Solid phase peptide synthesis. I. The synthesis of a tetrapeptide, Journal of American Chemical Society 85 (1963) 2149–2154.
[71] G. Sabatino, A.M. Papini, Advances in automatic, manual and microwave-assisted solid-phase peptide synthesis, Current Opinion in Drug Discovery 11 (2008) 762–770.
[72] D.F.H. Winkler, Automated solid-phase peptide synthesis, Methods in Molecular Biology 2103 (2020) 59–94.
[73] G. Wang, J.L. Narayana, B. Mishra, Y. Zhang, F. Wang, C. Wang, et al., Design of antimicrobial peptides: progress made with human cathelicidin LL-37, Advances in Experimental Medicine and Biology 1117 (2019) 215–240.
[74] G. Wang, M.L. Hanke, B. Mishra, T. Lushnikova, C.E. Heim, V.C. Thomas, et al., Transformation of human cathelicidin LL-37 into selective, stable, and potent antimicrobial compounds, ACS Chemical Biology 9 (2014) 1997–2002.

[75] C.H. Chen, C.G. Starr, E.T.G. Wiedman, W.C. Wimley, J.P. Ulmschneider, M.B. Ulmschneider, Simulation-guided rational de novo design of a small pore-forming antimicrobial peptide, Journal of American Chemical Society 141 (2019) 4839–4848.
[76] S.J. Son, R. Huang, C.J. Squire, I.K.H. Leung, MCR-1: a promising target for structure-based design of inhibitors to tackle polymyxin resistance, Drug Discovery Today 24 (2019) 206–216.
[77] P.R. Puentes, M.C. Henao, C.E. Torres, S.C. Gómez, L.A. Gómez, J.C. Burgos, et al., Design, screening, and testing of nonrational peptide libraries with antimicrobial activity: in silico and experimental approaches, Antibiotics (Basel), 9, 2020, p. 854.
[78] M. Ashby, A. Petkova, J. Gani, R. Mikut, K. Hilpert, Use of peptide libraries for identification and optimization of novel antimicrobial peptides, Current Topics in Medicinal Chemistry 17 (2017) 537–553.
[79] S.A. Guralp, Y.E. Murgha, J.-M. Rouillard, E. Gulari, From design to screening: a new antimicrobial peptide discovery pipeline, PLoS One 8 (2013) e59305.
[80] B. Mishra, S. Reiling, D. Zarena, G. Wang, Host defense antimicrobial peptides as antibiotics: design and application strategies, Current Opinion in Chemical Biology 38 (2017) 87–96.
[81] Y. Liu, D. Shi, J. Wang, X. Chen, M. Zhou, X. Xi, et al., Amphibian antimicrobial peptide, phylloseptin-PV1, exhibits effective anti-*staphylococcal* activity without inducing either hepatic or renal toxicity in mice, Frontiers in Microbiology 11 (2020) 565158.
[82] C.N. Pedron, I. Araújo, P.I. da Silva Junior, F.D. da Silva, M. Der, T. Torres, et al., Repurposing the scorpion venom peptide VmCT1 into an active peptide against Gram-negative ESKAPE pathogens, Bioorganic Chemistry 90 (2019) 103038.
[83] Q. Lin, B. Deslouches, R.C. Montelaro, Y.P. Di, Prevention of ESKAPE pathogen biofilm formation by antimicrobial peptides WLBU2 and LL37, International Journal of Antimicrobial Agents 52 (2018) 667–672.
[84] C. Brunati, T.T. Thomsen, E. Gaspari, S. Maffioli, M. Sosio, D. Jabes, et al., Expanding the potential of NAI-107 for treating serious ESKAPE pathogens: synergistic combinations against gram-negatives and bactericidal activity against non-dividing cells, Journal of Antimicrobial Chemotherapy 73 (2018) 414–424.
[85] H. Du, S. Puri, A. McCall, H.L. Norris, T. Russo, M. Edgerton, Human salivary protein histatin 5 has potent bactericidal activity against ESKAPE pathogens, Frontiers in Cellular and Infection Microbiology 7 (2017) 41.
[86] Y. Su, H. Wang, B. Mishra, J.L. Narayana, J. Jiang, D.A. Reilly, et al., Nanofiber dressings topically delivering molecularly engineered human cathelicidin peptides for the treatment of biofilms in chronic wounds, Molecular Pharmaceutics 16 (2019) 2011–2020.
[87] J.L. Narayana, B. Mishra, T. Lushnikova, R.M. Golla, G. Wang, Modulation of antimicrobial potency of human cathelicidin peptides against the ESKAPE pathogens and in vivo efficacy in a murine catheter-associated biofilm model, Biochimica et Biophysica Acta (BBA)-Biomembranes 1861 (2019) 1592–1602.
[88] J. Menousek, B. Mishra, M.L. Hanke, C.E. Heim, T. Kielian, G. Wang, Database screening and in vivo efficacy of antimicrobial peptides against methicillin-resistant *Staphylococcus aureus* USA300, International Journal of Antimicrobial Agents 39 (2012) 402–406.
[89] B. Mishra, G. Wang, Ab initio design of potent anti-MRSA peptides based on database filtering technology, Journal of American Chemical Society 134 (2012) 12426–12429.
[90] Y. Dong, T. Lushnikova, R.M. Golla, X. Wang, G. Wang, Small molecule mimics of DFTamP1, a database designed anti-staphylococcal peptide, Bioorganic and Medicinal Chemistry 25 (2017) 864–869.

[91] M.A. Scorciapino, I. Serra, G. Manzo, A.C. Rinaldi, Antimicrobial dendrimeric peptides: structure, activity and new therapeutic applications, International Journal of Molecular Sciences 18 (2017) 542.

[92] R.W. Scott, G.N. Tew, Mimics of host defense proteins; strategies for translation to therapeutic applications, Current Topics in Medicinal Chemistry 17 (2017) 576–589.

[93] W.S. Horne, S.H. Gellman, Foldamers with heterogeneous backbones, Account of Chemical Research 41 (2008) 1399–1408.

[94] J.A. Patch, A.E. Barron, Mimicry of bioactive peptides via non-natural, sequence-specific peptidomimetic oligomers, Current Opinion in Chemical Biology. 6 (2002) 872–877.

[95] N.C.S. Costa, J.P. Piccoli, N.A. Santos-Filho, L.C. Clementino, A.M. Fusco-Almeida, S.R.D. Annunzio, et al., Antimicrobial activity of RP-1 peptide conjugate with ferrocene group, PLoS One 15 (2020) e0228740.

[96] P. Kumar, R.A. Shenoi, B.F.L. Lai, M. Nguyen, J.N. Kizhakkedathu, S.K. Straus, Conjugation of aurein 2.2 to HPG yields an antimicrobial with better properties, Biomacromolecules 16 (2015) 913–923.

[97] A.A. David, S.E. Park, K. Parang, R.K. Tiwari, Antibiotics-peptide conjugates against multidrug-resistant bacterial pathogens, Current Topics in Medicinal Chemistry 18 (2018) 1926–1936.

[98] A. Brezden, H. Mohammad, J. Chmielewski, M.N. Seleem, Targeting biofilms and persisters of ESKAPE pathogens with P14KanS, a kanamycin peptide conjugate, Biochimica et Biophysica Acta (BBA)-General Subjects 1861 (2017) 848–859.

[99] T. Rounds, S.K. Straus, Lipidation of antimicrobial peptides as a design strategy for future alternatives to antibiotics, International Journal of Molecular Sciences 21 (2020) 9692.

[100] M.L. Mangoni, Y. Shai, Short native antimicrobial peptides and engineered ultrashort lipopeptides: similarities and differences in cell specificities and modes of action, Cellular and Molecular Life Sciences 68 (2011) 2267–2280.

[101] E. Kamysz, E. Sikorska, M. Jaśkiewicz, M. Bauer, D. Neubauer, S. Bartoszewska, et al., Lipidated analogs of the LL-37-derived peptide fragment KR12-structural analysis, surface-active properties and antimicrobial activity, International Journal of Molecular Sciences 21 (2020) 887.

[102] J.L. Narayana, R. Golla, B. Mishra, X. Wang, T. Lushnikova, Y. Zhang, et al., Short and robust anti-infective lipopeptides engineered based on the minimal antimicrobial peptide KR12 of human LL-37, ACS Infectious Diseases 7 (2021) 1795–1808.

[103] S. Huo, C. Chen, Z. Lyu, S. Zhang, Y. Wang, B. Nie, et al., Overcoming planktonic and intracellular *Staphylococcus aureus*-associated infection with a cell-penetrating peptide-conjugated antimicrobial peptide, ACS Infectious Diseases 6 (2020) 3147–3162.

[104] U.H. Dürr, U.S. Sudheendra, A. Ramamoorthy, LL-37, the only human member of the cathelicidin family of antimicrobial peptides, Biochimica et Biophysica Acta 1758 (2006) 1408–1425.

[105] G. Wang, B. Mishra, R.F. Epand, R.M. Epand, High-quality 3D structures shine light on antibacterial, anti-biofilm and antiviral activities of human cathelicidin LL-37 and its fragments, Biochimica et Biophysica Acta 1838 (2014) 2160–2172.

[106] D. Xhindoli, S. Pacor, M. Benincasa, M. Scocchi, R. Gennaro, A. Tossi, The human cathelicidin LL-37–a pore-forming antibacterial peptide and host-cell modulator, Biochimica et Biophysica Acta (BBA) - Biomembranes 2016 (1858) 546–566.

[107] G. Wang, Human antimicrobial peptides and proteins, Pharmaceuticals (Basel)., 7, 2014, pp. 545–594.

[108] P.K. Singh, B.F. Tack, P.B. McCray Jr, M.J. Welsh, Synergistic and additive killing by antimicrobial factors found in human airway surface liquid, American Journal of Physiology-Lung Cellular and Molecular Physiology 279 (2000) L799–L805.

[109] K. Bitschar, B. Sauer, J. Focken, H. Dehmer, S. Moos, M. Konnerth, et al., Lugdunin amplifies innate immune responses in the skin in synergy with host- and microbiota-derived factors, Nature Communications 10 (2019) 2730.

[110] T. Nakatsuji, T.H. Chen, S. Narala, K.A. Chun, A.M. Two, T. Yun, et al., Antimicrobials from human skin commensal bacteria protect against *Staphylococcus aureus* and are deficient in atopic dermatitis, Science Translational Medicine 9 (2017) eaah4680.

[111] T.B. Bedran, M.P. Mayer, D.P. Spolidorio, D. Grenier, Synergistic anti-inflammatory activity of the antimicrobial peptides human beta-defensin-3 (hBD-3) and cathelicidin (LL-37) in a three-dimensional co-culture model of gingival epithelial cells and fibroblasts, PLoS One 9 (2014) e106766.

[112] S. Wongkaewkhiaw, S. Taweechaisupapong, S. Thanaviratananich, J.G.M. Bolscher, K. Nazmi, C. Anutrakunchai, et al., D-LL-31 enhances biofilm-eradicating effect of currently used antibiotics for chronic rhinosinusitis and its immunomodulatory activity on human lung epithelial cells, PLoS One 15 (2020) e0243315.

[113] S. Sadeghi, H. Bakhshandeh, R.A. Cohan, A. Peirovi, P. Ehsani, D. Norouzian, Synergistic anti-staphylococcal activity of niosomal recombinant lysostaphin-LL-37, International Journal of Nanomedicine 14 (2019) 9777–9792.

[114] M.S. Zharkova, D.S. Orlov, O.Y. Golubeva, O.B. Chakchir, I.E. Eliseev, T.M. Grinchuk, et al., Application of antimicrobial peptides of the innate immune system in combination with conventional antibiotics-a novel way to combat antibiotic resistance? Frontiers in Cellular Infection Microbiology 9 (2019) 128.

[115] B. Mishra, G. Wang, Individual and combined effects of Engineered peptides and antibiotics on *Pseudomonas aeruginosa* biofilms, Pharmaceuticals (Basel), 10, 2017, p. 58.

[116] J. Portelinha, A.M. Angeles-Boza, The antimicrobial peptide Gad-1 clears *Pseudomonas aeruginosa* biofilms under cystic fibrosis conditions, Chembiochem: a European Journal of Chemical Biology 22 (2021) 1646–1655.

[117] R. Sabzevari, A.M. Roushandeh, A. Mehdipour, M. Alini, M.H. Roudkenar, SA/G hydrogel containing hCAP-18/LL-37-engineered WJ-MSCs-derived conditioned medium promoted wound healing in rat model of excision injury, Life Sciences 261 (2020) 118381.

[118] K. Niemirowicz, B. Durnaś, G. Tokajuk, E. Piktel, G. Michalak, X. Gu, et al., Formulation and candidacidal activity of magnetic nanoparticles coated with cathelicidin LL-37 and ceragenin CSA-13, Scientific Reports 7 (2017) 4610.

[119] Y. Su, V.L. Mainardi, H. Wang, A. McCarthy, Y.S. Zhang, S. Chen, et al., Dissolvable microneedles coupled with nanofiber dressings eradicate biofilms via effectively delivering a database-designed antimicrobial peptide, ACS Nano 14 (2020) 11775–11786.

[120] Y. Palmowski, J. Bürger, A. Kienzle, A. Trampuz, Antibiotic treatment of postoperative spinal implant infections, Journal of Spine Surgery 6 (2020) 785–792.

[121] S.W. Lee, K.S. Phillips, H. Gu, M. Kazemzadeh-Narbat, D. Ren, How microbes read the map: effects of implant topography on bacterial adhesion and biofilm formation, Biomaterials 268 (2021) 120595.

[122] W. Yin, S. Xu, Y. Wang, Y. Zhang, S.-H. Chou, M.Y. Galperin, et al., Ways to control harmful biofilms: prevention, inhibition, and eradication, Critical Reviews in Microbiology 47 (2021) 57–78.

[123] M. Kazemzadeh-Narbat, H. Cheng, R. Chabok, M.M. Alvarez, C.D.L. Fuente-Nunez, K.S. Phillips, et al., Strategies for antimicrobial peptide coatings on medical devices: a review and regulatory science perspective, Critical Reviews in Biotechnology 41 (2021) 94–120.
[124] B. Mishra, G. Wang, Titanium surfaces immobilized with the major antimicrobial fragment FK-16 of human cathelicidin LL-37 are potent against multiple antibiotic-resistant bacteria, Biofouling 33 (2017) 544–555.
[125] F. Jahanmard, M. Croes, M. Castilho, A. Majed, M.J. Steenbergen, K. Lietaert, et al., Bactericidal coating to prevent early and delayed implant-related infections, Journal of Controlled Release 326 (2020) 38–52.
[126] G. Wang, B. Mishra, K. Lau, T. Lushnikova, R. Golla, X. Wang, Antimicrobial peptides in 2014, Pharmaceuticals (Basel), 8, 2015, pp. 123–150.
[127] G. Wang, R.F. Epand, B. Mishra, T. Lushnikova, V.C. Thomas, K.W. Bayles, et al., Decoding the functional roles of cationic side chains of the major antimicrobial region of human cathelicidin LL-37, Antimicrobial Agents and Chemotherapy 56 (2012) 845–856.
[128] W. Xiuqing, J. Bozelli Junior, B. Mishra, T. Lushnikova, R.M. Epand, G. Wang, Arginine-lysine positional swap of the LL-37 peptides reveals evolutional advantages of the native sequence and leads to bacterial probes, Biochimica et Biophysica Acta (BBA)-Biomembranes 1859 (2017) 1350–1361.
[129] C.I. von Deuster, V. Knecht, Competing interactions for antimicrobial selectivity based on charge complementarity, Biochimica et Biophysica Acta (BBA) - Biomembranes 2011 (1808) 2867–2876.
[130] G. Wang, Determination of solution structure and lipid micelle location of an engineered membrane peptide by using one NMR experiment and one sample, Biochimica et Biophysica Acta (BBA) - Biomembranes 2007 (1768) 3271–3281.
[131] G. Wang, Structures of human host defense cathelicidin LL-37 and its smallest antimicrobial peptide KR-12 in lipid micelles, Journal of Biological Chemistry 283 (2008) 32637–32643.
[132] R.M. Epand, Anionic lipid clustering model, Advances in Experimental Medicine and Biology 1117 (2019) 65–71.
[133] R.F. Epand, G. Wang, B. Berno, R.M. Epand, Lipid segregation explains selective toxicity of a series of fragments derived from the human cathelicidin LL-37, Antimicrobial Agents and Chemotherapy 53 (2009) 3705–3714.
[134] Y. Shai, Mode of action of membrane active antimicrobial peptides, Biopolymers 66 (2002) 236–248.
[135] K. Matsuzaki, Membrane permeabilization mechanisms, Advances in Experimental Medicine and Biology 1117 (2009) 9–16.
[136] S.J. Ludtke, K. He, W.T. Heller, T.A. Harroun, L. Yang, H.W. Huang, Membrane pores induced by magainin, Biochemistry 35 (1996) 13723–13728.
[137] N. Salinas, E. Tayeb-Fligelman, M.D. Sammito, D. Bloch, R. Jelinek, D. Noy, et al., The amphibian antimicrobial peptide uperin 3.5 is a cross-/cross- chameleon functional amyloid, Proceedings of the National Academy of Sciences of the United States of America 118 (2021) e2014442118.
[138] C. Aisenbrey, M. Amaro, P. Pospíšil, M. Hof, B. Bechinger, Highly synergistic antimicrobial activity of magainin 2 and PGLa peptides is rooted in the formation of supramolecular complexes with lipids, Scientific Reports 10 (2020) 11652.
[139] G. Wang, The antimicrobial peptide database provides a platform for decoding the design principles of naturally occurring antimicrobial peptides, Protein Science 29 (2020) 8–18.

[140] H.E. Hasper, N.E. Kramer, J.L. Smith, J.D. Hillman, C. Zachariah, O.P. Kuipers, et al., An alternative bactericidal mechanism of action for lantibiotic peptides that target lipid II, Science (New York, N.Y.) 313 (2006) 1636–1637.

[141] G. Kragol, S. Lovas, G. Varadi, B.A. Condie, R. Hoffmann, L. Otvos Jr., The antibacterial peptide pyrrhocoricin inhibits the ATPase actions of DnaK and prevents chaperone-assisted protein folding, Biochemistry 40 (2001) 3016–3026.

[142] M. Mardirossian, R. Sola, M. Degasperi, M. Scocchi, Search for shorter portions of the proline-rich antimicrobial peptide fragment Bac5(1–25) that retain antimicrobial activity by blocking protein synthesis, ChemMedChem 14 (2019) 343–348.

[143] L. Kolano, D. Knappe, D. Volke, N. Sträter, R. Hoffmann, Ribosomal target-binding sites of antimicrobial peptides Api137 and Onc112 are conserved among pathogens indicating new lead structures to develop novel broad-spectrum antibiotics, Chembiochem: A European Journal of Chemical Biology 21 (2020) 2628–2634.

[144] R. Sola, M. Mardirossian, B. Beckert, L.S. De Luna, D. Prickett, A. Tossi, et al., Characterization of cetacean proline-rich antimicrobial peptides displaying activity against ESKAPE pathogens, International Journal of Molecular Sciences 21 (2020) 7367.

[145] W. Li, M.-A. Sani, E. Jamasbi, L. Otvos Jr., M.A. Hossain, J.D. Wade, et al., Membrane interactions of proline-rich antimicrobial peptide, Chex1-Arg20, multimers, Biochimica et Biophysica Acta (BBA)-Biomembranes 1858 (2016) 1236–1243.

[146] G. Wang, Improved methods for classification, prediction, and design of antimicrobial peptides, Methods in Molecular Biology 1268 (2015) 43–66.

# Antimicrobial peptide resistance and scope of computational biology in antimicrobial peptide research

C.K.V. Ramesan[1,]*, N.V. Vinod[2,]* and Sinosh Skariyachan[2,]*
[1]Department of Microbiology, Sree Narayana College, Kannur, Kerala, India,
[2]Department of Microbiology, St. Pius X College, Kasaragod, Kerala, India

## 14.1 Introduction

One of the potential issues overcomes at present is the emerging patterns of antimicrobial resistance. The uncontrolled growth and the explosion of antimicrobial resistance demand an urgent need for an alternative. The best solution to overcome this situation is small peptides having antimicrobial activities, that is, antimicrobial peptides (AMPs) [1]. In recent years AMPs are getting high interest among scientists and pharmaceutical fields because of their high therapeutic properties [2]. AMPs are large classes of low molecular weight protein molecules which are part of the innate immunity of all types of organisms [3]. They are having a broad spectrum of antibacterial, antifungal, antiviral, and antiinflammatory activities [4]. They are highly specific and possess immune-modulatory activities [2]. Because AMPs make bacteria develop low or no resistance, they are highly promising compounds that can be used as an effective alternative to antibiotics [5]. In many bacterial species, one of the major mechanisms of antibiotic resistance is the formation of biofilm. It is found that biofilm formation can be effectively hindered by many of the AMPs. AMPs exhibit broad-spectrum activity against most strains of gram-positive and gram-negative bacteria, including multidrug-resistant strains. Studies of different animal models found that AMPs are also effective in neutralizing toxins [6].

AMPs can be categorized into various classes that include thionins, snakins, defensins, glycine-rich proteins, lipid transferases, cyclotides, and hevein proteins [7]. There are various types of AMPs isolated and characterized from various sources such as plants, animals, and microbes [8]. AMPs are found abundantly in plants and can be extracted from all parts of plants like root, stem, seeds, leaves, and different organs of the plants [9]. In plants, AMPs not only protect plants from microbial diseases but also help in the growth and development of the plant [7]. Another important source of AMPs is insects [10]. A diverse type of AMPs is also

---

* These authors have equally contributed.

Antimicrobial Peptides. DOI: https://doi.org/10.1016/B978-0-323-85682-9.00007-6
© 2023 Elsevier Inc. All rights reserved.

found in the secretions of the skin of many amphibians [11]. AMPs like defensins and cathelicidins are the major AMPs found in various cell types of humans which link the innate and adaptive branches of immunity together. AMPs are also found in some foods and secretions (milk and lysozyme in animals). One of the important AMP classes of bacterial origin, Lantibiotics, is a new class having anti-infective properties [12].

The major antimicrobial mechanism of AMPs is considered to be able to target negatively charged membranes of bacteria [13]. They are usually positively charged proteins which are having evolutionarily conserved peptides with hydrophobic and hydrophilic regions. Due to this, they can easily get solubilized in lipid-rich membranes of bacteria making their entry easy. Once they reach inside a target cell, they kill the cells by several mechanisms [14]. The main target of the AMP is the microbial cell membrane. The specificity and the action of these small peptides are influenced by the interaction with the target membrane [15]. The peptide backbone can be modified to enhance the antimicrobial activity and also for the stabilization of the proteins. These modifications can also be used to reduce the toxic nature of some AMPs [16].

The major concerns of the large-scale production of AMPs are scarcity of sources, instability, toxicity, and bioavailability. We can improve the properties of the synthesized AMPs by modifying and designing artificial peptides, which are the main focus of the AMPs research. Almost half of the currently isolated AMPs are naturally modified. There are different methods of modifying AMPs, which mainly include biochemical and chemical modifications. These modification processes are complicated and costly when compared to the production of antibiotics. Due to the advancement of the protein modification enzyme, the biochemical modification of the antimicrobial peptide became an easy approach in AMPs research [17].

With the alarming increase of infections caused by pathogenic multidrug-resistant bacteria over the last decades, AMPs have been investigated as a potential treatment for those infections, directly through their lytic effect or indirectly, due to their ability to modulate the immune system. There are still concerns regarding the use of such molecules in the treatment of infections, such as cell toxicity and host factors that lead to peptide inhibition. To overcome these limitations, different approaches like peptide modification to reduce toxicity and peptide combinations to improve therapeutic efficacy are being tested [18].

Though AMPs have varieties of the mechanism of action against a diverse group of microorganisms, biochemical distinctions among the peptides themselves, target versus host cells, and the microenvironments in which these counterparts convene, likely provide for varying degrees of selective toxicity among diverse AMP types. In balance, successful microbial pathogens have evolved multifaceted and effective countermeasures to avoid exposure to and subvert mechanisms of AMPs. A clearer recognition of these opposing themes will significantly advance our understanding of how AMPs function in defense against infection. Furthermore, this understanding may provide new models and strategies for developing novel antimicrobial agents, that may also augment immunity, restore potency or amplify the mechanisms of conventional antibiotics, and minimize antimicrobial resistance mechanisms among pathogens [8,19].

Cationic antimicrobial peptides (CAMPs) contribute a very important role in the defensive mechanism against many bacterial infections in animals. They have direct microbicidal activity. However, the problem is many pathogenic bacteria can resist the action of these peptides. Among these Streptococcus (Group A and B) and *Streptococcus pneumoniae* are among the most common human pathogens, display a lot of adaptations to resist CAMPs. The common mechanism of resistance exhibited by these pathogens is repulsion, sequestration, export, and destruction. Among these different pathogens exhibit different mechanisms [19]. Many bacteria have developed several mechanisms to prevent AMP-mediated destruction. This resistance may be broad-range or specific resistance to AMPs. Specific mechanisms are against one type of AMP, whereas broad resistance mechanisms protect the bacterium from a large group of AMPs that have similar characteristics. Gram-positive bacteria are the largest AMP-producing groups, but the mechanism of resistance is well studied in gram-negative bacteria than gram-positive bacteria [20]. The genome of some bacterial groups has specific genes which can deal with AMP resistance by modifying lipopolysaccharide (LPS) [21].

## 14.2 Antimicrobial peptide resistance in gram-positive bacteria

### 14.2.1 Bacterial cell surface—cell wall and cell membrane

The major target for the bactericidal activities of AMPs is the bacterial cell surface [22]. Bacteria modify cell surface components frequently to resist the effects of AMPs by altering the net negative charge of the cell, changing membrane fluidity, or reorganizing AMP targets [23–25].

#### 14.2.1.1 Repulsion of antimicrobial peptides

Electrostatic interactions are the major force involved in targeting the bacterial cell by AMPs [22,26,27]. The anionic components such as phospholipids and teichoic acids present in the cell wall and cell membrane of bacteria contribute to the net charge of the cell [28–30]. The oppositely charged AMPs are more easily attracted toward the bacterial cell surface. Hence, a universal strategy of resistance to positively charged AMPs is the alteration in the net negative charge of the cell, thereby decreasing the affinity to positively charged AMPs. A different mechanism for the reduction in the net negative charge of bacterial cell surface includes masking of the negatively charged phosphatidylglycerol via the addition of a positively charged amino acid by the multipeptide resistance factor protein, MprF [31]. MprF confers resistance to positively charged AMPs in many gram-positive bacteria [32–34].

The addition of D-alanine to the backbone of teichoic acids can mask the negative charge present along these glycopolymers, thereby leading to increased surface charge and lower attraction of positively charged antimicrobials [35].

### 14.2.1.2 Target modification

A common target in gram-positive bacteria is the cell wall for antimicrobial compounds including AMPs. Lysozyme, an antimicrobial enzyme that is an integral part of host innate immunity is cationic at physiological pH, which interacts with negatively charged bacterial surfaces. The muramidase activity of lysozyme hydrolyzes the $\beta$-1, 4 linkages between N-acetyl glucosamine and N-acetylmuramic acid of peptidoglycan, leading to the breakdown of the peptidoglycan and causing lysis of the cell [36–38]. Several peptidoglycan-modifying enzymes contribute to AMP resistance in some gram-positive bacteria. The enzymes activity leads to steric hindrance between AMPs and the cell surface, thereby limiting the interaction between the cell surface and AMPs [39].

### 14.2.1.3 Alterations to membrane composition

Besides AMP repulsion and target modifications as resistance mechanisms, changes in membrane composition can also reduce the bactericidal activity of AMP. One example of this mechanism is the saturation of membrane fatty acids. Investigations into the cell membrane components of nisin-resistant *Listeria monocytogenes* showed that some resistant strains contained a higher proportion of saturated fatty acids versus unsaturated fatty acids [40,41]. The studies suggest that higher concentrations of saturated fatty acids lead to a decrease in cell membrane fluidity. The decrease in membrane fluidity increases nisin resistance by blocking nisin insertion into the bacterial membrane [42–44].

## 14.2.2 Extracellular mechanism of antimicrobial peptide resistance

### 14.2.2.1 Extracellular proteases

The most common mechanism is by producing extracellular proteases. Several bacteria like *Staphylococcus aureus*, *Enterococcus faecalis*, and *Staphylococcus epidermidis* produce proteases that degrade AMPs [45–47].

### 14.2.2.2 Protein-mediated sequestration

Some gram-positive bacteria produce extracellular or surface-linked proteins that directly bind to AMPs and block access to the cell membrane. Mechanisms of protein-mediated AMP sequestration vary between species and strains. Proteins that inhibit AMP activity through binding can be secreted into the extracellular environment to inhibit contact of bactericidal peptides with the cellular surface. For example, the Streptococcal inhibitor of complement (SIC) produced by *Streptococcus pyogenes* is a hydrophilic, secreted protein that sequesters many AMPs, thereby preventing them from reaching cell-surface targets [48].

### 14.2.3 Inhibition of antimicrobial peptide activity by surface-associated polysaccharides

Extracellular polysaccharide production has long been recognized as a factor that promotes both virulence and host colonization by many bacteria [49–51]. Extracellular polysaccharides are composed of structurally diverse polymers that are enzymatically produced by some gram-positive species. Extracellular polysaccharides that are attached to the cellular surface through covalent linkages with the cell wall are known as capsules, while loosely attached polymers are referred to as exopolysaccharides, or EPS [52–54]. Polysaccharide-mediated AMP resistance is thought to occur by shielding the bacterial membrane via binding or electrostatic repulsion of AMPs [55].

## 14.3 Mechanisms of antimicrobial peptides resistance in gram-negative bacteria

CAMPs are known to have microbicidal properties against a variety of pathogens including gram-negative bacteria. They are a large group of different types of peptides produced by many organisms ranging from prokaryotes to eukaryotes [56]. These peptides contain similar amino acid sequence regions and they maintain certain key characters. Most commonly they are cationic, amphipathic, and relatively hydrophobic in nature [57]. These attributes help them to allow CAMPs to interact with bacterial membranes, which contain anionic head groups and hydrophobic fatty acid chains. The intense pressure is exerted on bacteria by CAMP leading to killing bacteria. They involve in pore formation in the membrane which leads to the destabilization of the bacterial membranes and ultimately end up in cell lysis [58]. Some CAMPs may also target some intracellular sites and some involved in direct induction. The intracellular site targets lead to inhibition and disruption of the cell wall, protein and nucleic acid synthesis occurs. [59].

The CAMPs can kill a variety of pathogens. Gram-negative bacteria represent a major target. Some species of bacteria have evolved mechanisms to resist the action of these antimicrobials. Resistance to CAMPs is a significant negative effect on the ability of the host to prevent bacterial infections, and also affects the utility of polymyxins in the treatment. Unfortunately, due to the increased use of polymyxins recently, the resistance to polymyxins is on verge of the rise [60]. The development of CAMP resistance of gram-negative bacteria contributes to avoiding killing by both the host immune system and polymyxin antibiotics. Different mechanisms of resistance involved in gram-negative bacteria against CAMPs are described in the following sections and these mechanisms are depicted in Fig. 14.1.

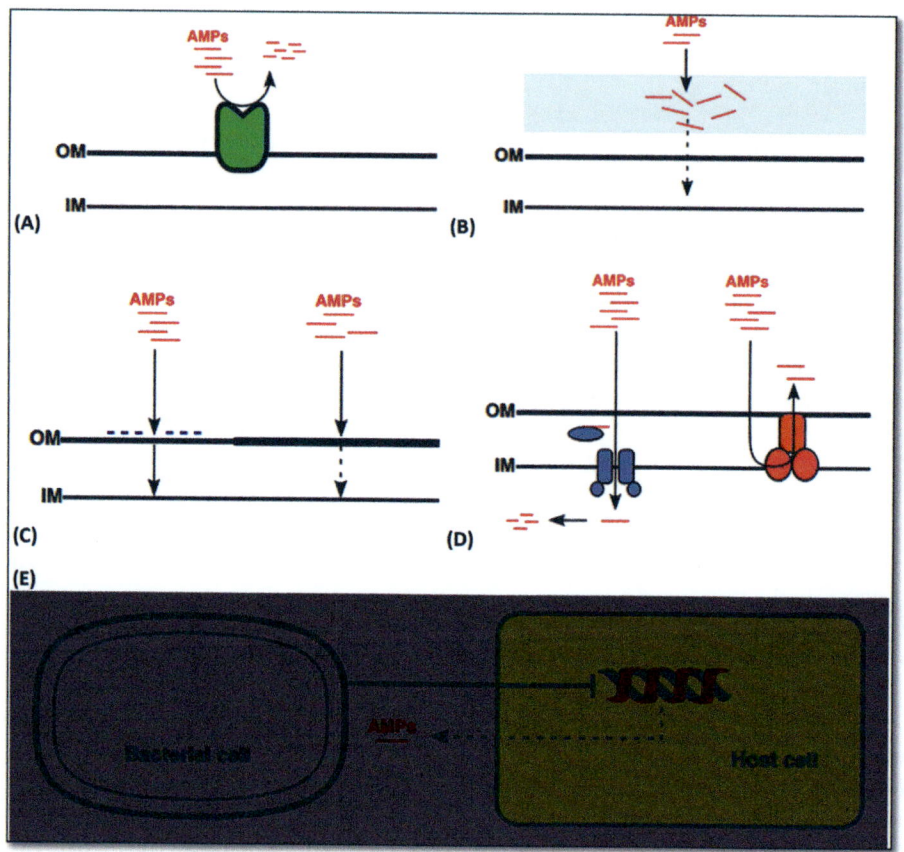

**Figure 14.1** AMPs resistant mechanisms in gram-negative bacteria. (A) Proteolytic degradation of AMPs. (B) Shielding of bacterial surface. (C) Modification of bacterial outer membrane. (D) Pumping AMPs in and out of the cell. (E) Downregulating the expression of AMPs. *AMPs*, antimicrobial peptides.

## 14.3.1 Modifications in the bacterial outer membrane

The outer membrane of gram-negative bacteria can act as the barrier against the action of CAMPs and is therefore quite often modified to enhance resistance. One of the main ways in which killing by CAMPs can be avoided is through an increase in bacterial surface charge. Host CAMPs contain a region of high positive charge and are attracted to negatively charged molecules, such as the surface of many bacteria. Thus increasing the surface charge of bacterial cell membrane prevents access to CAMPs [59]. The surface of gram-negative bacteria is largely composed of the glycolipid LPS. LPS is a major constituent of the outer membrane phospholipid bilayer, which envelops the peptidoglycan containing periplasm and the inner membrane. It specifically comprised the hydrophobic lipid and it can serve as one of the initial barriers against extracellular stresses including the action of CAMPs.

In particular, the lipid A and core oligosaccharides often contain multiple negatively charged residues, such as phosphate groups that contribute to the resistance mechanism of bacteria.

### 14.3.1.1 Lipopolysaccharide modifications

To alleviate the negative charge of LPS, numerous species of bacteria add positive residues to the lipid A structure. The amine-containing sugar amino arabinose is added to a lipid A phosphate group in *Pseudomonas aeruginosa* as well as *Salmonella typhimurium*, resulting in increased survival chances of both bacteria in the presence of polymyxin B [61,62]. *Francisella novicida* adds another amine-containing sugar called galactosamine, to its single lipid A phosphate group, and promotes polymyxin B resistance. Increased CAMP resistance is also linked to cationic sugar addition in *Bordetella pertussis* and *Bordetella bronchiseptica*, in which glucosamine groups are added to both lipids A phosphates [63,64].

The amine-containing moieties, such as amino acids, are also added to the lipid A component of LPS to counteract its negative charge. For example, specific strains of *Vibrio cholerae* add glycines to their lipid A [65]. Hankins et al. have shown that the O1 El Tor strain of *V. cholerae* adds glycine and diglycine amino acid residues to lipid A acyl chains, increasing the net positive charge of LPS and the bacterial cell surface [65,66].

Phosphoethanolamine is another amine-containing group that can be added to lipid A, as is the case in *Neisseria gonorrhoeae*. This phosphoethanolamine is required for its ability to resist CAMP-mediated killing [67]. Importantly, this increased resistance to CAMPs facilitates the establishment of a more severe disseminated form of gonorrheal infection [68]. Phosphoethanolamine addition to LPS also occurs in *S. typhimurium* and in colistin-resistant strains of *A. baumannii* [69], where it increases resistance to polymyxin B [70].

In some bacteria, it removes negative residues as an alternative mechanism of limiting the overall negative charge. The anionic phosphate groups of lipid A are major negative residues on LPS and are thus targets for removal. In *Francisella tularensis*, the 4' lipid A phosphate is removed by the phosphatase LpxF, leaving only one phosphate group on lipid A [71,72]. The lpxF gene expressed in *E. coli* containing two phosphate groups normally becomes modified *E. coli* lacking a 4' lipid A phosphate and consequently shows an increase in polymyxin minimum inhibiting concentration (MIC) [73]. The negatively charged phosphate groups on lipid A are a target for removal in many other pathogenic bacterial species, including *Porphyromonas gingivalis* [74], *Bacteroides fragilis* [75], and *Helicobacter pylori* [76]. Together these data demonstrate that CAMP resistance can be induced by the removal of negatively charged lipid A residues.

Another mechanism involved in generating CAMP resistance is the increase in the hydrophobicity of LPS. Hydrophobic lipid chains, added to lipid A phosphates, the glucosamine backbone, acyl chains, serve to increase LPS saturation and decrease overall permeability and prevent CAMPs from inserting into the membrane [77]. In *Salmonella*, acyl chains are added to the glucosamine backbone and phosphates of lipid A by PagP [78,79]. Enhanced acylation of lipid A also occurs in *E. coli* and *Yersinia enterocolitica* [80].

### 14.3.1.2 Phospholipid modifications

Phospholipids are the other major component of the gram-negative outer membrane. Similar to LPS, phospholipids in the outer membrane can also be modified to increase CAMP resistance. *S. typhimurium* uses PhoPQ system to modify phospholipids that reside in the outer membrane. PhoPQ-activated PagP adds palmitoyl groups to phospholipids. This leads to an increase in the levels of palmitoylated phosphatidylglycerols within the outer membrane, which is highly hydrophobic in nature than many other phospholipids in the outer membrane. Increased hydrophobicity in the outer membrane may decrease permeability, similar to the effect in lipid A palmitoylation [81]. Therefore localizing these modified phospholipids results in increased CAMP resistance. The inner membrane may also be modified to increase CAMP resistance. The addition of lysine to phospholipids within the plasma membrane increases the charge of anionic phosphatidyl glycerol to a cationic form and helps in repel cationic CAMPs and reducing their binding to the membrane. Similar to *S. aureus*, these lysylated phospholipids are also present in gram-negative species [82] including *Rhizobium tropici* [27] and *Caulobacter crescentus* [83]. It has been hypothesized that one of the reasons that CAMPs do not properly damage eukaryotic host membranes is that eukaryotic cells have a much more robust form of membrane repair than bacteria [84]. Thus it is possible that bacteria with increased membrane repair capacity could survive higher concentrations of CAMPs, simply repairing the membrane as it is damaged. Dalebroux et al. suggest that proteins involved in membrane repair are prime candidates for the investigation of microbial resistance [85]. The bacterial capsule is a protective layer external to the outer membrane that acts as an additional barrier and is comprised primarily of long-chained repeating polysaccharides [86]. *Klebsiella pneumoniae* capsule provides increased resistance against cationic defensins, lactoferrins, and polymyxins. Furthermore, there is a direct correlation between higher amounts of capsular polysaccharides, decreased levels of CAMPs binding to the outer membrane, and increased resistance to polymyxins [87]. It is likely critical for bacterial virulence during in vivo infection as a capsular mutant was unable to cause pneumonia in a mouse model [88]. In *Neisseria meningitidis*, capsule production was shown to increase resistance to the human CAMP LL-37 [89]. Furthermore, upon exposure to sublethal levels of LL-37, the capsule biosynthetic genes siaC and siaD were upregulated and contributed to increased capsule production [90]. *P. aeruginosa* has also been shown to use its capsule to resist CAMPs [91].

### 14.3.2 Biofilm formation

Bacteria can further resist the action of CAMPs through their organization into specialized structures known as biofilms. These structures consist of a group of bacteria adhering to a surface in a highly organized manner that allows for the circulation of nutrients and information [55]. Bacteria in a biofilm often secrete a slimy extracellular matrix that both aids in adherence to surfaces and acts as a barrier to outside stresses. This extracellular matrix can be composed of various compounds including cellulose, teichoic acids, proteins, lipids, and nucleic acids [37]. Biofilms can develop on any surface including on hospital equipment, allowing the populations to persist, and contributing to the growing problems of hospital-acquired infections. Biofilms are also able to form on biological surfaces such as the oral cavity or the respiratory tract, often facilitating some of the chronic infections [92].

The extracellular components of bacteria offered by the biofilm structure contribute to its protection. As a biofilm matures, it progresses from a thin homogeneous structure to a thicker heterogeneous form that contains many substructures that can increase the thickness of membranes [93]. In *P. aeruginosa*, this factor of biofilms displays exceptional resistance to CAMPs and antibiotics, in some cases over 1000 times as great as their planktonic form [94]. *Pseudomonas* biofilms contain a high level of the polysaccharide alginate, which is known to cause significant alterations to biofilm structure. A strain that overproduces alginate-formed biofilms that were much thicker and more structurally heterogeneous acts as a barrier to CAMPs [95]. Additionally, expression of *Pseudomonas* biofilm genes in *E. coli*, whose biofilms are normally flat and unstructured, resulted in the formation of biofilms with more complex architecture, correlating with increased resistance to the polymyxin antibiotic colistin. This increased resistance was not observed against other antibiotics such as ciprofloxacin, indicating that this protection may be specific to CAMPs [96]. Biofilm structure can vary greatly across different species and strains, which may account for some of the differences in CAMP susceptibility in various biofilms.

Specific components of the extracellular matrix have been shown to impart resistance to CAMPs. Anionic alginate in *P. aeruginosa* not only contributes to biofilm structure but can also bind to and induce conformational changes in invading CAMPs [97]. The CAMP-alginate complexes then oligomerize, hindering their ability to enter the biofilm [98]. Further, polysaccharides from biofilms of *K. pneumoniae* and *Burkholderia pyrrocinia* can bind and withdraw CAMPs [99]. Adding these polysaccharides to *E. coli* increased its inhibitory concentrations to CAMPs LL-37 and human beta-defensin 3.

The extracellular DNA also forms an integral component of *P. aeruginosa* [100] and *S. typhimurium* [101] biofilms, and can also contribute to CAMP resistance. The negative charge of DNA allows it to bind and sequester cations from the surrounding environment. This results in an environment with a low concentration of cations, which is an activating signal for the PhoPQ system that results in the activation of CAMP resistance genes via PhoPQ lead to LPS and other modifications [102].

Biofilms have several other inducible defenses against CAMPs. *P. aeruginosa* encodes the inducible biofilm gene psrA, which has greatly increased levels of CAMP resistance [103]. This gene was upregulated threefold in the presence of the CAMP indolicidin. Deletion mutants lacking psrA were less able to form biofilms and showed significantly increased killing when challenged with indolicidin or polymyxin B. Pamp et al. have shown that tolerance to colistin in *Pseudomonas* biofilms is due to metabolically active cells within the biofilm. While the less metabolically active cells in the biofilm were killed by colistin, a spatially distinct subset of more active cells was able to resist killing. These cells were able to upregulate PmrAB-regulated resistance genes responsible for lipid A modification [104]. Overall, biofilms confer bacteria with the ability to form a hardy structure that can withstand and resist destruction by high concentrations of CAMPs, as well as many other types of antimicrobials.

## 14.3.3 Efflux pumps

Efflux pumps are complexes of mostly membrane-bound proteins that move toxic compounds out of cells. Bacterial efflux pumps are active transporters, either directly requiring adenosine triphosphate (ATP) or using an existing electrochemical potential gradient. These complexes play an important role in antibiotic resistance, as many bacteria use them to resist major classes of antibiotics, including fluoroquinolones, macrolides, tetracyclines, glycylcyclines, beta-lactams, and aminoglycosides [105]. In addition, bacterial efflux pumps contribute to colonization and persistence, likely in part by defending against host antimicrobials such as CAMPs [106]. Generally, gram-negative bacteria that use efflux pumps to increase the chance for survival and virulence in vivo as in the case seen in *S. typhimurium* [107,108], *S. enteritidis* [53], *Enterobacter cloacae* [109], *Borrelia burgdorferi* [110], *P. aeruginosa* [111], *K. pneumoniae* [81], *V. cholerae* [112], and *N. gonorrhoeal* [79]. *K. pneumoniae* uses the AcrAB-TolC efflux pump system, known to mediate resistance against fluoroquinolones, to resist CAMPs. When the AcrB component of the efflux pump system was removed, mutant bacteria exhibited increased sensitivity to fluoroquinolones as well as polymyxin B [81].

The acrB mutant also exhibited 10-fold lower survival in bronchoalveolar lavage fluid, which contains many CAMPs, and specifically displayed increased sensitivity to the human alpha defensin HNP-1 as well as human beta-defensins HBD-1 and HBD-2 [79]. A human gut pathogen, *Y. enterocolitica*, has a high level of resistance to human CAMPs, at least in part due to the action of the RosAB efflux pump system. A rosAB deletion mutant was more sensitive than wild-type to the CAMPs polymyxin B, cecropin P1, and melittin [113]. This pump acts as a potassium antiporter, using a potassium gradient that pumps K+ ions into the cell as it pumps out harmful CAMPs. Interestingly, the RosAB pump is activated at 37°C and in the presence of CAMPs, similar to conditions encountered within the host during infection [113]. Under these conditions, pathogenic *Y. enterocolitica* strains are more resistant to CAMPs than nonpathogenic strains [91], the data suggest that RosAB-mediated CAMP resistance is likely important for maintaining pathogenicity in *Y. enterocolitica*. *N. gonorrhoeae* possess the Mtr (multiple transferrable resistances) efflux pump which facilitates resistance to numerous antimicrobials. This three-protein system has been shown to pump out various hydrophobic compounds, such as bile salts and fatty acids, which can cause membrane damage. This pump also confers resistance to CAMPs as well [114]. The MIC of the human CAMP LL-37 and horseshoe crab-derived tachyplesin-1 was also reduced in the Mtr deletion mutant. Thus, the Mtr efflux pump can recognize a variety of CAMP structures and remove them from the bacterial cell [115]. Gonococci lacking Mtr were completely outcompeted by the wild-type strain in a competitive infection of the mouse genital tract [59] and this was correlated with the levels of CAMP resistance in vitro [116]. The closely related *N. meningitidis*, which can cause meningitis in humans, also expresses the Mtr efflux pump and it was similarly shown to contribute to CAMP resistance [117].

The resistance nodulation division (RND) family of efflux pumps in *Vibrio* species has a similar activity in mediating resistance to polymyxins and bile acids. *V. cholerae* encode RND family proteins, including the VexB protein which can mediate CAMP resistance. When this protein is deleted from a virulent strain, the mutant bacteria exhibit increased susceptibility to polymyxin B as well as bile acids, which are found in the gastrointestinal tract that *V. cholerae* infects. Further, this deletion mutant was unable to effectively colonize the gut of mice when compared to the wild-type strain [118]. The closely related *Vibrio vulnificus*, which can cause wound infections and sepsis, encodes a different efflux pump, TykA, which is responsible for resistance to the CAMPs protamine and polymyxin B [119].

Efflux pumps are important for resistance to a wide range of antibiotics and there has been much interest in using efflux pump inhibitors to enhance antibiotic treatment [120]. However, the extensive evidence that these pumps can enhance CAMP resistance and play a role in virulence suggests that efflux pump inhibitors may also be used therapeutically to sensitize bacteria to innate immune defenses. Inactivating bacterial efflux pumps responsible for CAMP resistance could enhance the ability of the host CAMPs to clear infections, while at the same time increasing sensitivity to antibiotics.

### 14.3.4 Binding and sequestering cationic antimicrobial peptides

Another method of resistance is seen when bacteria are confronted with a large concentration of CAMPs, which can bind and sequester these peptides so they cannot reach the bacterial membrane. One method for binding external CAMPs is through the release of negatively charged molecules that will attract these amphipathic antimicrobials. Negatively charged proteoglycans are found in abundance on the surface of fibroblasts and epithelial cells and can be cleaved and released by bacterial enzymes. For example, the connective tissue proteoglycan decorin is one of the major secreted products of human fibroblasts [45], and when incubated with *P. aeruginosa* or *P. mirabilis*, it is cleaved to release several products, including dermatan sulfate. This degradation occurs in the presence of bacteria-conditioned media, purified *P. aeruginosa* elastase, or alkaline proteinase, even in the absence of fibroblast enzymes. This released dermatan sulfate was able to efficiently bind neutrophil-derived α-defensin, unlike the full-length uncleaved decorin molecule. This free and soluble dermatan sulfate was able to completely inhibit killing by defensins at concentrations 10 times above the MIC for *P. aeruginosa* [121].

Similarly, *P. aeruginosa* takes advantage of the release of the cell surface heparin sulfate proteoglycan syndecan-1. This proteoglycan is found on the surface of epithelial cells and is shed during tissue injury as a soluble ectodomain. Incubating epithelial cells with cell culture supernatants from *P. aeruginosa* led to cleavage of syndecan-1 and release of its soluble ectodomain [82]. This activity was found to be dependent on the *P. aeruginosa* protein LasA, which is a known virulence factor and has been previously shown to modify other proteins. Shredded ectodomains of syndecan-1 can bind and interfere with the antimicrobial activity of CAMPs, specifically those that are Pro/Arg-rich like cathelicidins [82], likely due to charge-based interactions [83]. The virulence of the pathogen was dependent on syndecan-1 shedding, as there was a three-log decrease in virulence if syndecan-1 was absent or rendered resistant to shedding [83]. Syndecan-1 ectodomains not only bind to CAMPs but can also bind and interfere with a range of other immune-signaling molecules [82] such as cytokines and matrix metalloproteases [122]. The immune modulation in addition leading to direct interference with CAMPs may together account for the observed virulence decrease [83].

The proteoglycans that can interfere with host CAMP activity suggest that the bacterial capsule, which is rich in polysaccharides, may also be able to capture and sequester CAMPs [59]. *K. pneumoniae* capsular mutants are more susceptible to α- and β-defensins [114]. This idea was further strengthened by evidence from Llobet et al., showing that the anionic polysaccharide component (CPS) of the bacterial capsule can impart CAMP resistance in *K. pneumoniae* and *P. aeruginosa* [102]. Purified CPS was able to increase the resistance of capsular mutants, and was shown to bind to soluble CAMPs in a charge-dependent manner. This resulted in fewer peptides reaching the surface of the bacteria. After exposure to CAMPs, these anionic polysaccharides are released by the bacteria to bind and sequester the antimicrobials [102]. It is possible that other encapsulated bacteria can use this mechanism to enhance CAMP resistance as well.

Another important component of the bacterial cell membrane that is released to trap CAMPs is a form of enclosed vesicles secreted out from the surface known as outer membrane vesicles (OMVs). OMV release is a normal part of bacterial cell growth [102] and may be used for a variety of processes including toxin delivery [123]. In *E. coli* the accumulation of proteins in the outer membrane causes membrane stress inducing an increase in OMV formation. As the targets of CAMPs are bacterial membranes, CAMPs can be bound and sequestered in these vesicles, diverting them from the membranes of living bacteria. The mutants that overproduce OMVs are sixfold more resistant to killing by polymyxin B, while a mutant lacking vesicle release was 10-fold more susceptible [124]. *V. cholerae* has adapted its OMV response to aid in CAMP resistance as well. In the presence of sublethal concentrations of polymyxin B, it was noted that OMVs released from the bacteria were larger and had altered protein content [125]. These OMVs were better able to protect against CAMPs, as coincubating bacteria with them doubled the level of protection against LL-37 when compared to OMVs produced by bacteria in the absence of polymyxin B. Thus OMV release can act as an inducible defense against CAMPs that can significantly increase levels of resistance.

## 14.3.5 Proteolytic degradation of antimicrobial peptides

The bacterial enzymes especially proteases secreted or localized at the outer membrane can act on active AMPs and be degraded into inactive fragments. A study reveals [126] that elastase from *P. aeruginosa* and a protease from *P. mirabilis*, inactivated LL-37. The *P. mirabilis* protease was identified as the ZapA and it is a zinc-metalloprotease and confirmed to cleave human LL-37 and b-defensin 1, but not b-defensin 2 [45]. Although these proteases have broad-spectrum activity against various peptides and proteins, they show strict substrate specificity. For example, the ZmpA and ZmpB zinc-metalloproteases from *Burkholderia cenocepacia* cleaved LL-37 and b-defensin 1, respectively [127]. Several proteases secreted by oral cavity bacteria have also been implicated in AMP resistance. For example, *P. gingivalis*, a pathogen mostly associated with chronic periodontal disease is highly proteolytic and secretes proteases known as gingipains cleave substrates. Degradation and inactivation of LL-37 and b-defensin 3 by gingipains were reported [128,129]. Many gram-negative pathogens, mainly of the *Enterobacteriaceae* family, rely on proteases found at the outer membrane to inactivate AMPs. These proteases, exemplified by *E. coli* OmpT, belong to the omptin family [128,130]. Omptins possess a unique active site that combines elements of both serine and aspartate proteases, and interaction with LPS is critical for activity. Omptins impact bacterial virulence by degrading several host proteins or peptides [131]. *E. coli* K12 OmpT was reported to efficiently degrade the AMP protamine [132]. Other studies have shown that *S. typhimurium* PgtE and *Yersinia pestis* Pla cleave a-helical AMPs such as C18G and human LL-37 [133,134]. CroP, the omptin of the murine enteric pathogen *C. rodentium*, was shown to degrade a-helical AMPs, including mCRAMP [135]. CroP-mediated degradation of AMPs occurred before they reached the periplasmic space and triggered a PhoPQ-mediated adaptive response. OmpT of enterohemorrhagic *E. coli* (EHEC) was shown to inactivate human LL-37 by cleaving it twice at dibasic sites [136].

## 14.3.6 Modulation of cationic antimicrobial peptide expression

CAMPs are present more or less at all mucosal sites of the body to prevent infection by invading microorganisms. However, in the context of an infection, CAMPs can also be upregulated to help fight the invading microbes. Stimulation of increased CAMP production occurs mainly through innate immune recognition of microbes [27] and subsequent signaling by pattern recognition receptors (PRRs). For example, the host PRR toll-like receptor 4 (TLR4) recognizes and signals for a proinflammatory response, including CAMP induction, in response to LPS.

The modulation of microbial components like LPS, such that they cannot be detected by host PRRs, can lead to decreased PRR signaling and thus lower levels of CAMPs. Modification of lipid A structure can also have a significant effect on TLR4 signaling. In *P. gingivalis*, removal of a single phosphate group from lipid A results in both increased resistance to polymyxin B [137] and reduced activation of TLR4 [138]. In addition, incorporation of a seventh acyl chain in *Salmonella* lipid

A results in increased CAMP resistance and decreased TLR4 signaling [139]. These examples highlight that modification of LPS structure can both directly repel or prevent binding of CAMPs and indirectly by reducing induction of CAMPs lead to CAMP resistance.

Interestingly, LPS modifications that increase CAMP resistance in some bacteria increase detection by TLR4 [140]. For example, palmitoylation of lipid A in *P. aeruginosa* results in increased resistance to CAMPs but a more inflammatory LPS [141], indicating that CAMP resistance and evasion of TLR4 signaling do not always occur in tandem. Those modifications that are able to provide both increased CAMP resistance and evasion of TLR4 signaling may be more beneficial to bacteria than modifications that provide only one of these attributes. Alternatively, increased inflammation may promote pathogenesis by some bacteria, and in these cases increased CAMP resistance, as well as increased TLR4 signaling, may be beneficial to the pathogen.

The link between tandem alterations of CAMP resistance and inflammatory signaling extends to other bacterial components and host receptors as well. Similar to LPS and TLR4, bacterial lipoproteins (BLP) are recognized by the host PRR toll-like receptor 2 (TLR2), leading to the initiation of inflammatory signaling. In *Francisella novicida*, the CRISPR-Cas protein Cas9 plays a regulatory role in repressing the expression of a BLP [142]. This BLP repression leads to enhanced resistance to polymyxin B, as well as direct suppression of this TLR2 ligand and thus evasion of signaling (including CAMP induction) by this host receptor [143]. Many diverse PRRs detect bacteria, and avoiding recognition by these receptors could represent a broad and critical strategy to subverting the induction of CAMPs.

Another mechanism of limiting host inflammatory signaling, and thus CAMP induction, may be provided by the bacterial capsule. By facilitating resistance to CAMPs, the capsule prevents damage to bacterial membranes that contain activators of host signaling such as LPS and BLPs. As such, the capsule serves as a barrier to prevent the release of bacterial components that can be recognized by PRRs. This in turn prevents induction of higher levels of CAMPs. For example, in *P. aeruginosa*, the absence of capsules leads to an increased sensitivity to CAMPs as well as increased induction of beta-defensins during murine infection [143]. Even further, the capsule polysaccharides of *P. aeruginosa* also activate the anti-inflammatory immune receptors, CYLD and MKP-1, which results in the release of anti-inflammatory molecules that downregulate beta-defensin production [144], illustrating yet another indirect mechanism by which the capsule facilitates resistance to CAMPs.

*Shigella*, which causes varying degrees of bacterial dysentery in children and adults, can downregulate the CAMP response of their human hosts through an as-yet undetermined mechanism. Biopsies from *Shigella*-infected colons show that there is a significant downregulation of transcripts encoding LL-37 and HBD-1 in epithelial cells, corresponding with decreased LL-37 and HBD-1 protein in the majority of the infected biopsies [145]. This downregulation was detected up to day 30, after which LL-37 levels began to increase above healthy control levels. Inhibition of LL-37 production was observed during in vitro infections of macrophage and epithelial cell lines and was shown to be dependent upon *Shigella* plasmids within the host cells acting by an unknown mechanism [145]. It is known, however, that the MxiE protein controls the injection of plasmid-encoded effectors into the host cell leading to this CAMP inhibition. Downregulation of CAMPs early in the infection likely enhances the ability of *Shigella* to adhere to mucosal surfaces and infect host epithelial cells [146].

In addition to the examples cited above, bacteria have a wide range of mechanisms of altering or avoiding the host immune response (for a review of these, see Hornef et al. [147]). Many of these could result in decreased levels of CAMPs, which could aid in colonization and infection by bacteria. As CAMP levels are dynamic and intertwined with the overall immune response, it is likely that diverse strategies to limit immune signaling indirectly play important roles in CAMP resistance.

## 14.4 Scope of computational biology in antimicrobial peptide research

Bioinformatics and computational biology played a vital role in AMP research. Computational biology and bioinformatics are the major disciplines that employ a lot of computational approaches that included the sequence and structural alignment and interpretation of large sets of biological data, such as nucleotide and protein sequences from several samples to make novel predictions or identify new aspects in biology including AMP research [148–150]. At present, computational biology techniques such as molecular docking have been used to study the ligand–protein interactions and also used to calculate the binding energy during the docking simulation. Bioinformatics studies also aid to perform studies at the molecular level to understand the structure, sequences or gene and proteins, molecular markers and correlate and compare them with the previously known structures [150]. Computational biology studies have provided an essential framework for modeling a biological system (with nucleotide sequence) and docking peptides that help model and predict effective therapeutic strategies to combat the issue of AMP resistance that is becoming one of emerging public health concerns. Computational biology is probably used to find the link between the prediction and mathematical modeling of omics-based data along with specific aspects of the immune system [151]. Computational biology and bioinformatics played a vital role in designing novel AMPs and understanding their sequence to structural relationships, especially the mutational impact of the resistance mechanisms.

With the aids of computational biology, bioinformatics, and rational design, it is possible to develop several potential peptides that is seen in natural sources. This approach is assimilated by the data of several peptides obtained from both natural sources and that are identified by virtual screening of large constructed libraries. There are several methods available in these aspects such as ab initio or de novo design method known as "database filtering" [148,149] coupled with protein design and engineering and three-dimensional modeling and rational design, to develop potential AMPs toward prioritized molecular targets. The concept "database filtering" employed a library of known peptides with antimicrobial activities against the specific organisms to design a characteristic length of peptides, charge, hydrophobicity profile, and the major amino acids that code for a particular peptide [148,149]. The rational design mainly helps to prioritize the set of amino acids that enhances the amphipathic features of the peptide. The three-dimensional modeling is used to understand the major secondary structural conformations especially α-helices and the arrangement of amino acids that produce the amphipathic surface. Some studies revealed that these techniques were successfully employed against *Mycobacterium tuberculosis* that results from the design of novel AMPs with appropriate antibacterial activity against the *Mycobacterium* [149]. The peptides that were designed by this approach exhibited potential antibacterial activities with a MIC less than 4 μM toward *M. tuberculosis* and also showed broad-spectrum activity toward other gram-positive pathogenic bacteria such *Streptococcus* and gram-negative bacteria such as *E. coli*. The designed peptides exhibited ideal selectivity and less toxicity against lung epithelial cells and cultured macrophages. Thus these AMPs serve as a platform for the design of novel antibacterial agents and for understanding the quantitative structure—activity relationships (QSAR) in the case of the biomembrane in *M. tuberculosis*.

The advanced approaches of computational design have been used to develop novel and powerful AMPs and study their resistance mechanism [152]. The computational design of novel lead molecules has become one of the potential areas in medicinal chemistry, which aim at creating potential pharmaceutical agents with high specificity toward microorganisms with reduced side effects [153]. Thus several bioinformatics tools were prioritized to design the variants of AMPs. The major such computational biology methods are empirical, machine learning, and stochastic approaches, focusing on the standardization of peptides via random approaches [153]. Machine learning models are one of the best and efficient screening models which help to optimize the small number of sequences validate them experimentally. The major type of machine learning model is the QSAR model [154], which uses several molecular descriptors based on the physiochemical properties to model the biological properties of AMPS from their basic sequences of amino acids [155].

Further, ab initio or de novo computational approaches are used to generate the sequences of AMPs based on the position and frequency which can produce characteristics features such as amphipathicity, load, and structure [156]. This approach is allowed to generate multiple sequences of amino acids with a great diversity in their composition, 3D structures, and mode of action [157]. Several de novo computational methods are available, linguistic models are one such tool. AMP design can be performed by a formal language that includes the rules (e.g., amino acid patterns) and vocabulary (e.g., amino acid residues), using this model, it is suggested that AMPs can act specifically by spotting the intracellular targets or straightly on bacterial membranes. Recently, this model was utilized by associating the patterns in the amino acid composition in databases in the public domain, followed by their integration into a peptide unit (AMP or not) focusing on the production of optimized AMPs [156]. In addition, a genetic algorithm is another important approach to designing the novel AMPs and study of their resistance mechanism. The phylogenetic models depend on genetic algorithms to generate the successive combination of insertion and deletions in a target AMP sequence to enhance fitness and estimate the determinants that possess antibacterial activity by the prediction methods [152]. The sequences are evaluated in each generation and the sequence with lower fitness values are segregated out from the candidate sequences and generate more specific candidates for the well-defined function [152].

The major computational approaches applied for designing AMPs are shown in Fig. 14.2. There are five different approaches for computationally designing AMPs which included QSAR, de novo, linguistic, pattern insertion, and evolutionary and genetic algorithms. The computational design of AMPs is starting from de novo methods where no sequence seeds are required, or based on known AMPs focusing on producing optimized analogs. Based on the approach, various parameters will influence the design that included the activity and molecular descriptors, 3D structures, linguistic rules, pattern or motifs detection, and fitness functions. Following the initial step, diverse sequence candidates are obtained and submitted to the prediction of structures and screening for antibacterial and hemolytic potential. The

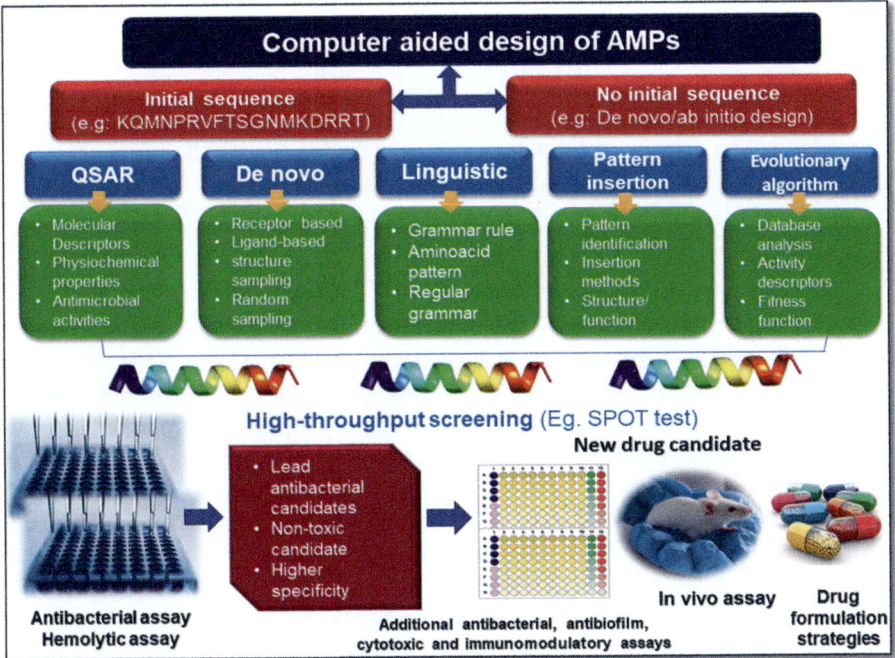

**Figure 14.2** The major computational biology approaches are used for the design of antimicrobial peptides. The major approaches included quantitative structure—activity relationships, de novo, linguistic, pattern insertion, and evolutionary and genetic algorithms. The major utilities of each method are illustrated in the figure.

ideal candidates are further submitted to the detailed structural and functional analyses that comprised of the immunomodulatory, antibiofilm, antibacterial, and in vivo assays. Thus the amalgamation of several AMP-generating strategies is examined and optimized for the evaluation of these AMP-based molecules for further clinical trials [158].

### 14.4.1 Antimicrobial peptide databases

There are several types of AMP databases available to study the various aspects of AMPs and their potential applications and utilities. The resources obtained from these databases are also helpful in understanding AMP-based resistance and its resistance mechanisms. The various types of AMP databases are briefly outlined in the following sections.

#### 14.4.1.1 Data repository of antimicrobial peptides

Data repository of antimicrobial peptides (DRAMP) (URL: http://dramp.cpu-bioinfor.org/) is a comprehensive open-access database comprised of several kinds of general, patent, and clinical AMPs. The current version 2.0 contains approximately 19,899 entries, which includes 5084 general entries, 14,739 patent entries, and 76 clinical entries. The updated version contains certain new entries, annotations, classifications, structures, and downloads. DRAMP comprised 14,040 nonoverlapping sequences when compared to the other popular AMP databases. To facilitate the search, the PubMed ID of references contains the activity information. The data can be downloaded by activity and dataset, and the website source code is also available for downloading the resources. Although there are thousands of AMPs reported in this database, only a few of them have undergone clinical trials [159].

#### 14.4.1.2 Dragon antimicrobial peptide database

Dragon antimicrobial peptide database (DAMPD) (URL: http://apps.sanbi.ac.za/dampd) is a manually curated database that contains 232 AMPs. It is an updated version of the ANTIMIC database. This database contains an integrated interface that allows the query searches based on AMP family, species, taxonomy, citation, keywords, and a combination of search terms and fields. Several important bioinformatics tools, such as Clustal W, Blast, HMMER, SignalP, Hydro calculator, AMP predictor, provide additional details about the results, these data are integrated into DAMPD to analyze the biological features of AMPs [160].

#### 14.4.1.3 The antimicrobial peptide database

The antimicrobial peptide database (APD) (URL: http://aps.unmc.edu/AP/) is a database originally developed in 2003. The revised versions are APD2 and APD3. The APD2 is developed in 2009 and it is updated and expanded into the APD3. Te3h major content of this database is natural AMPs with well-defined sequence and their activities. The databases comprise 2619 AMPs with 261, 4, 7, 13, 321, and 1972 from bacteria (bacteriocins), archaea, protists, fungi, plants, and animals. The APD3 consists of 2169, 172, 105, 959, 80, and 185 antibacterial, antiviral, anti-HIV, antifungal, antiparasitic, and anticancer peptides, respectively. There are several new AMPs with high-level annotation including antibiofilm, antiprotist, antimalarial, insecticidal, chemotactic, spermicidal, antioxidant, wound healing, and protease inhibiting features. The database also contains searchable annotations that included animal models, molecule-binding partners, target pathogens, and posttranslational modifications. The profile of amino acids or signatures of natural AMPs are vital for peptide prediction, classification, and their design [161].

### 14.4.1.4 Database of antimicrobial activity and structure of peptides

This database is manually curated which provides various analytical information to the scientific community for developing AMPs with a high therapeutic index. Database of antimicrobial activity and structure of peptides (DBAASP) provides information on detailed 3-D chemical structure and activity for peptides in which the antimicrobial activity against a particular target is experimentally evaluated. The database provides information on nonribosomal, ribosomal, and synthetic AMPs in the form of monomers, multimers, and multipeptides. It also provides data on synergistic activities that include the susceptibility of the particular strain toward certain kinds of peptides and in amalgamation with another antibiotic/peptide [162].

### 14.4.1.5 Collection of antimicrobial peptides

Collection of antimicrobial peptides (CAMPR3) is developed with the support of the Department of Science and Technology, India, and the Indian Council of Medical Research to expand and accelerate the studies based on various families of AMPs. These peptides contain the composition of a family-specific sequence of AMPs which can be retrieved to design new possible AMPs. This database consists of conserved sequence patterns of motifs in the forms of Hidden Markov Models (HMMs). This database is of 1386 AMPs with 45 families. The fundamental details for the sequence, accession numbers, protein definition, activity, target organisms, source organism, descriptions of the protein families, and external hyperlinks to databases such as UniProt, PubMed, and other AMPs databases. The database also gives tools for sequence alignment and comparison, creation of patterns, pattern, and HMM-based search [163,164].

### 14.4.1.6 A database linking antimicrobial peptides

Linking antimicrobial peptides (LAMP) is a comprehensive database (URL: http://biotechlab.fudan.edu.cn/database/lamp) of AMPs with descriptions about their antimicrobial potential and helps to enhance the process of discovering new AMPs with improved antimicrobial features and accelerate the discovery of new AMPs into preclinical or clinical trials. This linking AMP serves as a major platform to design AMPs as novel antimicrobial candidates. The preset version of LAMP consists of 5547 entries that included 3904 natural AMPs and 1643 synthetic AMPs. The database can be searched by simple keywords or combinations of keywords in searches. Along with the antimicrobial activity and cytotoxicity data, the cross-linking function of AMPs executed in LAMP help to enhance the present knowledge of AMPs and it probably accelerates the design of novel AMPs for several medical applications [165].

### 14.4.1.7 Yet another database of antimicrobial peptides

Yet another database of antimicrobial peptides (YADAMP) (URL: http://www.yadamp.unisa.it.) is primarily focused on bacteria, with a detailed description of 2133 peptides active against various bacteria. This database is created to simplify access to potential data on AMPs. The major difference between YADAMP and other AMP databases is the clear occurrence of antimicrobial potential toward most bacterial strains. Complex queries can be easily accessed through a web interface of this database. The details of the peptide can be retrieved based on the name of the peptide, net charge, hydrophobic percentage, number of amino acids, sequence motif, structure, and activity against bacteria. This database is ideal for understanding data on AMPs and for structure–activity analysis of various peptides [166].

### 14.4.1.8 Database of anuran defense peptide

Database of anuran defense peptide (DADP) is a comprehensive manually curated database that contains 2571 entries with 1923 various bioactive sequences in which 921 peptides have MIC against at least one microorganism. The database content is obtained from the UniProt-KB database and published manuscripts. The major specialties of this database are that the database is devoted to precursors and conserved signal peptides that are common to AMPs and related peptides with host defense activities, including defense from predators. This database is used to study the evolutionary origin and phylogenetic analysis of bioactive peptides present in the skin of toads and frogs. It is also used to find potential AMPs in expressed sequence tag (EST) databases of amphibians by the search for the peptides of conserved signals, and the presence of tripartite sequence arrangement—signal peptide followed by acidic pro-peptide and mature bioactive AMP. The MIC and HC50 data can be used for the studies of QSAR and related exercises [167].

### 14.4.1.9 Antimicrobial peptide scaffold by property alignment

The tool Antimicrobial PeptiDe scAffold by Property alignment (ADAPTABLE) (URL: http://gec.u-picardie.fr/adaptable) works based on the concept "property alignment" to produce families of property and sequence-related peptides (SR families). This feature helps to retrieve meaningful AMPs from more than 40,000 nonredundant sequences. The major selectable properties included the target organism and concentration of experimental activity, which allows the retrieval of peptides with simultaneous multiple actions. This tool not only combines AMP sequences databases but also merges their data, and optimizes the values, and handles amino acids with nonproteinogenic nature. In this tool, SR families permit the formation of peptide scaffolds based on the traits such as activity, source among the peptides [168].

### 14.4.1.10  Invertebrate antimicrobial peptide database

It is a comprehensive database of AMPs from invertebrates. The peptides hosted in this database are experimentally validated and they were manually annotated and curated from another public database and scientific literature. Currently, there are 702 peptides in this database. The description in this database included source (phylum and species), identification code, name, sequence, length, isoelectric point, secondary structure, charge, molar mass, aliphatic index, hydrophobicity, percentage of hydrophobic amino acids, Boman index, target organisms, experimental verification, and external literature to the link. Additional information about invertebrate antimicrobial peptide database (InverPep) can be found on the Help page [169].

### 14.4.1.11  Database of biofilm-active antimicrobial peptides

Biofilm-active antimicrobial peptides (BaAMPs) (URL: http://www.baamps.it) is a database consisting of information related to AMP sequence, associated with target organisms, chemical modifications, experimental methods, peptide concentration, and percentage of biofilm inhibition/reduction. These databases are manually extracted and curated from the literature. This database of AMPs is assayed specifically against microbial biofilms (BaAMPs). This database provides the information on a minimal standard set of required information such as details for each AMP, the microorganism and the tested biofilm conditions, and the specific assay and the concentration of the peptide used. The data will be given in an organized framework that would benefit antibiofilm-based research and the database support the design of novel AMPs active against bacterial biofilm.

### 14.4.1.12  Bacteria peptide database

Bacteria peptide database (BactPepDB) is a database that contains new bacterial peptides. It consists of the reannotation of the complete prokaryotic genomes present in RefSeq for the size of the peptide between 10 and 80 residues. These peptides are classified as previously identified in RefSeq, potential pseudogenes, intergenic and entity-overlapping. Further, the additional task is carried out included the similarity searches among genus, signal sequence or transmembrane segments searches, secondary structure, and disulfide bonds searches, and homologs protein search with a known 3D structure in the Protein Data bank. This database provides information on the peptides in complete prokaryote genomes, and details about their conservation and biological/structural features. Presently the database comprises 1,747,413 peptides from 557, 1226, and 2240 genera, species, and strains, respectively [170].

### 14.4.2  Detection of antimicrobial peptides and their resistance patterns by machine learning approach

Machine learning (ML) is one of the ideal methods for the detection of AMPs and their resistance patterns in the sequences. The ML methods and several algorithms have been effectively implemented for the analysis of novel AMPs. ML is an effective computational approach for predicting novel AMPs and it can give us potential candidates and provide insights for drug design and discovery. Recent studies suggested that the deep learning model was successfully used to identify several AMP sequences with the help of embedding layers and the multiscale convolutional network. The multiscale convolutional network is comprised of multiple layers of varying length of filters that utilizes all recent aspects obtained by the multiple convolutional layers [171].

Several methods are available for the prediction of AMPs in which combining the sequence alignment and the feature selection method are one of the major approaches. The prediction was performed by assigning peptide sequence to the group of peptide that showed greater sequence similarity to the query peptide sequence which is the main aspect of the sequence alignment method. The feature selection approach deals with the individual peptide unit encoded with 270 features such as amino acid composition, and pseudo-amino acid composition with the electrostatic charge, diversity of codon, polarity, molecular volume, and secondary structural features. Analogously, the feature selection and analysis approach, which includes the maximum relevance minimum redundancy method (mRMR) and the incremental feature selection (IFS) method, was utilized to select the ideal features for the prediction of AMPs and non-AMPs. The model for prediction was generated using the nearest neighbor algorithm (NNA) [172–174].

To recognize and categorize AMPs, several bioinformatics tools are available, that includes ADAM, AMPer, CAMP, CAMPR3, AntiBP, AVPpred, AntiBP2, EFC-FCBF, iAMP-2L, classAMP, and several other online AMP prediction tools. The majority of these tools work based on ML approaches. For example, random forest (RF), support vector machine (SVM), and artificial neural network (ANN) were used in CAMP. The application of the ML approach demand feature engineering and amino acid composition is the major feature. For example, basic amino acid counts instead of the full peptide features are employed in AntiBP. The method with pseudo-amino acid composition is also employed in several models [175].

Further, recent studies suggested that $^{13}$CNMR spectra of amino acid clusters were employed to build feature vectors for the sequences of AMPs based on the composition, distribution, and transition of cluster members. Along with the physicochemical properties of AMPs, these features were utilized to train mathematical models to predict AMPs from their basic amino acid sequences. The statistical approaches such as k-nearest neighbors (KNN), RF, Naïve Bayes (NB), SVM, and eXtreme Gradient Boosting (XGBoost) were used to make the classification system among the collected AMP datasets from several AMP datasets [175].

### 14.4.3 Recent perspectives on the scope of computational biology in antimicrobial peptide research

Based on the concept of AMP structural diversity, computational biology and bioinformatics along with data sciences help to perform computational enumeration of probable sequences of AMPs [176]. The databases that contain several AMP libraries and several computational biology tools based on ML and classification algorithms have been constructed recently to support the AMP design, which allows the screening and standardization of novel lead molecules in cost-effective approaches. The major applications of these tools are the usage of quantitative matrices (QM), ANN, and SVM to model novel antibacterial peptides and study the resistance patterns in AMPs. Computer-aided AMP design helps assess the anticipated biological activity from the primary structure of the peptide unit. Several computational prediction methods for AMPs are available that included prediction based on mature peptide sequences only, prediction based on precursor sequences only, prediction based on both mature and precursor sequences, prediction based on sequence similarity of the modified enzymes, and prediction based on available genomic data. The databases can also provide information such as antimicrobial activity, specific target, and cytotoxicity [177]. There are several online tools available for such kind of prediction. There are also several peptide modeling tools available to predict physicochemical features such as pharmacokinetic profiles and aggregation in aqueous media. The major types of computational biology tools for modeling included TANGO—this tool is used to predict amyloid cross-beta aggregation [178], ZipperDB—this tool is helpful in the evolution of the peptide fibrillogenic propensity profile [178], GROMACS—this is used for molecular dynamics simulations [179]; and SWISSADME [180], PreADMET, and DruLitoo—these tools are used to predict drug-likeness and medicinal chemistry friendliness and pharmacokinetic properties. ML approaches have been developed as a high-throughput molecular design platform for AMPs based on the detailed understanding of AMP activity at the molecular level such as membrane activity, and available molecular datasets developed by researchers [181]. These high-throughput platforms accelerate the development of lead molecules for clinical trials of rationally designed AMPs. Thus computational biology and bioinformatics provide profound scope in the design, development modeling of potential AMPs with diverse features and the data obtained from these exercises are crucial in understanding the resistance type and mechanisms of AMPs.

## 14.5 Conclusion

AMPs are a group of small peptide molecules that commonly exist in nature and they are important in the innate immune system of various organisms. AMPs possess a wide range of antimicrobial activities toward bacteria, fungi, viruses, and protozoa. The emergence of multidrug-resistant bacteria and the raising concerns about the usage of antibiotics resulted in the designing of novel AMPs, which have potential applications in pharmacy and healthcare, animal husbandry, food, aquaculture, and agriculture. They possess several inhibitory mechanisms against various bacteria and other microorganisms, however, recently reported the resistance mechanism of several bacteria toward AMPs. The major mechanism of AMP resistance in gram-positive bacteria related to the cell surfaces are repulsion of AMPs, target modification, and alterations of membrane composition. The extracellular mechanism of AMP resistance included the activity of extracellular proteases, protein-mediated sequestration, and surface-associated polysaccharides. The major mechanisms of AMPs resistance in gram-negative bacteria included bacterial outer membrane modification such as LPS and phospholipid modifications. In addition, the formation of biofilm and efflux pumps plays a vital role in the AMP resistances. Further, binding and sequestering CAMPs, proteolytic degradation of AMPs, and modulation of CAMP expression were considered to be the other mechanisms of AMP resistance. The recent advancement of computational biology, bioinformatics, and ML approaches provide significant insights into AMP research. There are several AMP databases available that provide up-to-date information on various kinds of AMPs, and these resources are important in the study of AMP resistance by several organisms. Computational biology and data sciences approach played a vital role in AMP research and provide a significant breakthrough for the future investigation of antimicrobial resistance exhibited by various microorganisms.

# References

[1] M. Magana, M. Pushpanathan, A.L. Santos, L. Leanse, M. Fernandez, A. Ioannidis, et al., The value of antimicrobial peptides in the age of resistance, The Lancet Infectious Diseases 20 (9) (2020) 216–230.

[2] J.K. Boparai, K.S. Pushpender, Mini review on antimicrobial peptides, sources, mechanism and recent applications, Protein & Peptide Letters 27 (1) (2020) 4–16.

[3] G. Annunziato, G. Costantino, Antimicrobial peptides (AMPs): a patent review (2015–2020), Expert Opinion on Therapeutic Patents 30 (12) (2020) 931–947.

[4] H. Kang, C. Kim, C.H. Seo, Y. Park, The therapeutic applications of antimicrobial peptides (AMPs): a patent review, Journal of Microbiology 55 (1) (2017) 1–12.

[5] P. Kumar, J.N. Kizhakkedathu, S.K. Straus, Antimicrobial peptides: diversity, mechanism of action and strategies to improve the activity and biocompatibility in vivo, Biomolecules 8 (1) (2018) 4.

[6] P.Y. Chung, R. Khanum, Antimicrobial peptides as potential anti-biofilm agents against multidrug-resistant bacteria, Journal of Microbiology, Immunology and Infection 50 (4) (2017) 405–410.

[7] J. Li, S. Hu, W. Jian, C. Xie, X. Yang, Plant antimicrobial peptides: structures, functions, and applications, Botanical Studies 62 (1) (2021) 5.

[8] M.R. Yeaman, N.Y. Yount, Mechanisms of antimicrobial peptide action and resistance, Pharmacological Reviews 55 (1) (2003) 27–55.

[9] S. Tang, Z.H. Prodhan, S.K. Biswas, C. Le, S.D. Sekaran, Antimicrobial peptides from different plant sources: isolation, characterisation, and purification, Phytochemistry 154 (2018) 94–105.

[10] H. Yi, M. Chowdhury, Y. Huang, X. Yu, Insect antimicrobial peptides and their applications, Applied Microbiology and Biotechnology 98 (13) (2014) 5807–5822.

[11] J. Wiesner, A. Vilcinskas, Antimicrobial peptides: the ancient arm of the human immune system, Virulence 1 (5) (2010) 440–464.

[12] T. Magrone, M.A. Russo, E. Jirillo, Antimicrobial peptides: phylogenic sources and biological activities. First of two parts, Current Pharmaceutical Design 24 (10) (2018) 1043−1053.
[13] T.V. Vineeth Kumar, G.A. Sanil, Review of the mechanism of action of amphibian antimicrobial peptides focusing on peptide-membrane interaction and membrane curvature, Current Protein & Peptide Science 18 (12) (2017) 1263−1272.
[14] A. Izadpanah, R.L. Gallo, Antimicrobial peptides, Journal of the American Academy of Dermatology 52 (3) (2015) 381−390.
[15] T. Lee, K. Hall, M. Aguilar, Antimicrobial peptide structure and mechanism of action: a focus on the role of membrane structure, Current Topics in Medicinal Chemistry 16 (1) (2016) 25−39.
[16] K. Browne, S. Chakraborty, R. Chen, M.D. Willcox, D.S. Black, W.R. Walsh, et al., New era of antibiotics: the clinical potential of antimicrobial peptides, International Journal of Molecular Sciences 21 (19) (2020) 704.
[17] Y. Gao, H. Fang, D. Liu, J. Liu, M. Su, Z. Fang, et al., The modification and design of antimicrobial peptide, Current Pharmaceutical Design 24 (8) (2018) 904−910. .
[18] L. Assoni, B. Milani, M.R. Carvalho, L.N. Nepomuceno, N.T. Waz, M.E.S. Guerra, et al., Resistance mechanisms to antimicrobial peptides in gram-positive bacteria, Frontiers in Microbiology 11 (2020) 593215.
[19] Y. Huan, Q. Kong, H. Mou, H. Yi, Antimicrobial peptides: classification, design, application and research progress in multiple fields, Frontiers in Microbiology 11 (2020) 582779. Available from: https://doi.org/10.3389/fmicb.2020.582779.
[20] C.N. LaRock, V. Nizet, Cationic antimicrobial peptide resistance mechanisms of streptococcal pathogens, Biochimica et Biophysica Acta 1848 (2015) 3047−3054.
[21] K.L. Nawrocki, E.K. Crispell, S.M. McBride, Antimicrobial peptide resistance mechanisms of gram-positive bacteria, Antibiotics 3 (4) (2014) 461−492.
[22] H.G. Boma, Peptide antibiotics and their role in innate immunity, Annual Review of Immunology 13 (1995) 61−92.
[23] C. Pandin, M. Caroff, G. Condemine, Antimicrobial peptide resistance genes in the plant pathogen *Dickeya dadantii*, Applied and Environmental Microbiology 82 (21) (2016) 6423−6430.
[24] A. Peschel, How do bacteria resist human antimicrobial peptides? Trends in Microbiology 10 (2002) 179−186.
[25] R.E. Hancock, A. Rozek, Role of membranes in the activities of antimicrobial cationic peptides, FEMS Microbiology Letters 206 (2002) 143−149.
[26] J.P. Powers, R.E. Hancock, The relationship between peptide structure and antibacterial activity, Peptides 24 (2003) 1681−1691.
[27] M. Zasloff, Antimicrobial peptides of multicellular organisms, Nature 415 (2012) 389−395.
[28] C. Weidenmaier, A. Peschel, Teichoic acids and related cell-wall glycopolymers in gram-positive physiology and host interactions, Nature Reviews. Microbiology 6 (2008) 276−287.
[29] H. Goldfine, Bacterial membranes and lipid packing theory, Journal of Lipid Research 25 (1984) 1501−1507.
[30] A. Wiese, M. Munstermann, T. Gutsmann, B. Lindner, K. Kawahara, U. Zahringer, et al., Molecular mechanisms of polymyxin B-membrane interactions: direct correlation between surface charge density and self-promoted transport, The Journal of Membrane Biology 162 (1998) 127−138.

[31] C.M. Ernst, P. Staubitz, N.N. Mishra, S.J. Yang, G. Hornig, H. Kalbacher, et al., The bacterial defensin resistance protein MprF consists of separable domains for lipid lysinylation and antimicrobial peptide repulsion, PLoS Pathogens 5 (2009) e1000660.
[32] S.A. Kristian, M. Durr, J.A. van Strijp, B. Neumeister, A. Peschel, MprF-mediated lysinylation of phospholipids in *Staphylococcus aureus* leads to protection against oxygen-independent neutrophil killing, Infection and Immunity 71 (2003) 546–549.
[33] Y. Bao, T. Sakinc, D. Laverde, D. Wobser, A. Benachour, C. Theilacker, et al., Role of mprF1 and mprF2 in the pathogenicity of *Enterococcus faecalis*, PLoS One 7 (2012) e38458.
[34] A.B. Hachmann, E.R. Angert, J.D. Helmann, Genetic analysis of factors affecting susceptibility of *Bacillus subtilis* to daptomycin, Antimicrobial Agents and Chemotherapy 53 (2009) 1598–1609.
[35] F.C. Neuhaus, J. Baddiley, A continuum of anionic charge: structures and functions of D-alanyl-teichoic acids in gram-positive bacteria, Microbiology and Molecular Biology Reviews 67 (2003) 686–723.
[36] K. Meyer, J.W. Palmer, R. Thompson, D. Khorazo, On the mechanism of lysozyme action, Journal of Biological Chemistry 113 (1936) 479–486.
[37] D.M. Chipman, N. Sharon, Mechanism of lysozyme action, Science (New York, N.Y.) 165 (1969) 454–465.
[38] J.A. Nash, T.N. Ballard, T.E. Weave, H.T. Akinbi, The peptidoglycan-degrading property of lysozyme is not required for bactericidal activity in vivo, Journal of Immunology 177 (2006) 519–526.
[39] L. Hebert, P. Courtin, R. Torelli, M. Sanguinetti, M.P. Chapot-Chartier, Y. Auffray, A. Benachour, *Enterococcus faecalis* constitutes an unusual bacterial model in lysozyme resistance, Infection and Immunity 75 (2007) 5390–5398.
[40] A.S. Mazzotta, T.J. Montville, Nisin induces changes in membrane fatty acid composition of *Listeria monocytogenes* nisin-resistant strains at 10 degrees C and 30 degrees C, Journal of Applied Microbiology 82 (1997) 32–38.
[41] A. Verheul, N.J. Russell, R. Van't Hof, F.M. Rombouts, T. Abee, Modifications of membrane phospholipid composition in nisin-resistant *Listeria monocytogenes* Scott A, Applied and Environmental Microbiology 63 (1997) 3451–3457.
[42] X.T. Ming, M.A. Daeschel, Nisin resistance of foodborne bacteria and the specific resistance responses of *Listeria monocytogenes* Scott A, Journal of Food Protection 56 (1993) 944–948.
[43] A.D. Crandall, T.J. Montville, Nisin resistance in *Listeria monocytogenes* ATCC 700302 is a complex phenotype, Applied and Environmental Microbiology 64 (1998) 231–233.
[44] N.N. Mishra, J. McKinnell, M.R. Yeaman, A. Rubio, C.C. Nast, L. Chen, et al., In vitro cross-resistance to daptomycin and host defense cationic antimicrobial peptides in clinical methicillin-resistant *Staphylococcus aureus* isolates, Antimicrobial Agents and Chemotherapy 55 (2011) 4012–4018.
[45] A. Schmidtchen, I.M. Frick, E. Andersson, H. Tapper, L. Bjorck, Proteinases of common pathogenic bacteria degrade and inactivate the antibacterial peptide LL-37, Molecular Microbiology 46 (2002) 157–168.
[46] A. Sabat, K. Kosowska, K. Poulsen, A. Kasprowicz, A. Sekowska, B. van Den Burg, et al., Two allelic forms of the aureolysin gene (aur) within *Staphylococcus aureus*, Infection and Immunity 68 (2000) 973–976.

[47] Y. Lai, A.E. Villaruz, M. Li, D.J. Cha, D.E. Sturdevant, M. Otto, The human anionic antimicrobial peptide dermcidin induces proteolytic defence mechanisms in staphylococci, Molecular Microbiology 63 (2007) 497−506.
[48] P. Akesson, A.G. Sjoholm, L. Bjorck, Protein SIC, a novel extracellular protein of *Streptococcus pyogenes* interfering with complement function, Journal of Biological Chemistry 271 (1996) 1081−1088.
[49] P.K. Peterson, B.J. Wilkinson, Y. Kim, D. Schmeling, P.G. Quie, Influence of encapsulation on staphylococcal opsonization and phagocytosis by human polymorphonuclear leukocytes, Infection and Immunity 19 (1978) 943−949.
[50] A.L. Nelson, A.M. Roche, J.M. Gould, K. Chim, A.J. Ratner, J.N. Weiser, Capsule enhances pneumococcal colonization by limiting mucus-mediated clearance, Infection and Immunity 75 (2007) 83−90.
[51] C.D. Ashbaugh, H.B. Warren, V.J. Carey, M.R. Wessels, Molecular analysis of the role of the group A streptococcal cysteine protease, hyaluronic acid capsule, and M protein in a murine model of human invasive soft-tissue infection, Journal of Clinical Investigation 102 (1998) 550−560.
[52] T. Candela, A. Fouet, *Bacillus anthracis* CapD, belonging to the gamma-glutamyltranspeptidase family, is required for the covalent anchoring of capsule to peptidoglycan, Molecular Microbiology 57 (2005) 717−726.
[53] L. Deng, D.L. Kasper, T.P. Krick, M.R. Wessels, Characterization of the linkage between the type III capsular polysaccharide and the bacterial cell wall of group B *Streptococcus*, Journal of Biological Chemistry 275 (2000) 7497−7504.
[54] D. Mack, W. Fischer, A. Krokotsch, K. Leopold, R. Hartmann, H. Egge, et al., The intercellular adhesin involved in biofilm accumulation of *Staphylococcus epidermidis* is a linear beta-1, 6-linked glucosaminoglycan: purification and structural analysis, Journal of Bacteriology 178 (1996) 175−183.
[55] M.A. Campos, M.A. Vargas, V. Regueiro, C.M. Llompart, S. Alberti, J.A. Bengoechea, Capsule polysaccharide mediates bacterial resistance to antimicrobial peptides, Infection and Immunity 72 (2004) 7107−7114.
[56] UNMC Department of Pathology and Microbiology, The antimicrobial peptide database, 2014. Available online: <http://aps.unmc.edu/AP/main.php> (accessed 2.7.14).
[57] T. Nakatsuji, R.L. Gallo, Antimicrobial peptides: old molecules with new ideas, Journal of Investigative Dermatology 132 (2012) 887−895.
[58] W.C. Wimley, Describing the mechanism of antimicrobial peptide action with the interfacial activity model, ACS Chemical Biology 5 (2010) 905−917.
[59] V. Nizet, Antimicrobial peptide resistance mechanisms of human bacterial pathogens, Current Issues Molecular Biology 8 (2006) 11−26.
[60] S. Biswas, J.M. Brunel, J.C. Dubus, M. Reynaud-Gaubert, J.M. Rolain, Colistin: an update on the antibiotic of the 21st century, Expert Review of Anti-Infective Therapy 10 (2012) 917−934.
[61] S.M. Moskowitz, R.K. Ernst, S.I. Miller, PmrAB, a two-component regulatory system of *Pseudomonas aeruginosa* that modulates resistance to cationic antimicrobial peptides and addition of aminoarabinose to lipid A, Journal of Bacteriology 186 (2004) 575−579.
[62] J.S. Gunn, K.B. Lim, J. Krueger, K. Kim, L. Guo, M. Hackett, et al., PmrA-pmrB-regulatedgenes necessary for 4-aminoarabinose lipid A modification and polymyxin resistance, Molecular Microbiology 27 (1998) 1171−1182.

[63] A.C. Llewellyn, J. Zhao, F. Song, J. Parvathareddy, Q. Xu, B.A. Napier, et al., NaxD is a deacetylase required for lipid A modification and *Francisella* pathogenesis, Molecular Microbiology 86 (2012) 611−627.

[64] N.R. Shah, R.E. Hancock, R.C. Fernandez, *Bordetella pertussis* lipid A glucosamine modification confers resistance to cationic antimicrobial peptides and increases resistance to outer membrane perturbation, Antimicrobial Agents and Chemotherapy 58 (2014) 4931−4934.

[65] J.V. Hankins, J.A. Madsen, D.K. Giles, J.S. Brodbelt, M.S. Trent, Amino acid addition to *Vibrio cholerae* LPS establishes a link between surface remodeling in gram-positive and gram-negative bacteria, Proceedings of the National Academy of Sciences of the United States of America 109 (2012) 8722−8727.

[66] J.B. Harris, R.C. LaRocque, F. Qadri, E.T. Ryan, S.B. Calderwood, Cholera, Lancet 379 (2012) 2466−2476.

[67] L.A. Lewis, B. Choudhury, J.T. Balthazar, L.E. Martin, S. Ram, P.A. Rice, et al., Phosphoethanolamine substitution of lipid a and resistance of *Neisseria gonorrhoeae* to cationic antimicrobial peptides and complement-mediated killing by normal human serum, Infection and Immunity 77 (2019) 1112−1120.

[68] L.A. Lewis, W.M. Shafer, T. Dutta Ray, S. Ram, P.A. Rice, Phosphoethanolamine residues on the lipid a moiety of *Neisseria gonorrhoeae* lipooligosaccharide modulate binding of complement inhibitors and resistance to complement killing, Infection and Immunity 81 (2013) 33−42.

[69] M.R. Pelletier, L.G. Casella, J.W. Jones, M.D. Adams, D.V. Zurawski, K.R. Hazlett, et al., Unique structural modifications are present in the lipopolysaccharide from colistin-resistant strains of *Acinetobacter baumannii*, Antimicrobial Agents and Chemotherapy 57 (2013) 4831−4840.

[70] H. Lee, F.F. Hsu, J. Turk, E.A. Groisman, The pmra-regulated pmrc gene mediates phosphoethanolamine modification of lipid A and polymyxin resistance in *Salmonella enterica*, Journal of Bacteriology 186 (2004) 4124−4133.

[71] X. Wang, A.A. Ribeiro, Z.Z. Guan, S.N. Abraham, C.R. Raetz, Attenuated virulence of a *Francisella* mutant lacking the lipid a 4'-phosphatase, Proceedings of the National Academy of Sciences of the United States of America 104 (2007) 4136−4141.

[72] E. Vinogradov, M.B. Perry, J.W. Conlan, Structural analysis of *Francisella tularensis* lipopolysaccharide, European Journal of Biochemistry 269 (2002) 6112−6118.

[73] B.O. Ingram, A. Masoudi, C.R. Raetz, *Escherichia coli* mutants that synthesize dephosphorylated lipid a molecule, Biochemistry 49 (2010) 8325−8337.

[74] H. Kumada, Y. Haishima, T. Umemoto, K. Tanamoto, Structural study on the free lipid A isolated from lipopolysaccharide of porphyromonas gingivalis, Journal of Bacteriology 177 (1995) 2098−2106.

[75] A. Weintraub, U. Zähringer, H.W. Wollenweber, U. Seydel, E.T. Rietschel, Structural characterization of the lipid a component of bacteroides fragilis strain nctc 9343 lipopolysaccharide, European Journal of Biochemistry 183 (1989) 425−431.

[76] A.X. Tran, J.D. Whittimore, P.B. Wyrick, S.C. McGrath, R.J. Cotter, M.S. Trent, The lipid A 1-phosphatase of helicobacter pylori is required for resistance to the antimicrobial peptide polymyxin, Journal of Bacteriology 188 (2006) 4531−4541.

[77] B.D. Needham, M.S. Trent, Fortifying the barrier: the impact of lipid A remodelling on bacterial pathogenesis, Nature Reviews. Microbiology 11 (2013) 467−481.

[78] L. Guo, K.B. Lim, C.M. Poduje, M. Daniel, J.S. Gunn, M. Hackett, et al., A acylation and bacterial resistance against vertebrate antimicrobial peptides, Cell 95 (1998) 189−198.

[79] R.P. Darveau, J. Blake, C.L. Seachord, W.L. Cosand, M.D. Cunningham, L. Cassiano-Clough, et al., Peptides related to the carboxyl terminus of human platelet factor IV with antibacterial activity, Journal of Clinical Investigation 90 (1992) 447−455.

[80] A.E. Jerse, N.D. Sharma, A.N. Simms, E.T. Crow, L.A. Snyder, W.M. Shafer, A gonococcal efflux pump system enhances bacterial survival in a female mouse model of genital tract infection, Infection and Immunity 71 (2003) 5576−5582.

[81] E. Padilla, E. Llobet, A. Doménech-Sánchez, L. Martínez-Martínez, J.A. Bengoechea, S. Albertí, *Klebsiella pneumoniae* AcrAB efflux pump contributes to antimicrobial resistance and virulence, Antimicrobial Agents and Chemotherapy 54 (2010) 177−183.

[82] P.W. Park, G.B. Pier, M.J. Preston, O. Goldberger, M.L. Fitzgerald, M. Bernfield, Syndecan-1shedding is enhanced by LasA, a secreted virulence factor of *Pseudomonas aeruginosa*, Journal of Biological Chemistry 275 (2000) 3057−3064.

[83] P.W. Park, G.B. Pier, M.T. Hinkes, M. Bernfield, Exploitation of syndecan-1 shedding by *Pseudomonas aeruginosa* enhances virulence, Nature 411 (2001) 98−102.

[84] J.A. Bengoechea, R. Díaz, I. Moriyón, Outer membrane differences between pathogenic and environmental Yersinia enterocolitica biogroups probed with hydrophobic permeants and polycationic peptides, Infection and Immunity 64 (1996) 4891−4899.

[85] Z.D. Dalebroux, S. Matamouros, D. Whittington, R.E. Bishop, S.I. Miller, PhoPQ regulates acidic glycerophospholipid content of the *Salmonella typhimurium* outer membrane, Proceedings of the National Academy of Sciences of the United States of America 111 (2014) 1963−1968.

[86] E. Cox, A. Michalak, S. Pagentine, P. Seaton, A. Pokorny, Lysylated phospholipids stabilize models of bacterial lipid bilayers and protect against antimicrobial peptides, Biochimica et Biophysica Acta 1838 (2014) 2198−2204.

[87] C. Sohlenkamp, K.A. Galindo-Lagunas, Z. Guan, P. Vinuesa, S. Robinson, T.-J. Oates, et al., The lipid lysyl-phosphatidylglycerol is present in membranes of *Rhizobium tropici* CIAT899 and confers increased resistance to polymyxin B under acidic growth conditions, Molecular Plant-Microbe Interactions 20 (2007) 1421−1430.

[88] D.E. Jones, J.D. Smith, Phospholipids of the differentiating bacterium *Caulobacter crescentus*, Canadian Journal of Biochemistry and Physiology 57 (1979) 424−428.

[89] T.S. Yokum, R.P. Hammer, M.L. McLaughlin, P.H. Elzer, Peptides with indirect *in vivo* activity against an intracellular pathogen: selective lysis of infected macrophages, The Journal of Peptide Research 59 (2002) 9−17.

[90] R.A. Dorschner, B. Lopez-Garcia, A. Peschel, D. Kraus, K. Morikawa, V. Nizet, et al., The mammalian ionic environment dictates microbial susceptibility to antimicrobial defense peptides, FASEB Journal 20 (2006) 35−42.

[91] L.M. Willis, C. Whitfield, . Structure, biosynthesis, and function of bacterial capsular polysaccharides synthesized by abc transporter-dependent pathways, Carbohydrate Research 378 (2013) 35−44.

[92] G. Cortés, N. Borrell, B. de Astorza, C. Gómez, J. Sauleda, S. Albertí, Molecular analysis of the contribution of the capsular polysaccharide and the lipopolysaccharide o side chain to the virulence of *Klebsiella pneumoniae* in a murine model of pneumonia, Infection and Immunity 70 (2002) 2583−2590.

[93] A. Jones, M. Geörg, L. Maudsdotter, A.B. Jonsson, Endotoxin, capsule, and bacterial attachmentcontribute to *Neisseria meningitidis* resistance to the human antimicrobial peptide LL-37, Journal of Bacteriology 191 (2009) 3861−3868.

[94] E. Llobet, J.M. Tomás, J.A. Bengoechea, Capsule polysaccharide is a bacterial decoy for antimicrobial peptides, Microbiology (Reading, England) 154 (2008) 3877−3886.

[95] J.W. Costerton, P.S. Stewart, E.P. Greenberg, Bacterial biofilms: a common cause of persistent infections, Science (New York, N.Y.) 284 (1999) 1318–1322.
[96] J. Wimpenny, W. Manz, U. Szewzyk, Heterogeneity in biofilms, FEMS Microbiology Reviews 24 (2000) 661–671.
[97] J.C. Nickel, I. Ruseska, J.B. Wright, J.W. Costerton, Tobramycin resistance of *Pseudomonasaeruginosa* cells growing as a biofilm on urinary catheter material, Antimicrobial Agents and Chemotherapy 27 (1985) 619–624.
[98] M. Hentzer, G.M. Teitzel, G.J. Balzer, A. Heydorn, S. Molin, M. Givskov, et al., Alginate overproduction affects *Pseudomonas aeruginosa* biofilm structure and function, Journal of Bacteriology 183 (2001) 5395–5401.
[99] A. Folkesson, J.A. Haagensen, C. Zampaloni, C. Sternberg, S. Molin, Biofilm induced tolerancetowards antimicrobial peptides, PLoS One 3 (2008) 1891.
[100] C. Chan, L.L. Burrows, C.M. Deber, Helix induction in antimicrobial peptides by alginate in biofilms, Journal of Biological Chemistry 279 (2004) 38749–38754.
[101] C. Chan, L.L. Burrows, C.M. Deber, Alginate as an auxiliary bacterial membrane: binding of membrane-active peptides by polysaccharides. The Journal of Peptide Research, 65 (2005) 343–351.
[102] M. Benincasa, M. Mattiuzzo, Y. Herasimenka, P. Cescutti, R. Rizzo, R. Gennaro, Activity of antimicrobial peptides in the presence of polysaccharides produced by pulmonary pathogens, Journal of Peptide Science 15 (2009) 595–600.
[103] H. Mulcahy, L. Charron-Mazenod, S. Lewenza, Extracellular DNA chelates cations and induces antibiotic resistance in *Pseudomonas aeruginosa* biofilms, PLoS Pathogens 4 (2008) e1000213.
[104] L. Johnson, S.R. Horsman, L. Charron-Mazenod, A.L. Turnbull, H. Mulcahy, M.G. Surette, et al., Extracellular DNA-induced antimicrobial peptide resistance in *Salmonella enterica* serovar Typhimurium, BMC Microbiology 13 (2013) e115.
[105] W.J. Gooderham, M. Bains, J.B. McPhee, I. Wiegand, R.E. Hancock, Induction by cationic antimicrobial peptides and involvement in intrinsic polymyxin and antimicrobial peptide resistance, biofilm formation, and swarming motility of PsrA in *Pseudomonas aeruginosa*, Journal of Bacteriology 190 (2008) 5624–5634.
[106] S.J. Pamp, M. Gjermansen, H.K. Johansen, T. Tolker-Nielsen, Tolerance to the antimicrobial peptide colistin in *Pseudomonas aeruginosa* biofilms is linked to metabolically active cells, and depends on the *pmr* and *mexAB-oprM* genes, Molecular Microbiology 68 (2008) 223–240.
[107] K. Poole, Efflux pumps as antimicrobial resistance mechanisms, Annals of Medicine 39 (2007) 162–176.
[108] L.J. Piddock, Multidrug-resistance efflux pumps—not just for resistance, Nature Reviews. Microbiology 4 (2006) 629–636.
[109] A.M. Buckley, M.A. Webber, S. Cooles, L.P. Randall, R.M. La Ragione, M.J. Woodward, et al., The acrab-tolc efflux system of *Salmonella enterica* serovar Typhimurium plays a role in pathogenesis, Cellular Microbiology 8 (2006) 847–856.
[110] K. Nishino, T. Latifi, E.A. Groisman, Virulence and drug resistance roles of multidrug efflux systems of *Salmonella enterica* serovar Typhimurium, Molecular Microbiology 59 (2006) 126–141.
[111] B.J. Stone, V.L. Miller, *Salmonella enteritidis* has a homologue of tolC that is required for virulence in BALB/c mice, Molecular Microbiology 17 (1995) 701–712.
[112] A. Pérez, M. Poza, A. Fernández, M.C. Fernández, S. Mallo, M. Merino, et al., Involvement of the AcrAB-TolC efflux pump in the resistance, fitness, and virulence

of *Enterobacter cloacae*, Antimicrobial Agents and Chemotherapy 56 (2012) 2084−2090.
[113] J.A. Bengoechea, M. Skurnik, Temperature-regulaed efflux pump/potassium antiporter system mediates resistance to cationic antimicrobial peptides in *Yersinia*, Molecular Microbiology 37 (2000) 67−80.
[114] W.M. Shafer, X. Qu, A.J. Waring, R.I. Lehrer, Modulation of *Neisseria gonorrhoeae* susceptibility to vertebrate antibacterial peptides due to a member of the resistance/nodulation/division efflux pump family, Proceedings of the National Academy of Sciences of the United States of America 95 (1998) 1829−1833.
[115] D.M. Warner, W.M. Shafer, A.E. Jerse, Clinically relevant mutations that cause derepression of the *Neisseria gonorrhoeae* MtrC-MtrD-MtrE efflux pump system confer different levels of antimicrobial resistance and *in vivo* fitness, Molecular Microbiology 70 (2008) 462−478.
[116] Y.L. Tzeng, K.D. Ambrose, S. Zughaier, X. Zhou, Y.K. Miller, W.M. Shafer, et al., Cationic antimicrobial peptide resistance in *Neisseria meningitidis*, Journal of Bacteriology 187 (2005) 5387−5396.
[117] X.R. Bia, D. Provenzano, N. Nguyen, J.E. Bina, *Vibrio cholerae* RND family efflux systems are required for antimicrobial resistance, optimal virulence factor production, and colonization of the infant mouse small intestine, Infection and Immunity 76 (2008) 3595−3605.
[118] Y.C. Chen, Y.C. Chuang, C.C. Chang, C.L. Jeang, M.C. Chang, A K+ uptake protein, TrkA, is required for serum, protamine, and polymyxin B resistance in *Vibrio vulnificus*, Infection and Immunity 72 (2004) 629−636.
[119] C. Kourtesi, A.R. Ball, Y.Y. Huang, S.M. Jachak, D.M. Vera, D.M.P. Khondkar, et al., Microbial efflux systems and inhibitors: approaches to drug discovery and the challenge of clinical implementation, The Open Microbiology Journal 7 (2013) 34−52.
[120] A. Zamfir, D.G. Seidler, H. Kresse, J. Peter-Katalinić, Structural investigation of chondroitin/dermatan sulfate oligosaccharides from human skin fibroblast decorin, Glycobiology 13 (2003) 733−742.
[121] M. Bernfield, M. Götte, P.W. Park, O. Reizes, M.L. Fitzgerald, J. Lincecum, et al., Functions of cell surface heparan sulfate proteoglycans, Annual Review of Biochemistry 68 (1999) 729−777.
[122] A.J. McBroom, A.P. Johnson, S. Vemulapalli, M.J. Kuehn, Outer membrane vesicle production by *Escherichia coli* is independent of membrane instability, Journal of Bacteriology 188 (2006) 5385−5392.
[123] M.J. Kuehn, N.C. Kesty, Bacterial outer membrane vesicles and the host-pathogen interaction, Genes & Development 19 (2005) 2645−2655.
[124] A.J. McBroom, M.J. Kuehn, Release of outer membrane vesicles by gram-negative bacteria is a novel envelope stress response, Molecular Microbiology 63 (2007) 545−558.
[125] M. Duperthuy, A.E. Sjöström, D. Sabharwal, F. Damghani, B.E. Uhlin, S.N. Wai, Role of the *Vibrio cholerae* matrix protein Bap1 in cross-resistance to antimicrobial peptides, PLoS Pathogens 9 (2003) e1003620.
[126] R. Belas, J. Manos, R. Suvanasuthi, *Proteus mirabilis* ZapA metalloprotease degrades a broad spectrum of substrates, including antimicrobial peptides, Infection and Immunity 72 (2004) 5159−5167.

[127] C. Kooi, P.A. Sokol, *Burkholderia cenocepacia* zinc metalloproteases influence resistance to antimicrobial peptides, Microbiology (Reading, England) 155 (2009) 2818–2825.

[128] M. Gutner, S. Chaushu, D. Balter, G. Bachrach, Saliva enables the antimicrobial activity of LL-37 in the presence of proteases of *Porphyromonas gingivalis*, Infection and Immunity 77 (2009) 5558–5563.

[129] G. Maisetta, F.L. Brancatisano, S. Esin, M. Campa, G. Batoni, Gingipains produced by Porphyromonas gingivalis ATCC49417 degrade human-beta-defensin 3 and affect peptide's antibacterial activity in vitro, Peptides 32 (2011) 1073–1077.

[130] V. Hritonenko, C. Stathopoulos, Omptin proteins: an expanding family of outer membrane proteases in gram negative *Enterobacteriaceae*, Molecular Membrane Biology 24 (2007) 395–406.

[131] J. Haiko, M. Suomalainen, T. Ojala, K. Lahteenmaki, T.K. Korhonen, Breaking barriers-attack on innate immune defences by omptin surface proteases of enterobacterial pathogens, Innate Immunity 15 (2009) 67–80.

[132] S. Stumpe, R. Schmid, D.L. Stephens, G. Georgiou, E.P. Bakker, Identification of OmpT as the protease that hydrolyzes the antimicrobial peptide protamine before it enters growing cells of *Escherichia coli*, Journal of Bacteriology 180 (1998) 4002–4006.

[133] E.M. Galvan, M.A. Lasaro, D.M. Schifferli, Capsular antigen fraction 1 and Pla modulate the susceptibility of *Yersinia pestis* to pulmonary antimicrobial peptides such as cathelicidin, Infection and Immunity 76 (2008) 1456–1464.

[134] T. Guina, E.C. Yi, H. Wang, M. Hackett, S.I. Miller, A PhoP-regulated outer membrane protease of *Salmonella enterica* serovar Typhimurium promotes resistance to alphahelical antimicrobial peptides, Journal of Bacteriology 182 (2000) 4077–4086.

[135] V. Le Sage, L. Zhu, C. Lepage, A. Portt, C. Viau, F. Daigle, et al., An outer membrane protease of the omptin family prevents activation of the *Citrobacter rodentium* PhoPQ two-component system by antimicrobial peptides, Molecular Microbiology 74 (2009) 98–111.

[136] J. Thomassin, J. Brannon, B.F. Gibbs, S. Gruenheid, H. Le Moual, OmpT outer membrane proteases of enterohemorrhagic and enteropathogenic *Escherichia coli* contribute differently to the degradation of human LL-37, Infection and Immunity 80 (2012) 483–492.

[137] S.R. Coats, T.T. To, T.S. Jain, P.H. Braham, R.P. Darveau, *Porphyromonas gingivalis* resistance to polymyxin B is determined by the lipid A 4'-phosphatase, PGN_0524, International Journal of Oral Science 1 (2009) 126–135.

[138] S.R. Coats, J.W. Jones, C.T. Do, P.H. Braham, B.W. Bainbridge, T.T. To, et al., Human Toll-like receptor 4 responses to *P. gingivalis* are regulated by lipid A 1- and 4'-phosphatase activities, Cellular Microbiology 11 (2009) 1587–1599.

[139] R.E. Bishop, The lipid A palmitoyltransferase PagP: molecular mechanisms and role in bacterial pathogenesis, Molecular Microbiology 57 (2005) 900–912.

[140] N. Maeshima, R.C. Fernandez, Recognition of lipid A variants by the TLR4-MD-2 receptor complex, Frontiers in Cellular and Infection Microbiology (Reading, England) 3 (2013).

[141] R.K. Ernst, E.C. Yi, L. Guo, K.B. Lim, J.L. Burns, M. Hackett, et al., Specific lipopolysaccharide found in cystic fibrosis airway *Pseudomonas aeruginosa*, Science (New York, N.Y.) 286 (1999) 1561–1565.

[142] T.R. Sampson, S.D. Saroj, A.C. Llewellyn, Y.L. Tzeng, D.S. Weiss, A CRISPR/C as system mediates bacterial innate immune evasion and virulence, Nature 497 (2013) 254–257.

[143] T.R. Sampson, B.A. Napier, M.R. Schroeder, R. Louwen, J. Zhao, C.Y. Chin, et al., CRISPR-C as system enhances envelope integrity mediating antibiotic resistance and inflammasome evasion, Proceedings of the National Academy of Sciences of the United States of America 111 (2014) 11163–11168.

[144] D. Moranta, V. Regueiro, C. March, E. Llobet, J. Margareto, E. Larrarte, et al., *Klebsiella pneumoniae* capsule polysaccharide impedes the expression of beta-defensins by airway epithelial cells, Infection and Immunity 78 (2010) 1135–1146.

[145] D. Islam, L. Bandholtz, J. Nilsson, H. Wigzell, B. Christensson, B. Agerberth, et al., Downregulation of bactericidal peptides in enteric infections: a novel immune escape mechanism with bacterial DNA as a potential regulator, Nature Medicine 7 (2001) 180–185.

[146] B. Sperandio, B. Regnault, J. Guo, Z. Zhang, S.L. Stanley, P.J. Sansonetti, P.J, et al., Virulent *Shigella flexneri* subverts the host innate immune response through manipulation of antimicrobial peptide gene expression, Journal of Experimental Medicine 205 (2008) 1121–1132.

[147] M.W. Hornef, M.J. Wick, M. Rhen, S. Normark, Bacterial strategies for overcoming host innate and adaptive immune responses, Nature Immunology 3 (2002) 1033–1040.

[148] B. Mishra, G. Wang, Ab initio design of potent anti-MRSA peptides based on database filtering technology, Journal of the American Chemical Society 134 (2012) 12426–12429.

[149] C.S. Pearson, Z. Kloos, B. Murray, E. Tabe, M. Gupta, J.H. Kwak, et al., Combined bioinformatic and rational design approach to develop antimicrobial peptides against *Mycobacterium tuberculosis*, Antimicrobial Agents and Chemotherapy 60 (5) (2016) 2757–2764. Available from: https://doi.org/10.1128/AAC.00940-15.

[150] H. Hemlata, A. Tiwari, Applications of bioinformatics tools to combat the antibiotic resistance, In: Proceedings of the International Conference on Soft Computing Techniques and Implementations, ICSCTI, 2016, pp. 96–98.

[151] N. Rapin, O. Lund., M. Bernaschi, F. Castiglione, Computational immunology meets bioinformatics: the use of prediction tools for molecular binding in the simulation of the immune system, PLoS One 4 (2010) e9862.

[152] C.D. Fjell, J.A. Hiss, R.E. Hancock, G. Schneider, Designing antimicrobial peptides: form follows function, Nature Reviews. Drug Discovery 11 (2012) 37–51. Available from: https://doi.org/10.1038/nrd3591.

[153] W. Porto, O.N. Silva, O.L. Franco, Prediction and rational design of antimicrobial peptides, in: E. Faraggi (Ed.), Protein Structure, InTech, London, 2012, pp. 377–396.

[154] J.B. Mitchell, Machine learning methods in chemoinformatics, Wiley Interdisciplinary Reviews: Computational Molecular Science 4 (2012) 468–481. Available from: https://doi.org/10.1002/wcms.1183.

[155] K. Hilpert, C.D. Fjell, A. Cherkasov, Short linear cationic antimicrobial peptides: screening, optimizing, and prediction. Available from: https://doi.org/10.1007/978-1-59745-419-3_8 in: L. Otvos (Ed.), Peptide-Based Drug Design, Springer, Berlin, 2008, pp. 127–159.

[156] W.F. Porto, I.C. Fensterseifer, S.M. Ribeiro, O.L. Franco, Joker: an algorithm to insert patterns into sequences for designing antimicrobial peptides antimicrobial peptides, Biochimica et Biophysica Acta General. Subjects 1862 (2018) 2043–2052. Available from: https://doi.org/10.1016/j.bbagen.2018.06.011.

[157] J.A. Hiss, M. Hartenfeller, G. Schneider, Concepts and applications of "natural computing" techniques in de novo drug and peptide design, Current Pharmaceutical Design 16 (2010) 1656–1665. Available from: https://doi.org/10.2174/138161210791164009.
[158] M.H. Cardoso, R.Q. Orozco, S.B. Rezende, G. Rodrigues, K.G.N. Oshiro, E.S. Cândido, et al., Computer-aided design of antimicrobial peptides: are we generating effective drug candidates? Frontiers in Microbiology 10 (2020) 3097. Available from: https://doi.org/10.3389/fmicb.2019.03097.
[159] X. Kang, F. Dong, C. Shi, S. Liu, J. Sun, J. Chen, et al., DRAMP 2.0, an updated data repository of antimicrobial peptides, Scientific Data 6 (1) (2019) 148. Available from: https://doi.org/10.1038/s41597-019-0154-y.
[160] V. Seshadri Sundararajan, M.N. Gabere, A. Pretorius, S. Adam, A. Christoffels, M. Lehväslaiho, et al., DAMPD: a manually curated antimicrobial peptide database, Nucleic Acids Research 40 (Database issue) (2012) D1108–D1112. Available from: https://doi.org/10.1093/nar/gkr1063.
[161] G. Wang, X. Li, Z. Wang, APD3: the antimicrobial peptide database as a tool for research and education, Nucleic Acids Research 44 (2016) (2016) D1087–D1093.
[162] M. Pirtskhalava, A. Gabrielian, P. Cruz, H.L. Griggs, R.B. Squires, D.E. Hurt, et al., DBAASP v.2: an enhanced database of structure and antimicrobial/cytotoxic activity of natural and synthetic peptides, Nucleic Acids Research 44 (D1) (2016) D1104–D1112.
[163] F.H. Waghu, R.S. Barai, P. Gurung, S. Idicula-Thomas, CAMPR3: a database on sequences, structures and signatures of antimicrobial peptides, Nucleic Acids Research (2015). gkv1051v1-gkv1051.
[164] F.H. Waghu, S. Idicula-Thomas, Collection of antimicrobial peptides database and its derivatives: applications and beyond, Protein Science 29 (1) (2020) 36–42. Available from: https://doi.org/10.1002/pro.3714.
[165] X. Zhao, H. Wu, H. Lu, G. Li, Q. Huang, LAMP: a database linking antimicrobial peptides, PLoS One 8 (6) (2013) e66557. Available from: https://doi.org/10.1371/journal.pone.0066557.
[166] S.P. Piotto, L. Sessa, S. Concilio, P. Iannelli, YADAMP: yet another database of antimicrobial peptides, International Journal of Antimicrobial Agents 39 (4) (2012) 346–351. Available from: https://doi.org/10.1016/j.ijantimicag.2011.12.003.
[167] M. Novković, J. Simunić, V. Bojović, A. Tossi, D. Juretić, DADP: the database of anuran defense peptides, Bioinformatics (Oxford, England) 28 (10) (2012) 1406–1407. Available from: https://doi.org/10.1093/bioinformatics/bts141.
[168] F. Ramos-Martín, T. Annaval, S. Buchoux, C. Sarazin, N. D'Amelio, ADAPTABLE: a comprehensive web platform of antimicrobial peptides tailored to the user's research, Life Science Alliance 2 (6) (2019) e201900512. Available from: https://doi.org/10.26508/lsa.201900512.
[169] E.A. Gómez, P. Giraldo, S. Orduz, InverPep: a database of invertebrates antimicrobial peptides, Journal of Global Antimicrobial Resistance 8 (2017) 13–17.
[170] M.D. Luca, G. Maccari, G. Maisetta, G. Batoni, BaAMPs: the database of biofilm-active antimicrobial peptides, Biofouling 31 (2) (2015) 193–199. Available from: https://doi.org/10.1080/08927014.2015.1021340.
[171] X. Su, J. Xu, Y. Yin, X. Quan, H. Zhang, Antimicrobial peptide identification using multi-scale convolutional network, BMC Bioinformatics 20 (2019) 730. Available from: https://doi.org/10.1186/s12859-019-3327-y.

[172] K.C. Chou, Prediction of protein cellular attributes using pseudo-amino acid composition, Proteins 43 (2001) 246–255.
[173] W.R. Atchley, J. Zhao, A.D. Fernandes, T. Druke, Solving the protein sequence metric problem, Proceedings of the National Academy of Sciences of the United States of America 102 (2005) 6395–6400.
[174] Y.D. Cai, K.C. Chou, Predicting membrane protein type by functional domain composition and pseudo-amino acid composition, Journal of Theoretical Biology 238 (2016) 395–400.
[175] K. Kavousi, M. Bagheri, S. Behrouzi, S. Vafadar, F.F. Atanaki, B.T. Lotfabadi, et al., IAMPE: NMR-assisted computational prediction of antimicrobial peptides, Journal of Chemical Information and Modeling 60 (10) (2020) 4691–4701.
[176] D. Nagarajan, T. Nagarajan, N. Roy, O. Kulkarni, S. Ravichandran, M. Mishra, et al., Computational antimicrobial peptide design and evaluation against multidrug-resistant clinical isolates of bacteria, Journal of Biological Chemistry 293 (2018) 3492–3509.
[177] R. Hammam, I. Fliss, Current trends in antimicrobial agent research: chemo- and bioinformatics approaches, Drug Discovery Today 15 (2010) 540–546.
[178] L. Goldschmidt, P.K. Teng, R. Riek, D. Eisenberg, The amylome, all proteins capable of forming amyloid-like fibrils, Proceedings of the National Academy of Sciences of the United States of America 107 (2010) 3487–3492.
[179] S. Pronk, S. Páll, R. Schulz, P. Larsson, P. Bjelkmar, R. Apostolov, et al., GROMACS 4.5: a high-throughput and highly parallel open-source molecular simulation toolkit, Bioinformatics (Oxford, England) (2013) 845–854.
[180] A. Daina, O. Michielin, V. Zoete, SwissADME: a free web tool to evaluate pharmacokinetics, drug-likeness and medicinal chemistry friendliness of small molecules, Scientific Reports 7 (2017) 42717.
[181] E.Y. Lee, G.C.L. Wong, A.L. Ferguson, Machine learning-enabled discovery and design of membrane-active peptides, Bioorganic & Medicinal Chemistry 6 (2008) 2708–2718.

# Recent advances and challenges in peptide drug development

**15**

N.K. Hemanth Kumar[1,2], K. Poornachandra Rao[3,4],
Rakesh Somashekaraiah[3], Shobha Jagannath[1] and M.Y. Sreenivasa[3]
[1]Department of Studies in Botany, University of Mysore, Mysuru, Karnataka, India,
[2]Department of Botany, Yuvarajas College, University of Mysore, Mysuru, Karnataka, India,
[3]Department of Studies in Microbiology, University of Mysore, Mysuru, Karnataka, India,
[4]Department of Analytical Food Microbiology, VIMTA Labs. Ltd., Hyderabad, Telangana, India

## 15.1 Introduction

Proteins are considered as the basic building blocks of amino acids. The amino acids were coded by genes resulting in the formation of the peptides, further these peptides form into various proteins. In addition, proteins also play a significant role in biological processes such as catalyzing the reactions, helps in transportation of molecules, signal transduction between cells, and mediate the immune reactions to the diverse pathogens. The fundamental basic biological processes like DNA replication, transcription, and translation depend on the functions of specific proteins. All these biological activities are regulated through the formation of protein complexes via protein–protein interactions [1,2].

The peptides are one of the distinctive classes of pharmaceutically interesting compounds, as vital signaling molecules for several functions. Peptides provide a way for therapeutic interference that is closely mimics the natural pathways. Certainly, many peptide drugs are fundamentally used in substitution therapy, where they provide the peptide hormone, where the endogenous levels of peptides are insufficient. This was first time exemplified by the isolation and use of insulin in the 1920s in diabetic condition where the persons who were not able to produce adequate amount of the hormone [3]. Conventionally peptides were considered as an unfortunate option for drug candidates due to their little bioavailability and high susceptibility to proteolysis enzymes [4]. However, the peptides were considered as significant natural mediators of biological processes [5] and around 60 wellbeing products from peptides have been approved by the Food and Drug Administration (FDA) with international sales estimated at US$14.7b in 2011 [6,7]. Action by highly active metathesis catalysts has found its use in the production of complex products [8] (Fig. 15.1). Substitution of amino acids is one of the mechanisms of protein delivery systems. Replacement of two extensively explored classes of acyclic β-amino acid into different β-hairpin loop prototypes within a β-sheet-rich protein tertiary structure, sequence structure stability relationships that can be generally extensible to prospect protein design efforts connecting these residues (Fig. 15.2) [9].

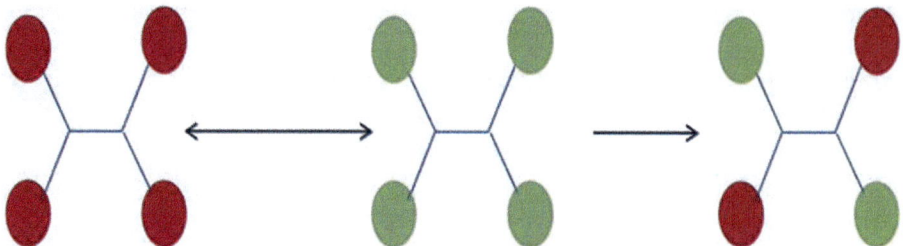

**Figure 15.1** Olefeinic replacement.

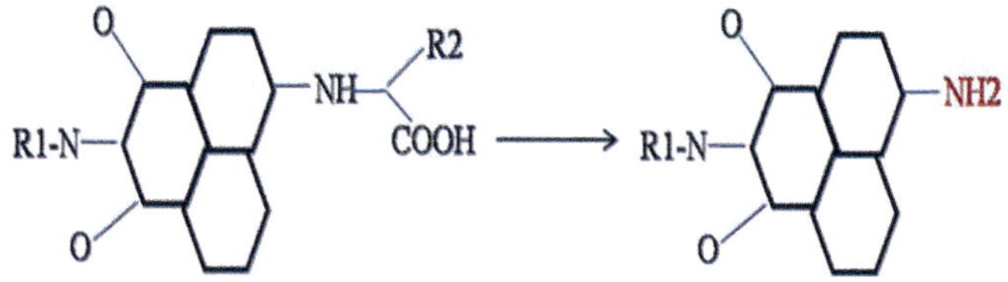

**Figure 15.2** Substitution of α-amino acid.

Antimicrobial peptides belong to a different group of molecules produced by cellular tissues in a wide array of organisms. These peptides demonstrate potent antimicrobial activity and can readily be mobilized to neutralize a wide range of microorganisms, including bacteria, viruses, protozoa, and fungi [10,11]. Besides, this class of antimicrobials has shown hopeful endotoxin neutralization properties [12]. Antimicrobial peptides are produced by multicellular organisms as a defense mechanism against pathogenic microorganisms. Revolutionary studies have led to the innovation of different types of hosts' defense peptides which includes defensins, magainins, cecropins, and cathelicidins with surprisingly diverse structures and bioactivity [13].

Nanotechnology employs two major approaches in encapsulating antimicrobial peptides, which includes direct delivery and nondirect delivery. Directed delivery is referred to as active targeting which involves the modification of the surface chemistry of the nanocarrier with ligands or other moieties to permit interaction of the nanocarrier and the intended site [14]. The nondirected one which involves conventional nanodelivery systems which do not own surface modification of their structure to guide the nanocarrier; this can be manipulated through controlling the nanocarrier's size and shape [15].

The exploitation of peptides as therapeutic agents has developed in over a period and thus continues to develop with modifications in drug design, development, and the way of treatments. Peptides which are isolated from the natural sources, like insulin and adrenocorticotrophic hormone (ACTH), offered life-protecting medicines during the initial period of the 20th century. When the elucidation of the peptides sequences and chemical synthesis were becoming possible in the 1950s, where the hormones oxytocin and vasopressin entered to the pharmaceutical field for clinical use [16]. Protein—protein interactions are considered as the basis for all the physiological processes which are frequently comprised of many activated receptors that are directly or indirectly control enzymatic activities [17]. The peptides are small molecules that obstruct with protein—protein interactions; therefore, they have high value in the field of pharmacology as therapeutic agents due to their capability to modify the protein interaction which are associated with the diseases [18]. The conjugation peptides of amphiphilic variants of elastin-like polypeptide possessed less water-soluble compounds like *p*-phenylenevinylene oligomers and polypyridine and metal complexes (Fig. 15.3) [19]. Bile acids have gained significant interest in the drug delivery tools due to their exclusive physicochemical properties. The major benefit of bile acids as drug delivery enhancers is their capability to act as drug-solubilizing agents (Fig. 15.4). Hence, bile acids improve the bioavailability of drugs with limiting absorption factors which include either poor aqueous solubility or the low membrane permeability [20]. Phonophoresis is another approach used for the improved transdermal delivery. In this mechanism, the ultrasound is applied through a coupling-contact agent to the target [21]. The drug absorption is enhanced through thermal effect of ultrasonic waves and subsequent temporary alterations in physical structure of skin (Fig. 15.5).

## 15.2 Historical overview of peptide drug development

The development of peptide based on chemical methods has over 100 years of history which has been presented in Table 15.1. Ever since, initial isolation and commercialization of insulin in 1920s, the peptide drugs have significantly revolutionized the modern pharmaceutical industry. As the progress in the field of recombinant DNA and protein

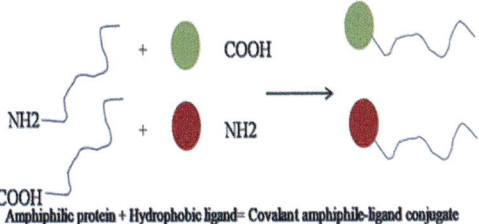

**Figure 15.3** Conjugation of amphiphilic proteins.

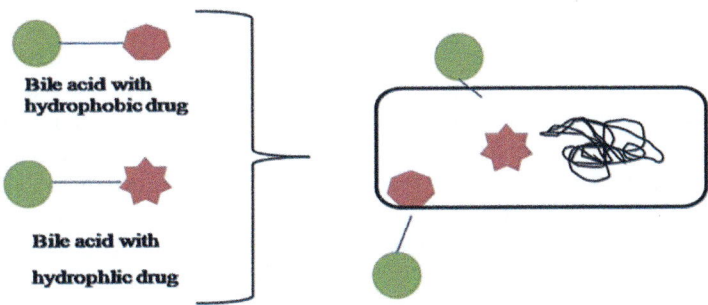

**Figure 15.4** Bile acid as drug delivery systems (enhancers).

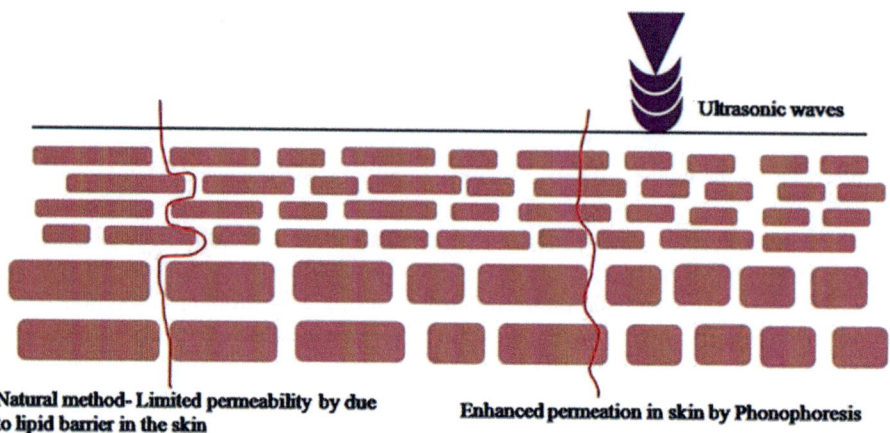

**Figure 15.5** Phonophoresis in drug delivery system.

**Table 15.1** Development of peptide-based drugs.

| | |
|---|---|
| 1881 | Invention of dipeptide by Theodor Curtius |
| 1901 | First report on synthesized dipeptide |
| 1932 | Invention of the peptide belongs to group Z |
| 1953 | First polypeptide oxytocin synthesizes by chemical method |
| 1955 | Invention of the $N,N'$-dicyclohexyl carbodiimide derivative (DCC) the coupling reagents |
| 1957 | Invention of the Boc protecting group |
| 1957–58 | Invention of the column chromatographic technique |
| 1963 | Invention of the solid phase peptide synthesis (SPPS) |
| 1970 | Invention of the Fmoc protecting group |
| 1970–73 | Discovery of resins butylated hydroxyanisole (BHA) and Wang |
| 1976 | Introduction of preparative high performance liquid chromatography (HPLC) |
| 1990 | Development of native chemical ligation method |
| 1999 | Synthesis of large scale of Enfuvirtide (Fuzeon or T-20) |

isolation and purification technology, the human recombinant insulin has been replaced with animal-based tissue-derived insulin product. From the last two decades, approximately 30 peptide drugs have been approved. Survey at global level regarding the peptide therapeutics estimated the annual growth rate of 9.1% from 2016 to 2024 and marketing of peptide drugs to go beyond 70 billion USD in 2019 [22]. The vigorous growth in the pharmaceutical industry is endorsed to the anticipated increasing in the prevalence of metabolic disorders. The peak selling peptide drugs for the treatment of metabolic disorders such as glucagon-like peptide both had at least 2 billion USD sales per annum [23]. Depo-Foam technologies consist of novel multivesicular liposomes characterized by their special structure of nonconcentric aqueous chambers encircled by a network of lipid membranes [24]. This technology might be used to construct sustained-release formulations of therapeutic proteins with maximum loading capacity (Fig. 15.6).

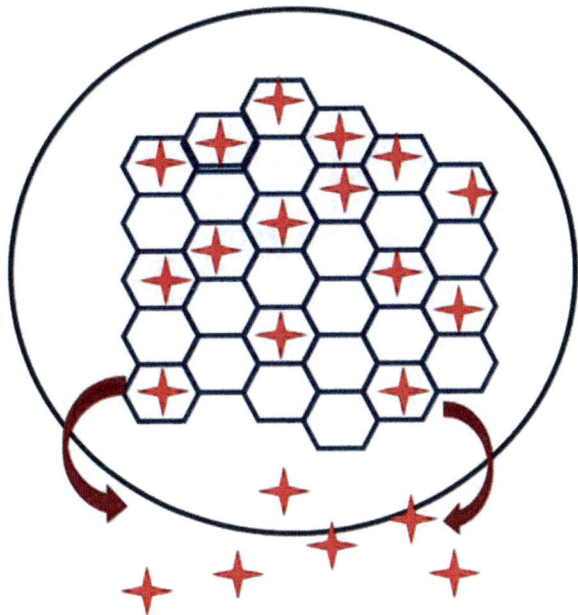

**Figure 15.6** Depo-Foam technology as drug delivery system.

## 15.3 Basic drawbacks of peptide drugs

Different peptide drugs are generally manufactured artificially, whereas the natural polypeptides like growth factors, hormones, and neurotransmitters are known to plant a key role in normal physiology and development. The peptide drugs have their own drawbacks, which includes in vivo instability and impermeability to the membrane and proteolytic deprivation of peptide in serum limits where the half-life of the drug decreases the bioavailable concentration of the drug. Regular dosing might be needed to sustain the drug at a clinically successful concentration. Several chemical modifications have been employed to avoid the proteolytic degradation and to enhance the in vivo half-life period of the peptide drugs [23].

## 15.4 Present approaches toward the discovery of protein–protein modulators

### 15.4.1 High-throughput screening

High-throughput screening (HTS) is a suitable method for the discovery of typical drug targets. Generally, it has been used to recognize the compounds that mark the hotspots of protein–protein interaction (PPI) interfaces. The compound library is used for the selection the predictable targets may not be appropriate for the selection of PPI modulators. However, HTS has been proved to be helpful in the recognition of molecules at the primary stage [25,26].

## 15.4.2 Fragment-based drug discovery

The objective of the fragment-based drug discovery (FBDD) is to recognize the molecular fragments from databases. The FBDD is a better approach when compared to HTS for PPIs modulators scheming since the PPI interface frequently consists of irregular hotspots. Several techniques can be employed for the discovery and validation of fragments which includes nuclear magnetic resonance (NMR), surface plasmon resonance (SPR), mass spectroscopy (MS), and X-ray crystallography [27–29]. Once the hits of the fragments were identified, the fragment optimization, linking, and self-gathering might be used to get the hits. The molecular weight of the fragments is very low, and the affinity is also comparatively low [30,31]. The techniques like NMR and X-ray crystallography provide structural information for the hit's optimization; therefore, the FBDD is relevant for the targets with unknown structures [32]. Therapeutic proteins have gained vital importance as they execute critical roles in different biological processes. The deliverance of therapeutic proteins to intention sites is, conversely, challenging due to their intrinsic sensitivity to diverse environmental conditions (Fig. 15.7). Polymeric nanoparticles can put forward not only physical protection from environmental but also targeted relief of such proteins to specific sites [33].

## 15.4.3 Structure-based design

Structure-based drug design is becoming a fundamental tool for the quicker and more cost-efficient method led to the discovery comparative to the traditional method. The genomic, proteomic, and structural studies have paved the way for hundreds of new target molecules and opportunity for prospect drug discovery [34]. Ever since most

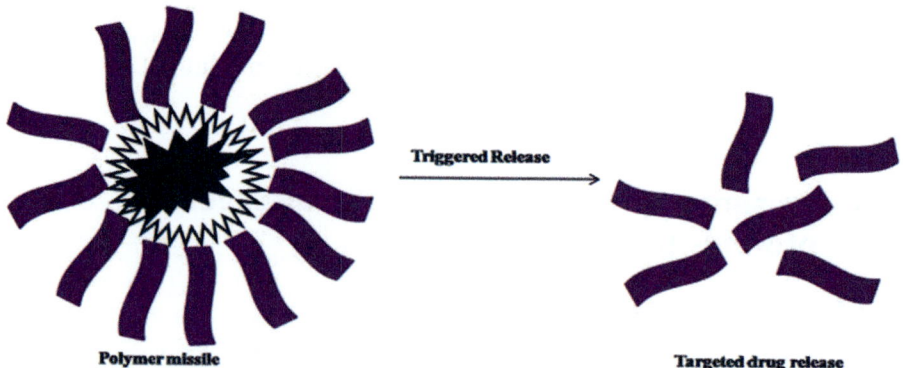

**Figure 15.7** Polymer-based protein delivery systems.

of the PPIs lack endogenous small signal molecule called ligands, it is difficult to design the connect PPI modulators. Nevertheless, the hotspots give significant structural data and a foundation for the rational design of PPI modulators. Currently, there are two designing techniques available for structure-based design in PPI modulators. The first one is based on the hotspots structure during bioisosterism, and the new modulators can be obtained [31]. The second is peptidomimetic design which mostly depends upon the phage display and computer modeling to simulate the secondary structure of the key peptides in PPIs. Additionally, some of the small molecules and binding peptides were synthesized depending on the stability of α-helix structure, is the most frequently identified secondary structure, several PPI modulators have been effectively developed based on the α-helix structure [35,36].

## 15.5 Peptides and protein–protein interactions

### 15.5.1 Potential developments for intrusive peptides

The interactions among the protein–protein complexes are considered as the central part of the normal and pathogenic cell biology and physiology. Irregular PPIs drive signaling changes that support some of the biological processes which include infection, neurodegeneration, chronic inflammation, and cardiovascular disease. Therefore the protein interaction surfaces are considered for the development of therapeutic targets as peptides molecules possess more benefits compared to smaller molecules particularly in this area. Several hopeful interfering peptides are presently under clinical trials [37]. A peptide drug (XG-102, Brimapitide) depending upon the N-terminal c-Jun sequence compete with endogenous c-Jun for interaction with c-Jun N-terminal kinases (JNK), repress JNK-driven inflammation is under a phase III trial [38]. Interfering peptides with helical structures bind to the protein interacting surfaces and they show hopeful interaction-blocking activity, possibly due to their superior stability and protease resistance [31]. Some of the peptide drugs structures were modified and these peptide drugs are called as peptidomimetics, and they are designed in such a way that they mainly interrupt the flat and large interfaces of their relevant targets [39].

### 15.5.2 Computational and experimental methods for determining protein–protein interactions

The PPIs have been determined experimentally by different biophysical techniques like NMR spectroscopy, X-ray crystallography, isothermal titration calorimetry (ITC), surface plasma resonance (SPR), spectrophotometric and fluorescence spectroscopy assay. The data which is obtained by these techniques have played a vital role in enhancement of our knowledge regarding the interaction and the secondary and tertiary structure of proteins. These analytical techniques, conversely, are often sustained and generally used to study one specific PPI at a time. X-ray diffraction (XRD) technique is certainly very powerful for the structural analysis of protein crystal. Several proteins are unable to crystallize well, or do not crystallize apart from as smaller protein domains. The NMR spectroscopy technique provides structure of protein complex in a lower resolution compared to the XRD. The calorimetric studies give the information regarding the amount energy and affinity properties, but are unable to identify a specific interaction the way NMR or XRD identifies. Several docking methods have been developed to obtain accurate predictions of surface charge and protein structure. This technique even can be applied for proteins even though the crystal structure of the protein is not available. These techniques are applied for the in silico identification of protein interaction surfaces and binding study of potential blocking peptides [23].

### 15.5.3 Computer-assisted docking strategies

The computational PPI docking studies provide the information for designing the drugs at the atomic level, as a few of the docking types can be developed in very short duration. This can be done by rigid-body docking technique, where two interacting proteins are considered for chemical and geometric orientation fit. Z-DOCK is a distinctive rigid-body protein docking software that normally produces precise predictions of PPI when suitable scoring scaffolds are provided [40]. Due to the accessibility and difficulty of a range of scoring parameters from mainly flexible docking methods, several docking programs have been introduced over the past decades. ATTRACT is a well-established PPI forecast server with authoritative toolkits [41].

### 15.5.4 Structural-based predictions

Several computers assisted docking techniques; principally the flexible body docking technique required structural information along with the number of hydrogen bonds and allosteric effects for determination of free energies to create more precise prediction on the binding affinities flanked by the interacting proteins [42]. On the other hand, the sequence-based strategies depend upon the functional information in widely accessible databases to produce predictions of binding affinity. PPA-Pred, for instance, developed a model that depended on sequence properties by classifying protein–protein complexes based on their biological functions [43]. In fact, the technique sequence-based strategies also make use of learning machines to improve their prophecy confidence over time [43].

## 15.6 Innovations and computational methods for peptide–protein interactions

### 15.6.1 Selection of preliminary peptide scaffolds

Analogous to PPIs, the available information on structural aspects for a drug target has limited prediction accuracy for peptide–protein interactions. Due to the nonavailability of protein costructures, several studies exploit accessible information from structural databases such as the Protein Data Bank (PDB) to determine sequence-binding motifs for designing the peptides [44]. Several well-characterized natural peptides had been isolated from proteins and established to conserve their unique functions as well as structural scaffolds. Repeated Arg-Gly-Asp (RGD) motifs were formed as a resultant of cell attachment domain of fibronectin that adheres to integrins a membrane-bound receptor protein and promotes the growth and differentiation [45]. The ability of RGD peptides in mimic the functions of their natural protein have served as a hopeful approach for structural and functional analyses of proteins. Recently in silico modeling-based design has emerged as a influential technique to new peptidic leads identification from natural proteins. Another exciting progress is the discovery of microtubule-binding peptides. Microtubules are hollow tubular proteins made up of $\alpha/\beta$ tubulin dimers with important nanodevice implications recognized to their participation in several eukaryotic cell functions along with tumor progression [46]. Another interesting innovation of a tetrapeptide Ser-Leu-Arg-Pro (SLRP) from a peptide database was revealed to disturb microtubule assembly function and lead to apoptosis of cancer cells with help of computation docking method Autodock Vina [47].

### 15.6.2 Molecular docking for peptide—protein interactions

The flourishing molecular docking of the structural pose of peptide—protein interactions has been mainly reliant on the number of structural scaffolds accessible regarding the communication complex. The spectacular augmentation in numbers of peptide—protein structures accessible in PDB database has really facilitate the progress of more powerful docking methods. The peptide—protein docking studies are generally divided into local or global docking depends on the degree of structural data provided as inputs.

### 15.6.3 Docking methods at local and global levels

#### 15.6.3.1 Local docking methods

Local docking is the usually used approach as it searches for a prospective binding pose for a peptide at a binding site in a determined structure of its target receptor. Several methods have the capacity to enhance the primary quality at atomic resolution within 1–2 Å root mean square deviations (RMSD) of the peptide conformation. Rosetta FlexPepDock, DynaRock, and PepCrawler are the majority accepted methods that give diverse approaches of important peptide-binding sites [48]. Rosetta FlexPepDock is a Monte Carlo-based method that reduces the optimization steps to obtain high-quality conformational sampling for binding motifs [49]. The sampling methods allow the achievement of near native peptide conformation turn into necessary before performing local docking [50]. In recent times, HADDOCK peptide docking method was used to theorize the unnecessity of previous backbone data in local docking by secondary structure [51].

### 15.6.3.2 Global docking methods

In contrast to local docking methods that searches for the peptide-binding site, these methods moreover look for the peptide-binding site at the targeted protein. This is frequently used when no previous information is accessible on binding sites. A spatial position specific scoring matrix (PSSM) was used in mounting the PepSite method to distinguish possible binding sites with predictable site for each residue [52]. The diverse degrees of peptide side-chain flexibility, however, provide flexible-body docking enormously ineffective. The well-established protocols for global peptide−protein docking consequently relay on rigid-body docking subsequent gaining of input peptide conformation. Many global docking methods are competent enough for predicting the peptide conformation from a specified query sequence. ClusPro and ATTRACT, for instance, use a established motif set of pattern conformations for threading query sequences [41,53]. Further global docking methods like PeptiMap and CABS-Dock also provide automatic docking simulation with diverse algorithms [29]. One more newly developed software, HPEPDOCK, used an assembly of peptide conformations for global docking and observed considerably elevated achievement rates as well as lesser simulation time necessary than ATTRACT [41,54].

### 15.6.4 Template-based docking method

Template-based docking methods are also called as comparative docking study. They are generally used to known structures as template scaffolds [55]. It is currently being considered as a novel group in peptide−protein docking due to the fast augmentation in the numeral of peptide−protein structures deposited in PBD, which significantly accelerate advances and designs in simulation algorithms. The GalaxyPepDock is a famous server that carry out such similarity based docking study by incisive templates of maximum similarity and constructs models using energy optimization to permit more precise predictions on structural elasticity among interacting complexes [56].

## 15.7 Conclusion

In present days, peptides have gained a lot of consideration and the many approved peptide biotherapeutics has been increased over the recent decades. Therefore this has been an attractive advance toward their ability to bind with larger peptide sites than small molecules. The widely available computer-assisted binding prediction software has led to effective coherent designs of novel peptide drugs of elevated therapeutic properties. Regardless of recent development in the field of various docking modeling protein−protein and peptide−protein structures, most important challenges are remained. Growing interests in peptide biotherapeutics have paved the way for rapid advances in both biocomputational and chemical methods.

# References

[1] S. Ferrari, F. Pellati, M.P. Costi, Protein–protein interaction inhibitors: case studies on small molecules and natural compounds, Disruption of Protein-Protein Interfaces, Springer, Berlin, Heidelberg, 2013, pp. 31–60.
[2] U. Stelzl, U. Worm, M. Lalowski, C. Haenig, F.H. Brembeck, H. Goehler, et al., A human protein-protein interaction network: a resource for annotating the proteome, Cell 122 (2005) 957–968.
[3] F.G. Banting, C.H. Best, J.B. Collip, W.R. Campbell, A.A. Fletcher, Pancreatic extracts in the treatment of diabetes mellitus, Canadian Medical Association Journal 12 (1922) 141.
[4] L. Otvos Jr, J.D. Wade, Current challenges in peptide-based drug discovery, Frontiers in Chemistry 2 (2014) 62.
[5] A.J. Kastin, Handbook of Biologically Active Peptides, Academic Press, Amsterdam; London, 2006.
[6] M. Hamzeh-Mivehroud, A.A. Alizadeh, M.B. Morris, W.B. Church, S. Dastmalchi, Phage display as a technology delivering on the promise of peptide drug discovery, Drug Discovery Today 18 (2013) 1144–1157.
[7] A.A. Kaspar, J.M. Reichert, Future directions for peptide therapeutics development, Drug Discovery Today 18 (2013) 807–817.
[8] H. Zhao, Z.Y. Lin, L. Yildirimer, A. Dhinakar, X. Zhao, J. Wu, Polymer-based nanoparticles for protein delivery: design, strategies, and applications, Journal of Materials Chemistry B 4 (2016) 4060–4071.
[9] M.S. Messina, H.D. Maynard, Modification of proteins using olefin metathesis, Materials Chemistry Frontiers 4 (2020) 1040–1051.
[10] Y. Shai, Mode of action of membrane active antimicrobial peptides, Peptide Science 66 (2002) 236–248.
[11] K.A. Brogden, Antimicrobial peptides: pore formers or metabolic inhibitors in bacteria? Nature Review Microbiology 3 (2005) 238–250.
[12] O. Fleitas Martinez, M.H. Cardoso, S.M. Ribeiro, O.L. Franco, Recent advances in anti-virulence therapeutic strategies with a focus on dismantling bacterial membrane microdomains, toxin neutralization, quorum-sensing interference, and biofilm inhibition, Frontiers in Cellular and Infection Microbiology 9 (2019) 74.
[13] R.E.W. Hancock, H.G. Sahl, Antimicrobial and host defense peptides as new anti-infective therapeutic strategies, Nature Biotechnology 24 (2006) 1551–1557.
[14] A.L.F. Lamprecht, N. Ubrich, H. Yamamoto, U. Schäfer, H. Takeuchi, P. Maincent, et al., Biodegradable nanoparticles for targeted drug delivery in treatment of inflammatory bowel disease, Journal of Pharmacology and Experimental Therapeutics 299 (2001) 775–781.
[15] G. Storm, S.O. Belliot, T. Daemen, D.D. Lasic, Surface modification of nanoparticles to oppose uptake by the mononuclear phagocyte system, Advanced Drug Delivery Reviews 17 (1995) 31–48.
[16] Y.H. Lau, P. De Andrade, Y. Wu, D.R. Spring, Peptide stapling techniques based on different macrocyclisation chemistries, Chemical Society Reviews 44 (2015) 91–102.
[17] Z. Yan, J. Wang, Specificity quantification of biomolecular recognition and its implication for drug discovery, Scientific Reports 2 (2012) 1–7.
[18] D. Thomas, A Big Year for Novel Drug Approvals, Biotechnology Innovation Organization, Washington, DC, 2013.

[19] D.E. Mortenson, D.F. Kreitler, N.C. Thomas, I.A. Guzei, S.H. Gellman, K.T. Forest, Evaluation of β-amino acid replacements in protein loops: effects on conformational stability and structure, Chembiochem: a European Journal of Chemical Biology 19 (2018) 604–612.
[20] A.M. Lillo, C.L. Lopez, T. Rajale, H.J. Yen, H.D. Magurudeniya, M.L. Phipps, et al., Conjugation of amphiphilic proteins to hydrophobic ligands in organic solvent, Bioconjugate Chemistry 29 (2018) 2654–2664.
[21] N. Pavlovic, S. Goločorbin-Kon, M. Đanić, B. Stanimirov, H. Al-Salami, K. Stankov, et al., Bile acids and their derivatives as potential modifiers of drug release and pharmacokinetic profiles, Frontiers in Pharmacology 9 (2018) 1283.
[22] T.M. Research, Global industry analysis, size, share, growth, trends and forecast, Peptide Mark (2016) 2016–2024.
[23] A.C.L. Lee, J.L. Harris, K.K. Khanna, J.H. Hong, A comprehensive review on current advances in peptide drug development and design, International Journal of Molecular Sciences 20 (2019) 2383.
[24] P. Jani, P. Manseta, S. Patel, Pharmaceutical approaches related to systemic delivery of protein and peptide drugs: an overview, International Journal of Pharmaceutical Sciences Review and Research 12 (2012) 42–52.
[25] J.G. Allen, M.P. Bourbeau, G.E. Wohlhieter, M.D. Bartberger, K. Michelsen, R. Hungate, et al., Discovery and optimization of chromenotriazolopyrimidines as potent inhibitors of the mouse double minute 2 − tumor protein 53 protein − protein interaction, Journal of Medicinal Chemistry 52 (2009) 7044–7053.
[26] B.L. Grasberger, T. Lu, C. Schubert, D.J. Parks, T.E. Carver, H.K. Koblish, et al., Discovery and cocrystal structure of benzodiazepinedione HDM2 antagonists that activate p53 in cells, Journal of Medicinal Chemistry 48 (2005) 909–912.
[27] P.J. Hajduk, J. Greer, A decade of fragment-based drug design: strategic advances and lessons learned, Nature Reviews. Drug Discovery 6 (2007) 211–219.
[28] T.V. Magee, Progress in discovery of small-molecule modulators of protein−protein interactions via fragment screening, Bioorganic and Medicinal Chemistry Letters 25 (2015) 2461–2468.
[29] M. Kurcinski, M. Jamroz, M. Blaszczyk, A. Kolinski, S. Kmiecik, CABS-dock web server for the flexible docking of peptides to proteins without prior knowledge of the binding site, Nucleic Acids Research 43 (2015) W419–W424.
[30] D.C. Rees, M. Congreve, C.W. Murray, R. Carr, Fragment-based lead discovery, Nature Reviews. Drug Discovery 3 (2004) 660–672.
[31] C. Sheng, G. Dong, Z. Miao, W. Zhang, W. Wang, State-of-the-art strategies for targeting protein−protein interactions by small-molecule inhibitors, Chemical Society Reviews 44 (2015) 8238–8259.
[32] C.W. Chung, A.W. Dean, J.M. Woolven, P. Bamborough, Fragment-based discovery of bromodomain inhibitor's part 1: inhibitor binding modes and implications for lead discovery, Journal of Medicinal Chemistry 55 (2012) 576–586.
[33] Q. Ye, J. Asherman, M. Stevenson, E. Brownson, N.V. Katre, DepoFoam™ technology: a vehicle for controlled delivery of protein and peptide drugs, Journal of Controlled Release 64 (2000) 155–166.
[34] M. Batool, B. Ahmad, S. Choi, A structure-based drug discovery paradigm, International Journal of Molecular Sciences 20 (2019) 2783.
[35] B.N. Bullock, A.L. Jochim, P.S. Arora, Assessing helical protein interfaces for inhibitor design, Journal of the American Chemical Society 133 (2011) 14220–14223.

[36] J.M. Mason, Design and development of peptides and peptide mimetics as antagonists for therapeutic intervention, Future Medicinal Chemistry 2 (2010) 1813–1822.
[37] M.A. Warso, J.M. Richards, D. Mehta, K. Christov, C. Schaeffer, L.R. Bressler, et al., A first-in-class, first-in-human, phase I trial of p28, a non-HDM2-mediated peptide inhibitor of p53 ubiquitination in patients with advanced solid tumours, British Journal of Cancer 108 (2013) 1061–1070.
[38] C. Chiquet, F. Aptel, C. Creuzot-Garcher, J.P. Berrod, L. Kodjikian, P. Massin, et al., Postoperative ocular inflammation: a single subconjunctival injection of XG-102 compared to dexamethasone drops in a randomized trial, American Journal of Ophthalmology 174 (2017) 76–84.
[39] A. Ellert-Miklaszewska, K. Poleszak, B. Kaminska, Short peptides interfering with signaling pathways as new therapeutic tools for cancer treatment, Future Medicinal Chemistry 9 (2017) 199–221.
[40] G. Sliwoski, S. Kothiwale, J. Meiler, E.W. Lowe, Computational methods in drug discovery, Pharmacological Reviews 66 (2014) 334–395.
[41] S.J. De Vries, J. Rey, C.E. Schindler, M. Zacharias, P. Tuffery, The pepATTRACT web server for blind, large-scale peptide–protein docking, Nucleic Acids Research 45 (2017) W361–W364.
[42] J.M. Choi, A.W. Serohijos, S. Murphy, D. Lucarelli, L.L. Lofranco, A. Feldman, et al., Minimalistic predictor of protein binding energy: contribution of solvation factor to protein binding, Biophysical Journal 108 (2015) 795–798.
[43] K. Yugandhar, M.M. Gromiha, Protein–protein binding affinity prediction from amino acid sequence, Bioinformatics (Oxford, England) 30 (2014) 3583–3589.
[44] H.M. Berman, T. Battistuz, T.N. Bhat, W.F. Bluhm, P.E. Bourne, K. Burkhardt, et al., The protein data bank, Acta Crystallographica Section D: Biological Crystallography 58 (2002) 899–907.
[45] S. Takahashi, M. Leiss, M. Moser, T. Ohashi, T. Kitao, D. Heckmann, et al., The RGD motif in fibronectin is essential for development but dispensable for fibril assembly, Journal of Cell Biology 178 (2007) 167–178.
[46] S. Mohapatra, A. Saha, P. Mondal, B. Jana, S. Ghosh, A. Biswas, et al., Synergistic anticancer effect of peptide-docetaxel nanoassembly targeted to tubulin: toward development of dual warhead containing nanomedicine, Advanced Healthcare Materials 6 (2017) 1600718.
[47] B. Jana, P. Mondal, A. Saha, A. Adak, G. Das, S. Mohapatra, et al., Designed tetrapeptide interacts with tubulin and microtubule, Langmuir: the ACS Journal of Surfaces and Colloids 4 (2018) 1123–1132.
[48] I. Antes, DynaDock: a new molecular dynamics-based algorithm for protein–peptide docking including receptor flexibility, Proteins: Structure, Function, and Bioinformatics 78 (2010) 1084–1104.
[49] B. Raveh, N. London, O. Schueler-Furman, Sub-angstrom modeling of complexes between flexible peptides and globular proteins, Proteins: Structure, Function, and Bioinformatics 78 (2010) 2029–2040.
[50] B. Raveh, N. London, L. Zimmerman, O. Schueler-Furman, Rosetta FlexPepDock ab-initio: simultaneous folding, docking and refinement of peptides onto their receptors, PLoS One 6 (2011) e18934.
[51] M. Trellet, A.S. Melquiond, A.M. Bonvin, A unified conformational selection and induced fit approach to protein-peptide docking, PLoS One 8 (2013) e58769.
[52] L.G. Trabuco, S. Lise, E. Petsalaki, R.B. Russell, PepSite: prediction of peptide-binding sites from protein surfaces, Nucleic Acids Research 40 (2012) W423–W427.

[53] K.A. Porter, B. Xia, D. Beglov, T. Bohnuud, N. Alam, O. Schueler-Furman, et al., ClusProPeptiDock: efficient global docking of peptide recognition motifs using FFT, Bioinformatics (Oxford, England) 33 (2017) 3299–3301.
[54] P. Zhou, B. Jin, H. Li, S.Y. Huang, HPEPDOCK: a web server for blind peptide–protein docking based on a hierarchical algorithm, Nucleic Acids Research 46 (2018) W443–W450.
[55] M.F. Lensink, S. Velankar, S.J. Wodak, Modeling protein–protein and protein–peptide complexes: CAPRI 6th edition, Proteins: Structure, Function, and Bioinformatics 85 (2017) 359–377.
[56] H. Lee, L. Heo, M.S. Lee, C. Seok, GalaxyPepDock: a protein–peptide docking tool based on interaction similarity and energy optimization, Nucleic Acids Research 43 (2015) W431–W435.

# Future perspective of peptide antibiotic market

## 16

B. Arun[1], E.P. Rejeesh[2] and N. Megha Rani[3]
[1]Department of Biotechnology and Microbiology, Kannur University, Kannur, Kerala, India, [2]Department of Pharmacology, Mount Zion Medical College, Adoor, Kerala, India, [3]Department of Pharmacology, Yenepoya Medical College, Mangalore, Karnataka, India

## 16.1 Introduction

Antimicrobial peptides (AMPs) have a wide range of applications in cosmetics, pharmaceuticals, biotech products, agriculture, feed additives, and others. The AMPs market size was US$ 5 million in 2020 and at a compound annual growth rate (CAGR) of 5.3% during 2020–26. Earlier it was predicted that by 2020 the peptide drugs market will reach 40 billion US dollars per year but the outcome was higher than these predictions [1–3], which is about 10% of the global pharmaceutical market.

Since the discovery of lysozyme in 1922 by Alexander Fleming, several AMPs and antibiotics have hit the market. The antimicrobial peptide database (APD3) lists around 3324 AMPs [4]. The US FDA database as of January 2020 shows that about 852 peptides and protein therapeutics are discovered out of which only 239 are validated for clinical use in humans. Among these 239 peptide drugs, 27 small peptides [4,5] are AMPs that are approved by the US FDA. These include gramicidin D, daptomycin (and its derivative cubicin), vancomycin, oritavancin, dalbavancin, and telavancin [5,6]. Colistin is an FDA-approved AMP not mentioned in the THPdb.

Peptide antibiotics are polypeptide chains that are anti-infective and antitumor antibiotics. They can be divided into ribosomally synthesized and nonribosomally synthesized peptide antibiotics [5–7]. Peptide drugs based on their molecular weight are also classified as traditional small molecules ($<500$ Da) and larger biologics ($>5000$ Da) [8–10]. The larger molecular weight makes it have poor oral bioavailability and hence not available orally. The increasing demand for peptide antibiotics is because of their better efficacy and specificity, low toxicity, and less prone to antibiotic resistance [4].

## 16.2 Global antimicrobial peptides market overview

In the present scenario globally, death due to drug-resistant diseases each year is around 700,000 which includes 230,000 deaths per year due to multidrug-resistant tuberculosis. Frequent diseases like sexually transmitted infections, respiratory tract infections, and urinary tract infections are becoming difficult to treat because of the emergence of multidrug-resistant strains of microorganisms and also medical procedures are becoming much more precarious, and our food systems are increasingly hazardous [11]. This global increase in the number of drug-resistant cases is calling for new antimicrobial agents.

Out of the AMPs used therapeutically, vancomycin is prescribed the most and occupies the major market share. The vancomycin market is predicted to grow to US$ 465.1 million by the year 2027 globally, which is estimated to be US$ 313.7 million in the year 2020, which is a 5.8% CAGR growth for 2020−27. The market in the United States has reached US$ 84.5 million, whereas the Chinese vancomycin market is predicted to grow at a 9.4% CAGR to reach US$ 102.3 million by 2027. Japanese and Canadian markets are predicted to grow at 3.1% and 4.6%, respectively during 2020−27. The German market is predicted to grow at a 4% CAGR [12].

## 16.3 Applications of antimicrobial peptide

### 16.3.1 Prospects in medicine

Pharmacotherapeutic uses of AMPs in healthcare like invasive procedures, and wound and ophthalmic applications are in the infancy stage now. Only a few AMPs are FDA-approved; like gramicidin, daptomycin, and colistin, but several AMPs are at various stages of drug discovery and trial pipeline. A few examples of therapeutic applications of AMPs under development are below.

AMP PXL150 when used in mice, exhibits pronounced efficacy in burn wounds as an anti-infective agent. AMP D2A21 also exhibits similar properties in burn wound infections and is in phase III of clinical trials. Protegrin-1 and lactoferricin B showed significant activity against ophthalmic pathogenic bacteria like *Pseudomonas aeruginosa*, *Staphylococcus aureus*, *Candida albicans*, *Aspergillus* spp., and *Streptococcus pneumoniae*. However, the clinical use is not confirmed and is at the theoretical stage.

Peptide ZXR-2 demonstrated antimicrobial activities against candidiasis, dental infections, endodontic infections, and periodontal infections caused by agents like *Porphyromonas gingivalis*, *Streptococcus mutans*, and *Streptococcus sobrinus*. PAC-113, a peptide, is used as an over-the-counter AMP for the treatment of oral candidiasis in Taiwan [13,14].

The therapeutic application of AMPs requires additional technical strategies like prodrug formulation to improve protease stability, reduce cytotoxicity, etc. Multiple trials are on combining AMPs with existing antibacterial agents and probiotics as vectors and other formulation strategies like gels, creams, ointments, nanoparticles, hydrogels, glutinous rice paper capsules (topical for wounds), sponges, modified starch envelope, and Thiol-PEG-Thiol immobilized covalently with AMP. Targeted formulations (pheromone-labeled AMPs, quantum dots, nanotubes, graphene, triggered drug release formulations like enzyme precursor, pH-activated release formulations, and metal nanoparticles) are potential methods to enhance the drug delivery of AMPs. Hybrid peptides are another approach to developing targeted peptides. For example, the peptide targeting *P. aeruginosa*, PA2 is combined with GNU7, which is AMP, with broad-spectrum activity to derive a hybrid that acts on OprF protein and yields bactericidal activity and specificity [15].

## 16.3.2 Food industry

The addition of preservatives for increasing the shelf life of food is always of safety concern. The use of AMPs as food preservatives are projected to be healthier alternatives with a low toxic propensity to currently used preservatives. AMPs have good antimicrobial activity, they are either acid/or alkali in nature, thermally stable and at the same time easily hydrolyzed by proteases in the human body. Nisin, an AMP from lactic acid bacteria, *Lactococcus lactis*, is used as a potential dairy preservative in many countries. Polysine and pedocin are other preservative AMPs in use. Pedocin inhibits *Listeria monocytogenes* and prevents meat deterioration. Enterocin AS-48 is used as a preservative in juices, cider, and soya milk.

New formulation techniques like liposomes and probiotic bioengineered vectors releasing the AMP are also tried to overcome problems like interactions with food and proteolytic degradations. Another futuristic application is AMP-added packaging methods, for example, ε-poly-L-lysine formulated with starch biofilms for effects on *Penicillium expansum* and *Aspergillus parasiticus*.

## 16.3.3 Animal husbandry and aquaculture

In the last decade, the application of animal growth promoters in livestock farms like poultry, pig, ruminants, and aquaculture was banned by most developed countries. AMPs have a promising market opportunity as new antibacterial and growth promoters since they can improve production performance, promote intestinal health, boost immunity, and prevent bacterial inflammation. For instance, SIAMP for the treatment of infectious bronchitis virus (IBV) in chicken, SIAMP was found to improve daily weight gain in broilers [16].

European sea bass dicentracin, Caerin 1.1, and NK-lysine peptides inhibit fish pathogens like *Nodavirus, Spring viremia carp, Septicemia hemorrhagic,* and *Infectious pancreatic necrosis virus. Bacillus subtilis* E20 mediated fermentation of soybean meal contains some AMPs that effectively inhibit *Vibrio alginolyticus and Vibrio parahaemolyticus*. When added to feeds, it improves *Litopenaeus vannamei* resistance level against *V. parahaemolyticus* [16]. For market analysis of peptide antibiotics, the usually adopted market segmentation is based on peptide source, nature, application or indication, geographic area, formulation, route of administration, etc. The analysis parameters are generally growth analysis and adoption rate. Various segments with forecasts and factors influencing the market growth are furnished in an ideal report. Several market analysis agencies publish yearly reports with projections for 5 years ahead.

## 16.4 Important parameters of market analysis

The important parameters required for a market analysis of AMP includes the following:

1. Segment wise and subsegment wise predictions of the global market with market structure.
2. Factors influencing the market growth.
3. Past, present, and future market segment wise revenues based on end-user, type, material, and design.
4. Region wise and country wise past, present, and future market segment and subsegment revenue predictions.
5. Key player profile in the market, based on market shares, competencies and also the competitive picture of the market.
6. Economic, technological, and market trend factors that influence the global market.

## 16.5 Drivers and restraints of the peptide antibiotics market

Antibiotic peptides have several identified drivers that boosted the market potential resulting in a high promotion of antibiotic peptide drug discovery and development. There are several restraints too, which are the major challenges faced by research and marketing of peptide antibiotics.

## 16.6 Conclusion

AMPs because of their immune-modulatory activities and direct antifungal, antibacterial, and antiviral activities have become a broad class and their high target specificity makes the preclinical to clinical access rate higher. AMPs boost anti-inflammatory cellular response and disintegrate cancerous cell membranes and hence increasing their market potential further. The growing need for drugs for the treatment of diabetes, HIV, and other infectious diseases by resistant microbes, etc., are the major driving factors for the AMPs market.

The track record of adverse effects on chronic use that appeared postmarketing that led to limited use or withdrawal from the market is a fear factor for the investors hampering the growth of the peptide antibiotics market. The major obstacle to the research and development of AMPs is the high investment and cost involved in manufacturing. This is compounded with the risk of emergence of adverse effects on chronic use and this is a major discouraging factor too. Insufficient financial aid for research in peptide therapeutics is a hindrance to the expansion of the AMP market globally. Additionally, environmental stability and half-life, bioavailability, and other pharmacokinetic features of AMPs are yet to be established. These factors also limit the market growth of AMPs.

AMPs use is mainly limited to the developed countries and higher socioeconomic patients, which forms only a small section of the global population narrowing the profitability as discussed earlier R&D is also facing challenges, which adds to the apprehension of the sponsors to invest in AMPs research. In the year 2012, United States and European governments joined together to encourage and promote the R&D process for the development of newer AMPs and launched a Transatlantic Task Force and the Infectious Diseases Society of America. Countries from the Asia-Pacific region are also collaborating in this area, for instance, Australian and New Zealand biotechnology organizations have joined together to finance the manufacturing of new peptides. In the year 2007, to facilitate the process of manufacturing new peptides and specialized small molecules being compliant with Good Manufacturing Practices (GMP), Mimotopes Pty Ltd. and GlycoSyn were financed with US$ 519.85 million [16].

Global collaborations for developing AMPs are also boosting the discovery and drug development in this field. One of the examples would be the collaboration between Symcel and Colzyx to develop antibacterial peptides. Application of AI and DNA analysis to discover peptides for healthcare applications is also a new trend [16].

During the COVID-19 pandemic, the vancomycin market around the world was estimated at US$ 313.7 million in 2020 and by 2027 the market is calculated to reach up to US$ 465.1 million, that is, a CAGR growth of 5.8% over 7 years. With this overall growth of AMP in the global market in the year 2020, the US vancomycin market was predicted at US$ 84.5 million. By the year 2027, China is predicted to touch a market size of US$ 102.3 million, which is a rise in CAGR of 9.4% over the 7 years (2020−27). Over the same period, Japan is predicted to grow 3.1%, Canada is predicted to grow 4.6%, and Germany is predicted to grow 4% CAGR, respectively [17].

The global animal antibacterial peptide market growth is estimated to rise from US$ 2195.80 million in 2020 to US$ 4591.36 million significantly by 2025. But the tight regulatory controls that are underway for use in livestock farming as antibiotics and growth promoters are because of the proven strong relationship between the emergence of microbial resistance and nonhuman applications. In 1986 Sweden for the first time in the world prohibited the addition of antibiotics in the animal feed for the purpose of promoting growth in animals. Later in 1995 and 1996 Denmark and Germany banned avoparcin, a glycopeptide antibiotic, because its use was suspected to contribute to the emergence of resistance to vancomycin in humans which is a very essential last-resort antibiotic [13]. These tight regulatory controls that are going to be implemented with immediate effect are going to have a catastrophic effect on the animal antibacterial peptide market.

Peptide antibacterial replaced the previously used animal antibacterial chemicals with an advantage of lower possibilities of toxicity from meat, making it safer for human consumption, but now because of threat from resistance development, this market which is in millions may largely fall in size (Figs. 16.1 and 16.2, Tables 16.1 and 16.2).

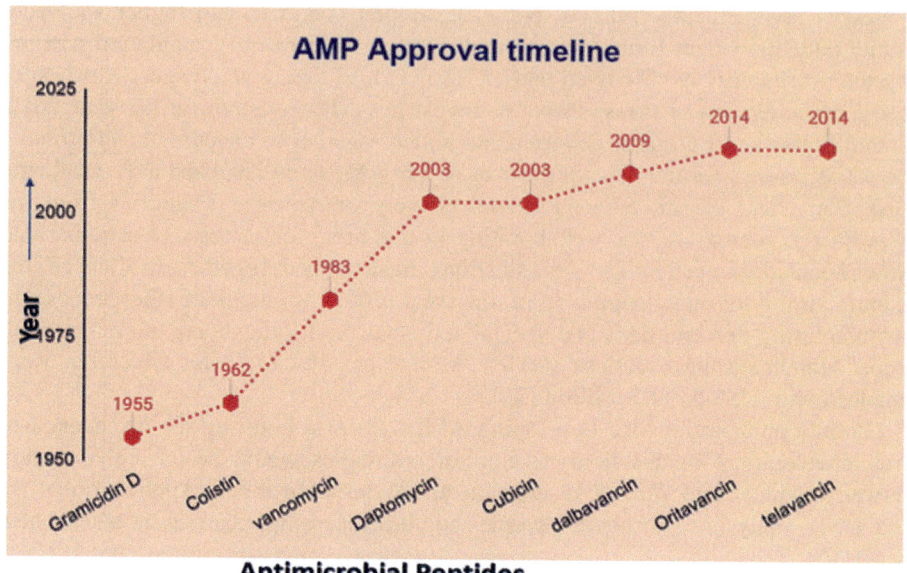

**Figure 16.1** Timeline of antimicrobial peptide approval.

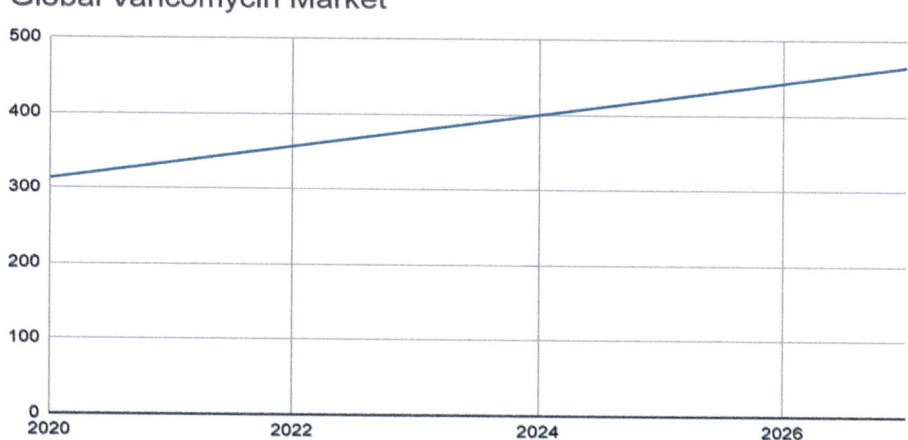

**Figure 16.2** Global vancomycin market to reach US$ 465.1 million by 2027.

**Table 16.1** Indications and current status of antimicrobial peptides.

| Name of the antimicrobial peptides | Indications | Status |
|---|---|---|
| **_Clinical applications_** | | |
| Dalbavancin | Acute bacterial skin infections | Approved |
| PAC-113, P-113 | Oral candidiasis | Under trials |
| Enfuvirtide | HIV-1 infection | Approved |
| Bacitracin | Localized skin and eye infections, wound infections | Approved |
| Vancomycin | Bacterial infections | Approved |
| Daptomycin | Gram-positive infections | Approved |
| Telavancin | Skin infections, nosocomial pneumonia | Approved |
| Polymyxin E | MDR infections caused by Gram-negative bacteria | Approved |
| D2A21—cationic cyclic decapeptide | Purulent skin disease | Under trials |
| PXL01—lactoferrin analog | Postsurgical adhesions | Under trials |
| Omiganan | Papulopustular rosacea | Under trials |
| **_Applications in the food industry_** | | |
| Nisin | Dairy (_Listeria monocytogenes_ and _Staphylococcus aureus_) | Approved |
| Polylysine | Sushi, boiled rice, noodles, meat, and drinks | Approved |
| **_Application in animal husbandry and aquaculture_** | | |
| SGAMP | Heat stress | Published |
| NKL-24 | _Vibrio parahaemolyticus_ infection in the scallop | Published |
| Caerin 1.1 | _Lactococcus garvieae_, porcine epidemic diarrhea virus (PEDV) | Published |
| Dicentracin | _L. garvieae_, viral hemorrhagic septicemia (VHSV), infectious pancreatic virus (IPNV) | Published |

**Table 16.2** Global antimicrobial peptides market segmentation.

| Basis | Types | Subtypes/examples | |
|---|---|---|---|
| Source | Plant | Thionins | |
| | | Defensins | |
| | | Snakins | |
| | Microbial | Nisin | |
| | | Gramicidin | |
| | Mammalian | Cathelicidins—cathelicidin LL-37 | |
| | | Defensins—α, β, θ defensins | |
| | | Lactoferricin B | |
| | | Casein 201 | |
| | | Histatins | |
| | | Granulysin | |
| | Amphibian | Frog—magainin | |
| | | Rana cancrivora—cancrin | |
| | Insects | Cecropin | |
| | | Jellein | |
| Activity | Antibacterial | Nisin | |
| | | Cecropins | |
| | | Defensins | |
| | | P5 and P9 | |
| | Antifungal | Ranatuerin | |
| | | Cecropins | |
| | Antiviral | Epi-1 | Dermaseptin-S4 |
| | | SIAMPs | Siamycin-I |
| | | a- and b- defensins | Siamycin-II |
| | | | RP 71955 |
| | | LL-37 | Enfuvirtide |
| | | Gramicidin D | Anticoronavirus |
| | | Caerin 1 | Peptide HR1, HR2 |
| | | Maximin 3 | Lipopeptide |
| | | Magainin 2, | EK1C4 |
| | | Dermaseptin-S1 | Temporin |
| | Antiparasitic | Cathelicidin | |
| | | Temporins-SHd | |
| | | Epi-1 | |
| | | Jellein | |
| Amino acid-rich species | Proline-rich (PrAMPs) | Tur1A+++++++ | |
| | | pPR-AMP1 | |
| | Tryptophan and arginine-rich | Indolicidin | |
| | | Triptrpticin | |
| | Histidine-rich | HV2 | |
| | | L4H4 | |
| | Glycine-rich | Attacins | |
| | | Diptericins | |
| | | GG3 | |

*(Continued)*

**Table 16.2** (Continued)

| Basis | Types | Subtypes/examples |
|---|---|---|
| Products | Natural antimicrobial peptides | |
| | Artificial antimicrobial peptides | |
| Chemical type | Cationic peptides versus anionic peptides | |
| | Linear versus helical | |
| Application | Personal care and cosmetics industry | |
| | Healthcare and pharmaceuticals industry | |
| | Biotechnology industry | |
| | Agriculture/farming | |

# References

[1] Peptide therapeutics market size worth $48.04 billion by 2025. https://www.grandviewresearch.com/press-release/global-peptide-therapeutics-market, (2020) (accessed 04.03.21).
[2] US$ 65 Billion opportunity in the global peptide therapeutics . . .. 31 Jan. 2020, https://www.prnewswire.com/news-releases/us65-billion-opportunity-in-the-global-peptide-therapeutics-market-clinical-patent-insight-on-197-marketed-peptides-300996774.html (2020) (accessed 04.03.21).
[3] K. Fosgerau, T. Hoffmann, Peptide therapeutics: current status and future directions, Drug Discovery Today 20 (2015) 122−128.
[4] S.S. Usmani, G. Bedi, J.S. Samuel, S. Singh, S. Kalra, P. Kumar, et al., HPdb: database of FDA-approved peptide and protein therapeutics, PLoS One 12 (2017) e0181748.
[5] N. Singh, J. Abraham, Ribosomally synthesized peptides from natural sources, The Journal of Antibiotics 67 (2014) 277−289.
[6] R. Kolter, R.F. Moreno, Genetics of ribosomally synthesized peptide antibiotics, Annual Review of Microbiology 46 (1992) 141−161.
[7] X. Zhao, Z. Li, O.P. Kuipers, Mimicry of a non-ribosomally produced antimicrobial, brevicidine, by ribosomal synthesis and post-translational modification, Cell Chemical Biology 27 (2020) 1262−1271.
[8] D.J. Craik, D.P. Fairlie, S. Liras, D. Price, The future of peptide-based drugs, Chemical Biology & Drug Design 81 (2013) 136−147.
[9] P. Xu, H. Zhang, R. Dang, P. Jiang, Peptide and low molecular weight proteins-based kidney targeted drug delivery systems, Protein and Peptide Letters 25 (2018) 522−527.
[10] J.L. Lau, M.K. Dunn, Therapeutic peptides: historical perspectives, current development trends, and future directions, Bioorganic & Medicinal Chemistry 26 (2018) 2700−2707.
[11] H. Kim, J.H. Jang, S.C. Kim, J.H. Cho, Development of a novel hybrid antimicrobial peptide for targeted killing of *Pseudomonas aeruginosa*, European Journal of Medicinal Chemistry 185 (2020) 111814.
[12] Global Antimicrobial Peptides Market Report: 2017−2030, September 2020. https://www.goldsteinresearch.com/report/anti-microbial-peptides-market-outlook-2024-global-opportunity-and-demand-analysis-market-forecast-2016-2024 (2020) (accessed 10.03.21).

[13] Y. Huan, Q. Kong, H. Mou, H. Yi, Antimicrobial peptides: classification, design, application and research progress in multiple fields, Frontiers in Microbiology 11 (2020) 582779.
[14] C.H. Chen, T.K. Lu, Development and challenges of antimicrobial peptides for therapeutic applications, Antibiotics 9 (2020) 24.
[15] M. Ibrahim, F. Ahmad, B. Yaqub, A. Ramzan, A. Imran, M. Afzaal, et al., Current trends of antimicrobials used in food animals and aquaculture, Antibiotics and Antimicrobial Resistance Genes in the Environment (2020) 39−69.
[16] Center for Disease Dynamics, Economics & Policy, 2015. State of the World's Antibiotics, 2015. CDDEP: Washington, DC. https://cddep.org/wp-content/uploads/2015/09/the-state-of-the-worlds-antibiotics-_2015.pdf (accessed 15.03.21).
[17] https://www.globenewswire.com/news-release/2021/04/19/2212568/0/en/Global-Vancomycin-Market-to-Reach-465-1-Million-by-2027.html.

# Index

*Note*: Page numbers followed by "*f*" and "*t*" refer to figures and tables, respectively.

## A

*Acinetobacter baumannii*, 237
Aculeacin A, 109
Adepantins, 37–39
Affinity chromatography, 70
Alternaramide, 206
Amino acid analysis, 73–74
Amolopin, 52
AMP D2A21, 312
AMP PXL150, 312
Amphibian host defense peptides
   antimicrobial peptide from, 140–149
   innate immune system, 139
Antibiotic resistance of ESKAPE pathogens
   direct drug interaction, 238–239
   indirect drug resistance, 239–240
Antibiotics, 81
Antibiotics against multidrug-resistant ESKAPE pathogens
   advantages and disadvantages of cationic AMPs, 241–243
   animal models and clinical use of AMPs, 249
   antimicrobial peptides to stop ESKAPE pathogens, 243–246
   bacterial killing by AMPs, 247–248
   resistance of, 238–240
   strategies to combat, 240–241
Antimicrobial lipopeptides of bacterial origin
   antiadhesion and antibiofilm activities, 90
   lipopeptides, 82–91
     amphisin, 88
     amphomycin (Amp), 87
     bogorol cationic peptides, 86
     circulocins, 87
     fengycin, 85–86
     gramicidins, 87
     iturin, 85
     kurstakin, 87
     surfactin, 83–84
     syringomycin, 88
     tridecaptins, 86
     viscosin, 87–88
   mechanism of action of, 89–90
   natural role of, 90
   structure-activity relationship of, 88–89
   treatment of multidrug-resistant infections, 90–91
Antimicrobial peptides (AMPs), 81
   from amphibians
     in Asia, 145–149
     in Europe, 145
     isolated from African frogs, 140–141
     in North America, 141–142
     in South America, 142–144
   applications of
     animal husbandry and aquaculture, 313
     food industry, 313
     prospects in medicine, 312–313
   in Asia
     Western Ghats, 146–149, 147*f*
   bacterial outer membrane
     lipopolysaccharide modifications, 267–268
     phospholipid modifications, 268
   binding and sequestering cationic AMPs, 271–272
   biofilm formation, 268–270
   biosynthesis and regulation, 41–43
   cell wall and cell membrane
     alterations to membrane composition, 264
     repulsion of, 263–264
     target modification, 264
   clinical applications of, 186*t*

Antimicrobial peptides (AMPs) (*Continued*)
  common families of, 43–47
  derived from different plant families
    cyclotides, 158
    defensins, 159–160
    heveins and hevein-like peptides, 160–162
    snakins, 160
    thionins, 159
  discovery, 243
  efflux pumps, 270–271
  features of, 35–40
  of fungal origin
    biotechnological applications, 108–110
    detection methods of antimicrobial peptides, 103–106
    fungal peptides, 101–102
    fungi-producing, 100
    mechanisms of synthesis, 103
    mode of action and biological activities, 102
    peptide databases, 107–108
  historical perspective, 34–35
  indications and current status of, 317*t*
  in innate immunity, 18–21, 19*t*
  limitations and challenges, 53–54
  from marine environment
    from algae, 207
    from invertebrates, 198–203
    from microorganisms, 203–206
    from vertebrates, 207
  from marine invertebrates
    from ascidians, 201
      tunicates, 201
    from Cnidaria, 203
    from crustaceans, 201–202
    from Echinodermata, 203
    from marine molluscs, 198–201
    from marine sponges, 198
      and activity, 199*t*
    from marine worms, 202–203
  from marine microorganisms
    from actinomycetes, 204–206
    from marine bacteria, 203–204
      antimicrobial NRPs, 205*t*
      nonribosomal antimicrobial peptides, 204
      ribosomal antimicrobial peptides, 204
    from marine fungi, 206
  from marine vertebrates
    from marine fishes, 207
  mechanism of, 264–265
    protein-mediated sequestration, 264–265
  modes of action, 48–51
  modulation of cationic, 273–275
  proteolytic degradation of, 273
  relationship of structure with function, 48
  resistance in gram-negative bacteria
    AMPs resistant mechanisms in, 266*f*
    bacterial outer membrane, 266–268
  resistance in gram-positive bacteria
    cell wall and cell membrane, 263–264
  to stop ESKAPE pathogens
    combined treatment, 245–246
    formulated antimicrobial peptides, 246
    library-based search and peptide mimetics, 244
    peptide conjugates, 244–245
    structure-based design, 243–244
    surface immobilized, 246
  by surface-associated polysaccharides, 265
Antimicrobial peptide database (APD), 241
APB-13, 227
Apidaecin, 37–39, 47
Arminins, 11
*Aspergillus parasiticus*, 313
*Aspergillus* spp., 312
Attacins, 21, 120

**B**
Bacillomycin L, 85
*Bacillus subtilis* E20, 313
Bactenecins, 45
Bactenins, 37–39, 47
Bacterial antibiotic resistance, 237–238
Bacterial killing by antimicrobial peptides
  bacterial membranes, 247–248
  bacterial ribosomes, 248
  cell wall, 248
  mechanisms of, 247–248
Bacteriocins, 47, 204
B aminolipopeptides, 206
Barrel-stave model, 50
Basic drawbacks of peptide drugs, 301
Biofilms, 239–240, 268–270
Biosynthesis and regulation, 41–43

# Index

Biotechnological applications, 108–110
Brevinins, 20–21, 145–146

## C

Caerins, 144
*Candida albicans*, 312
Capsules, 265
Cathelicidins, 17–18, 20, 25–26, 45, 177, 177f
Cationic antimicrobial peptides (CAMPs), 241–243, 245–246, 263, 265–267
  binding and sequestering, 271–272
  expression. modulation of, 273–275
Cationic host defense peptides (CHDPs), 139
Cecropins, 21, 45, 120–126, 171–172
Cefepime, 241
Classification of mammalian antimicrobial peptides
  based on activity, 179–182
    antibacterial, 179
    anticancer, 180–181
    antifungal, 180
    antiparasitic, 180
    antiviral, 180
    in fertility, 182
    immunomodulatory and chemotactic, 181
    in ophthalmology, 182
    in tissue regeneration and wound healing, 181–182
  based on amino acid sequence, 173–174
    glycine-rich antimicrobial peptides, 174
    histidine-rich peptides, 174
    proline-rich peptides, 173–174
    tryptophan and arginine-rich, 174
  based on structure, 175–179
    alpha, beta, and theta, 176f
    cathelicidins, 177, 177f
    defensins, 175–177
    histatins, 178, 178f
    neutrophil-activating peptide-2, 179f
    thrombocidin, 178–179
Colistin, 241, 245, 249
Common families of antimicrobial peptides
  cathelicidins, 45
  defensins, 45–46
  rich in specific amino acids, 46–47

    histatins, 47
    proline-rich, 47
    ribosomally synthesized antimicrobial peptides, 47
    tryptophan-rich, 46–47
  thionins, 46
Comparative docking study, 306
Compound annual growth rate (CAGR), 311
Computational biology in antimicrobial peptide research
  antimicrobial peptide databases, 278–281
    activity and structure of peptides, 279
    antimicrobial peptide database, 278–279
    of anuran defense peptide, 280
    bacteria, 281
    collection of, 279
    data repository of, 278
    dragon, 278
    invertebrate, 281
    linking, 279–280
    scaffold by property alignment, 280
    yet another database of, 280
  computational biology in, 275–283
  design of, 277f
  resistance in gram-negative bacteria, 265–275
  resistance in gram-positive bacteria, 263–265
  resistance patterns by machine learning approach, 281–282
  scope of, 275–283
Computational biology tools, 282–283
Computational methods for peptide-protein interactions
  docking methods at local and global levels, 305–306
    global, 306
    local, 305
    template-based, 306
  innovations and, 304–306
  molecular docking for peptide-protein interactions, 305
  preliminary peptide scaffolds, 304–305
Cyclotides, 158

## D

Dahleins, 144–145
Dalbavancin, 311

Damicornin, 203
Daptomycin, 82, 89, 245, 249, 311
Database filtering, 275–276
Defensins, 17–18, 20–21, 45–46, 102, 126–127, 159–160, 175–177
Depo-Foam technology, 301f
Dermaseptins, 20–21, 51–52, 142–143
Detection methods of antimicrobial peptides
　cellular MALDI-TOF-mass spectrum of Trichoderma isolate, 104f, 106t
　　amino acid sequence of selected peptaibols, 107t
　　(KUMB 9) producing 17- and 19-residue peptaibol, 105f
　　(KUMB 14) producing 14- and 17-residue peptaibol, 105f
　　(KUMB 59) producing 12-, 14-, and 17-residue peptaibol, 106f
　　(KUMB 553) producing 17-residue peptaibol, 104f
Discovery of protein–protein modulators
　approaches toward, 301–303
　fragment-based drug discovery, 302
　high-throughput screening, 301–302
　polymer-based protein delivery systems, 302f
　structure-based design, 302–303
Distinctin, 142–143
Drosocin, 47
Drosomycin, 37, 41
Dybowskin, 146
Dynein, 227

**E**
Early endosomes, 221
EBOV-7, 225
Edeines, 86
Edman procedure, 74
*Enterobacter* species, 237, 239
Enterocin AS-48, 313
*Enterococcus faecium*, 237
Epinecidin-1, 207
*Escherichia coli*, 239
ESKAPE pathogens, 237–238
　AMPs to stop
　　combined treatment, 245–246
　　formulated antimicrobial peptides, 246
　　library-based search and peptide mimetics, 244

　　peptide conjugates, 244–245
　　structure-based design, 243–244
　　surface immobilized, 246
　antibiotic resistance of, 238–240
　strategies to combat
　　antibiotic derivatives, 241
　　antimicrobial peptides, 241
　　phage therapy, 240–241
　　vaccines, 240

**F**
Fallaxin, 143
FDA-approved synthetic and modified peptides, 4t
Features of antimicrobial peptides
　αβ-antimicrobial peptides, 35, 37, 42t
　α-helical antimicrobial peptides, 37, 38t
　　PDB structures of, 39f
　β-sheet, 37
　　selected, 40t
　cationicity and amphipathicity, 35–36
　diversity, 35
　four major structural classes of, 36f
　non-αβ, 37–40, 44t, 45f
　structure, 36–40
Fungal peptides, 101–102
Fusion inhibitors, 224

**G**
Galactosamine, 267
Gel permeation chromatography, 69–70
Gingipains cleave substrates, 273
Gloverins, 127
Glycopeptide antibiotics (GPAs), 22–24
Gramicidin D, 311

**H**
Heveins and hevein-like peptides, 160–162
High-performance liquid chromatography (HPLC), 71–73
　cation-exchange chromatography, 72
　high-performance gel permeation chromatography, 71–72
　reversed-phase HPLC, 72–73
Histatins, 37–39, 41, 47, 178, 178f
Historical developments of antimicrobial peptide research
　AMPs as host innate defense barricade, 4–5

Index 325

AMPs modification for medical
   application, 9–10
chemical developments in AMPs, 7–8, 8t
current timeline of AMP approvals, 6–7
database, 5–6
for industrial applications, 10–11
interdisciplinary upgrade to AMPs, 11–12
Host defense peptides (HDPs).
   *See* Antimicrobial peptides
   (AMPs)

**I**
Identification of plant antimicrobial peptides
   analytical methods, 164t
   extraction and, 163–164
   extraction, purification and sequencing, 163f
Indolicidin, 37–39, 46–47
*Infectious pancreatic necrosis virus*, 313
Innate immunity, antimicrobial peptides in, 18–21, 19t
Insect peptides with antimicrobial effects, 121t
   antimicrobial resistance mechanism, 119f
   classification of insect peptides, 119–128
      attacins, 120
      cecropins, 120–126
      defensins, 126–127
      gloverins, 127
      lebocins, 127–128
      moricins, 128
   mode of action, 128–129
   need for AMPs, 118–119
Ion-exchange chromatography, 68–69

**K**
Kalata B1, 158
Kannurin, 84
*Klebsiella pneumonia*, 237
*Klebsiella pneumoniae*, 239

**L**
*Lactococcus lactis*, 313
Lactoferricins, 37–39
Lantibiotics, 47
Late endosomes, 221
Lebocins, 127–128
Lichenysin, 84
Ligands, 302–303

Limitations antimicrobial peptides
   aggregation propensity, 54
   and challenges, 53–54
   salt sensitivity, 54
   stability, 53–54
   toxicity, 54
Linear peptides, 242–243
Lipid clustering, 247–248
*Listeria monocytogenes*, 313
LL-37, 45, 52
LP-19, 225
Lysosomes, 221

**M**
Machine learning, 281–282
Maculatins, 144
Magainins, 20–21, 34–35, 45
Maginin I and II, 20–21
Mammalian antimicrobial peptides
   challenges in, 188–189
   classification of, 173–182
   clinical applications of, 185–188
   databases, 172t
   as first-line defense against invading microbes, 172–173
   history of, 171–172
Mass spectrometry, 74–75
Mathermycin, 206
Maximins, 45–46
Mechanism of action of mammalian antimicrobial peptides
   cell wall-targeting mechanism, 184
   immunomodulatory mechanism, 185
   membrane-targeting mechanism, 182–184, 183f
   targeting intracellular processes, 184–185
Mechanism of inhibition
   endosomal acidification inhibitors, 226–227
   plasma membrane and viral fusion inhibitors, 224–226
   replication and translation inhibitors, 227–228
   viral attachment inhibitors, 223–224
Meliacine, 227
Membrane filtration, 70–71
Modes of action
   membrane-independent/ nonmembrane-disruptive mechanism, 51
   membrane-mediated action, 49–50

Modes of action (*Continued*)
  microbial cell membrane, 49*f*
    barrel-stave model, 50
    carpet model, 50
    detergent model, 50
    toroidal pore model, 50
Moricins, 128
Multifaceted roles of antimicrobial peptides
  anticancer, 51–52
  antidiabetogenic, 52
  antiinflammatory and immunomodulatory, 52–53
  spermicidal peptides, 53
  wound-healing, 52
Mycosubtilin, 85

# N
Nisin, 47
*Nodavirus*, 313
Nonribosomally generated peptides (NRPs), 21–24
Nonribosomal peptides, 22–24
Nonribosomal peptide synthetase (NRPS), 18, 22–24
NOVEL-2, 226

# O
OC34-HR2P, 224–225
Ofloxacin, 241
Oritavancin, 311–312
Outer membrane vesicles (OMVs), 272

# P
P9, 226
Pattern recognition receptors (PRRs), 273–274
Penicillin-binding proteins (PBPs), 239
*Penicillium expansum*, 313
Peptaibiotics, 101, 103
Peptaibols, 101, 103–104, 106*t*, 107*t*, 108
Peptide antibiotic market
  applications of AMP, 312–313
  drivers and restraints of, 314
  global AMPs market overview, 311–312
  parameters of market analysis, 314
Peptide-based database, 5–6
Peptide databases
  for antimicrobial peptides from fungi, 108*t*
  peptaibol, 108

Peptide drug development
  basic drawbacks of, 301
  bile acid as drug delivery systems, 299*f*
  conjugation of amphiphilic proteins, 299*f*
  development of, 300*t*
  discovery of protein-protein, 301–303
  drawbacks of, 301
  historical overview of, 299–300
  innovations and computational methods for, 304–306
  olefeinic replacement, 298*f*
  overview of, 299–300
  peptides and protein-protein interactions, 303–304
  phonophoresis in drug delivery system, 300*f*
  substitution of á-amino acid, 298*f*
Peptides and protein-protein interactions
  computational and experimental methods for, 303–304
  computer-assisted docking strategies, 304
  potential developments for intrusive, 303
  structural-based predictions, 304
Peptides as therapeutics, 228
Peptides with antiviral activities
  challenges and future scope, 228–229
  mechanism of inhibition, 223–228
  as therapeutics, 228
  viral life cycle, 220–222
Peptidomimetics, 303
Phage therapy, 240–241
Phagocytic cells, 17–18
Phosphoethanolamine, 267
Phylloseptins, 143–144
Phylloxin, 142–143
Plant-derived antimicrobial peptides
  antimicrobial activity of, 159*f*
  derived from different plant families, 158–162
  extraction and identification of, 163–164
  technological and therapeutic applications, 164–165
Plectasin, 102
Polymyxin, 83, 89
Polyphemusin I, 21
Polyselective efflux pump, 238–239
PopuDef, 146
*Porphyromonas gingivalis*, 312
Posttranslationally modified peptides, 24–26

PR-39, 37–39, 47
Proline-rich peptides, 21
Proline-rich AMPs (PrAMPs), 173–174, 248
*Pseudomonas aeruginosa*, 237, 239, 312
*Pseudomonas* antimicrobial peptides, 87–88
Purification and characterization of
    antimicrobial peptides
  characterization techniques, 73–75
    amino acid analysis, 73–74
    Edman procedure, 74
    fragmentation pattern of peptide during
      mass spectrometry (MS/MS), 76f
    mass spectrometry, 74–75
    sequence by tandem mass spectrometry, 75
    two dimensional-poly acrylamide gel
      electrophoresis, 74
  techniques, 67–73
    affinity chromatography, 70, 70f
    high-performance liquid
      chromatography, 71–73
    ion-exchange chromatography, 68–69, 68f
    membrane filtration, 70–71
    principle of gel permeation
      chromatography, 69f
    solid-phase extraction on C18 column, 68
Pyrrhocoricin, 37–39

## R

Ranacyclins E and T, 145
*Rana temporalis*, 148
Resistance mechanisms, 1
Resistance-nodulation division (RND), 238–239
Ribosomally synthesized and
    posttranslationally modified
    peptides (RiPPs), 24–25
Ribosomally synthesized peptides (RPs), 21–22, 24–26
Ribosomal peptides, 21–26
Rifampicin, 241

## S

Sahyadri Hills, 146–147
SAH-RSVFBD, 225–226
Sclerotides A and B, 206
Scopularides A and B (cyclodepsipeptides), 206
*Septicemia hemorrhagic*, 313
Snakins, 160
*Spring viremia carp*, 313
*Staphylococcus aureus*, 237, 312
Statistical distribution of AMPs in
    THPdb, 6f
*Streptococcus mutans*, 312
*Streptococcus pneumoniae*, 312
*Streptococcus sobrinus*, 312
Structure-based design, 243
*Sylvirana temporalis*, 148

## T

Tandem mass spectrometry(MS/MS), 75
Tat-HA2Ec3, 226
Technological and therapeutic applications, 164–165
Telavancin, 311
Temporins, 145
Tetracycline, 241
Thionins, 46, 159
Tracheal antimicrobialpeptide (TAP), 20
Trichoderins, 206

## U

Urumin, 226

## V

Vancomycin, 241, 311–312
*Vibrio alginolyticus*, 313
*Vibrio parahaemolyticus*, 313

## W

WAP-8294A2 (WAP), 86
Western Ghats, 146–149

CPI Antony Rowe
Eastbourne, UK
November 28, 2022